Nanofiltration Membranes: Recent Advances and Environmental Applications

Nanofiltration Membranes: Recent Advances and Environmental Applications

Editor

Mohammad Peydayesh

MDPI • Basel • Beijing • Wuhan • Barcelona • Belgrade • Manchester • Tokyo • Cluj • Tianjin

Editor
Mohammad Peydayesh
Department of Health
Sciences and Technology
ETH Zürich
Zürich
Switzerland

Editorial Office
MDPI
St. Alban-Anlage 66
4052 Basel, Switzerland

This is a reprint of articles from the Special Issue published online in the open access journal *Membranes* (ISSN 2077-0375) (available at: www.mdpi.com/journal/membranes/special_issues/nanofiltration_membranes).

For citation purposes, cite each article independently as indicated on the article page online and as indicated below:

LastName, A.A.; LastName, B.B.; LastName, C.C. Article Title. *Journal Name* **Year**, *Volume Number*, Page Range.

ISBN 978-3-0365-4336-9 (Hbk)
ISBN 978-3-0365-4335-2 (PDF)

© 2022 by the authors. Articles in this book are Open Access and distributed under the Creative Commons Attribution (CC BY) license, which allows users to download, copy and build upon published articles, as long as the author and publisher are properly credited, which ensures maximum dissemination and a wider impact of our publications.

The book as a whole is distributed by MDPI under the terms and conditions of the Creative Commons license CC BY-NC-ND.

Contents

About the Editor vii

Mohammad Peydayesh
Nanofiltration Membranes: Recent Advances and Environmental Applications
Reprinted from: *Membranes* **2022**, *12*, 518, doi:10.3390/membranes12050518 1

Carlyn J. Higgins and Steven J. Duranceau
Removal of Enantiomeric Ibuprofen in a Nanofiltration Membrane Process
Reprinted from: *Membranes* **2020**, *10*, 383, doi:10.3390/membranes10120383 5

Dayuan Zheng, Dan Hua, Yiping Hong, Abdul-Rauf Ibrahim, Ayan Yao and Junyang Pan et al.
Functions of Ionic Liquids in Preparing Membranes for Liquid Separations: A Review
Reprinted from: *Membranes* **2020**, *10*, 395, doi:10.3390/membranes10120395 19

Anna Malakian, Zuo Zhou, Lucas Messick, Tara N. Spitzer, David A. Ladner and Scott M. Husson
Understanding the Role of Pattern Geometry on Nanofiltration Threshold Flux
Reprinted from: *Membranes* **2020**, *10*, 445, doi:10.3390/membranes10120445 47

Thomas J. Ainscough, Darren L. Oatley-Radcliffe and Andrew R. Barron
Groundwater Remediation of Volatile Organic Compounds Using Nanofiltration and Reverse Osmosis Membranes—A Field Study
Reprinted from: *Membranes* **2021**, *11*, 61, doi:10.3390/membranes11010061 59

Mujahid Aziz and Godwill Kasongo
The Removal of Selected Inorganics from Municipal Membrane Bioreactor Wastewater Using UF/NF/RO Membranes for Water Reuse Application: A Pilot-Scale Study
Reprinted from: *Membranes* **2021**, *11*, 117, doi:10.3390/membranes11020117 91

Antónia Macedo, David Azedo, Elizabeth Duarte and Carlos Pereira
Valorization of Goat Cheese Whey through an Integrated Process of Ultrafiltration and Nanofiltration
Reprinted from: *Membranes* **2021**, *11*, 477, doi:10.3390/membranes11070477 105

Tawsif Siddique, Naba K. Dutta and Namita Roy Choudhury
Mixed-Matrix Membrane Fabrication for Water Treatment
Reprinted from: *Membranes* **2021**, *11*, 557, doi:10.3390/membranes11080557 117

Zoulkifli Amadou-Yacouba, Julie Mendret, Geoffroy Lesage, François Zaviska and Stephan Brosillon
Impact of Pre-Ozonation during Nanofiltration of MBR Effluent
Reprinted from: *Membranes* **2022**, *12*, 341, doi:10.3390/membranes12030341 153

Benyamin Bordbar, Arash Khosravi, Ali Ahmadi Orkomi and Mohammad Peydayesh
Life Cycle Assessment of Hybrid Nanofiltration Desalination Plants in the Persian Gulf
Reprinted from: *Membranes* **2022**, *12*, 467, doi:10.3390/membranes12050467 169

Nur Sena Yüzbasi, Paweł A. Krawczyk, Kamila W. Domagała, Alexander Englert, Michael Burkhardt and Michael Stuer et al.
Removal of MS2 and fr Bacteriophages Using MgAl$_2$O$_4$-Modified, Al$_2$O$_3$-Stabilized Porous Ceramic Granules for Drinking Water Treatment
Reprinted from: *Membranes* **2022**, *12*, 471, doi:10.3390/membranes12050471 185

About the Editor

Mohammad Peydayesh

Dr. Mohammad Peydayesh received his Ph.D. degree in chemical engineering from Iran University of Science and Technology in 2018. He was awarded as a first-ranked graduate of the department in all B.Sc, M.Sc, and Ph.D. programs. Then, he joined ETH Zurich as a postdoctoral fellow under the supervision of Prof. Dr. Mezzenga. Since September 2021, he has been working as a senior assistant at the Food and Soft Material Laboratory, ETH Zürich. His research interests are soft matter, self-assembly phenomena, amyloid fibrils, advanced sustainable materials, waste valorization, water purification, and environmental engineering.

Editorial

Nanofiltration Membranes: Recent Advances and Environmental Applications

Mohammad Peydayesh

Department of Health Sciences and Technology, ETH Zurich, 8092 Zurich, Switzerland; mohammad.peydayesh@hest.ethz.ch

Nanofiltration (NF) is a cutting-edge filtration technology that may be considered a true paradigm shift in membrane science. NF can be used for a wide range of applications due to its unique properties to address major global issues in a sustainable and green way. Furthermore, NF technology has strong potential for supporting the UN Sustainable Development Goals (SDGs), particularly SDG 6: Clean Water and Sanitation. However, the successful industrial application of NF depends on developing efficient and reliable membranes, optimization of processes, understanding of the separation mechanisms, and justifications in terms of energy and economics.

This Special Issue, titled "Nanofiltration Membranes: Recent Advances and Environmental Applications" in the Membranes, aims to assess the recent developments and advances in all the aspects related to NF technology and its environmental applications.

There are ten contributions, including eight research articles and two reviews, in this Special Issue. Various topics are discussed, including fabrication of organic and inorganic NF membranes, tailoring NF membranes' surface properties, the application of NF in side-stream valorization, using NF technology in combination with other membrane technologies such as ultrafiltration (UF) and reverse osmosis (RO), wide-range applications of NF for removing organic and inorganic pollutants and viruses, and a detailed sustainability assessment of the use of NF technology via life cycle assessment (LCA).

Higgins et al. [1] studied the removal of Enantiomeric Ibuprofen in an NF membrane process. In addition to the rejection by the NF membrane itself, results show that the stainless steel equipment of a flat-sheet experimental unit could adsorb up to 23% of pharmaceuticals, primarily S-enantiomers of ibuprofen. This illustrates the importance of conducting initial mass balance experiments to determine possible losses of the compound in bench-scale membrane removal studies in order to determine the true removal capabilities of the membrane process.

Zheng et al. [2] reviewed the functions of ionic liquids (ILs) to prepare membranes for different filtration processes, including NF. The application of ILs as raw membrane materials, physical additives, chemical modifiers, and solvents was discussed. Moreover, related challenges and future perspectives of IL-assisted membranes were highlighted.

Colloidal fouling caused by the accumulation of particles such as colloidal silica, iron, aluminum, manganese oxides, and calcium carbonate precipitates on the membrane surface impedes NF operations by introducing additional hydraulic resistance that reduces water permeability. Malakian et al. [3] applied nanoscale line-and-groove patterns on commercial NF membranes using thermal embossing to address this issue. Experimental work combined with CFD simulations revealed that increasing the pattern ratio fraction leads to higher threshold flux, which is vital for the productivity of the NF process since it maintains low fouling rates.

Ainscough et al. [4] evaluated the performance of NF technology for removing chlorinated hydrocarbons such as trichloroethylene (TCE), tetrachloroethylene (PCE), cis-1,2-dichloroethylene (DCE), 2,2-dichloropropane (DCP), and vinyl chloride (VC) from groundwater. There are few reports on removing these compounds from water, which is considered

Citation: Peydayesh, M. Nanofiltration Membranes: Recent Advances and Environmental Applications. *Membranes* **2022**, *12*, 518. https://doi.org/10.3390/membranes12050518

Received: 28 April 2022
Accepted: 11 May 2022
Published: 13 May 2022

Publisher's Note: MDPI stays neutral with regard to jurisdictional claims in published maps and institutional affiliations.

Copyright: © 2022 by the author. Licensee MDPI, Basel, Switzerland. This article is an open access article distributed under the terms and conditions of the Creative Commons Attribution (CC BY) license (https://creativecommons.org/licenses/by/4.0/).

a particularly difficult separation. The results from the laboratory experiment showed high performance of the NF, where rejection rates of up to 93% for synthetic solutions and up to 100% for real groundwater samples were achieved. In site trials and due to the operational limitation, although the NF removed organic materials, it failed to remove volatile organic compounds (VOCs).

Aziz et al. [5] investigated the performance of UF, NF, and RO technologies for treating the effluent from a full-scale membrane bioreactor (MBR) municipal wastewater treatment plant. While the UF technology showed removal efficiencies of around 40% for chemical oxygen demand (COD), NH_4^{2+}, PO_4, and NO_3, NF and RO technologies exhibited excellent removal rates above 90%. The superiority of NF technology was further revealed by the fact that the energy consumption of RO was 1.4 times higher than that of NF.

NF can also be used to valorize industrial by-products, contributing to waste management and the circular economy. Macedo et al. [6] valorized goat cheese whey via an integrated process of UF and NF. The results showed that all the lactose could be retained using NF, allowing total whey recovery in cheese dairy plants.

Mixed matrix membranes (MMMs) are one of the primary membrane categories, combining the advantage of polymeric membranes and inorganic nanoparticles. Siddique et al. [7] reviewed the current state of the art of MMMs in different membrane processes, including NF. The focus here was on different fabrication approaches of MMMs, controlling parameters, fouling mitigation, and challenges and outlooks.

Amadou-Yacouba et al. [8] evaluated the impact of the pre-ozonation step during the NF of MBR effluent. Although the ozonation process had a low effect on the organic carbon mineralization, it decreased the COD and the specific UV absorbance by 50%, indicating the efficiency of ozonation in degrading a specific part of the organic matter fraction. The results showed that pre-ozonation during the use of NF could decrease the fouling and maintain the flux by partially mineralizing dissolved and colloidal organic matter.

Bordbar et al. [9] investigated and benchmarked hybrid NF desalination plants in terms of sustainability and the environmental footprint via LCA in the Persian Gulf region. The studied plants were multi-stage flash (MSF), hybrid RO-MSF, hybrid NF-MSF, RO, and hybrid NF-RO, and their impacts on climate change, ozone depletion, fossil depletion, human toxicity, and marine eutrophication were assessed. The LCA results demonstrated that hybrid NF-RO is a process with minimal environmental impacts for seawater desalination.

NF can be used to remove microorganisms and viruses in water sources prior to the consumption, addressing the problems associated with waterborne diseases. To do so, Yüzbasi et al. [10] developed two ceramic membranes composed of alumina platelets using spray granulation. They assessed the virus retention performance using two virus surrogates: MS2 and fr bacteriophages. The results showed complete virus removal from water, where the virus concentration decreased by 7 log10 reduction value (LRV), meeting the World Health Organization (WHO) criteria of LRV ≥ 4.

In conclusion, I would like to express my sincere appreciation to the authors, reviewers, and publisher for their outstanding work and contribution to the Special Issue "Nanofiltration Membranes: Recent Advances and Environmental Applications". I hope this collection will be useful for the membrane community worldwide, promoting the NF technology toward a more efficient, sustainable, and affordable process for environmental application.

Funding: This research received no external funding.

Acknowledgments: The guest editor is grateful to all the authors that contributed to this Special Issue.

Conflicts of Interest: The editor declares no conflict of interest.

References

1. Higgins, C.J.; Duranceau, S.J. Removal of Enantiomeric Ibuprofen in a Nanofiltration Membrane Process. *Membranes* **2020**, *10*, 383. [CrossRef] [PubMed]
2. Zheng, D.; Hua, D.; Hong, Y.; Ibrahim, A.-R.; Yao, A.; Pan, J.; Zhan, G. Functions of Ionic Liquids in Preparing Membranes for Liquid Separations: A Review. *Membranes* **2020**, *10*, 395. [CrossRef] [PubMed]
3. Malakian, A.; Zhou, Z.; Messick, L.; Spitzer, T.N.; Ladner, D.A.; Husson, S.M. Understanding the Role of Pattern Geometry on Nanofiltration Threshold Flux. *Membranes* **2020**, *10*, 445. [CrossRef] [PubMed]
4. Ainscough, T.J.; Oatley-Radcliffe, D.L.; Barron, A.R. Groundwater Remediation of Volatile Organic Compounds Using Nanofiltration and Reverse Osmosis Membranes—A Field Study. *Membranes* **2021**, *11*, 61. [CrossRef] [PubMed]
5. Aziz, M.; Kasongo, G. The Removal of Selected Inorganics from Municipal Membrane Bioreactor Wastewater Using UF/NF/RO Membranes for Water Reuse Application: A Pilot-Scale Study. *Membranes* **2021**, *11*, 117. [CrossRef] [PubMed]
6. Macedo, A.; Azedo, D.; Duarte, E.; Pereira, C. Valorization of Goat Cheese Whey through an Integrated Process of Ultrafiltration and Nanofiltration. *Membranes* **2021**, *11*, 477. [CrossRef] [PubMed]
7. Siddique, T.; Dutta, N.K.; Choudhury, N.R. Mixed-Matrix Membrane Fabrication for Water Treatment. *Membranes* **2021**, *11*, 557. [CrossRef] [PubMed]
8. Amadou-Yacouba, Z.; Mendret, J.; Lesage, G.; Zaviska, F.; Brosillon, S. Impact of Pre-Ozonation during Nanofiltration of MBR Effluent. *Membranes* **2022**, *12*, 341. [CrossRef] [PubMed]
9. Bordbar, B.; Khosravi, A.; Orkomi, A.A.; Peydayesh, M. Life Cycle Assessment of Hybrid Nanofiltration Desalination Plants in the Persian Gulf. *Membranes* **2022**, *12*, 467. [CrossRef]
10. Yüzbasi, N.S.; Krawczyk, P.A.; Domagała, K.W.; Englert, A.; Burkhardt, M.; Stuer, M.; Graule, T. Removal of MS2 and Fr Bacteriophages Using MgAl2O4-Modified, Al_2O_3-Stabilized Porous Ceramic Granules for Drinking Water Treatment. *Membranes* **2022**, *12*, 471. [CrossRef]

Article

Removal of Enantiomeric Ibuprofen in a Nanofiltration Membrane Process

Carlyn J. Higgins [1] and Steven J. Duranceau [2],*

1. Hazen and Sawyer, 1000 N. Ashley Dr. Suite 1000, Tampa, FL 33602, USA; chiggins@hazenandsawyer.com
2. Department of Civil, Environmental and Construction Engineering, University of Central Florida, 4000 Central Florida Blvd., Orlando, FL 32816, USA
* Correspondence: steven.duranceau@ucf.edu; Tel.: +1-407-823-1440

Received: 16 October 2020; Accepted: 26 November 2020; Published: 30 November 2020

Abstract: A study of the behavior of R- and S-enantiomers of ibuprofen (R-IBU and S-IBU) in aqueous solution by nanofiltration (NF) membranes revealed that up to 23% of the pharmaceutical was adsorbed onto the stainless steel equipment of a flat-sheet experimental unit. Mass balances disclosed that IBU's S-enantiomer was primarily responsible for the adsorption onto the equipment. Additional IBU adsorption was also experienced on the NF membrane coupons, verified by increased contact angle measurements on the surfaces. The IBU-equipment adsorptive relationship with and without the membrane coupon were best described by Freundlich and Langmuir isotherms, respectively. At a feed water pH of 4.0 units and racemic µg/L IBU concentrations, NF removal ranged from 34.5% to 49.5%. The rejection of S-IBU was consistently greater than the R-enantiomer. Adsorption onto the surfaces influenced NF rejection by 18.9% to 27.3%. The removal of IBU displayed a direct relationship with an increase in feed water pH. Conversely, the adsorption of IBU exhibited an indirect relationship with an increase in feed water pH.

Keywords: nanofiltration; ibuprofen; adsorption; enantiomer; chirality; removal

---

1. Introduction

The existence and subsequent discovery of chemicals of emerging concern (CECs) in aquatic environments has sparked interest in determining removal capabilities of specific water treatment technologies. Nanofiltration (NF) is a promising pressure-driven semipermeable membrane technology that can be employed to remove CECs from aqueous streams [1–5]. In a membrane process, the extent of solute removal is dependent on chemical properties, feed water matrix composition, membrane characteristics, and operational variables [1,2,6–8]. A strong research effort has attempted to elucidate the impact of CEC properties on solute removal through NF processes. It is widely accepted that molecular weight is an important parameter in the prediction of non-charged and non-polar compound rejection [1,9–11]. However, other solute characteristics such as chemical properties, solute geometry, and functional groups can also affect rejection of CECs [1,6,7]. In a membrane process, correlations between CEC removal and hydrophobicity [12], membrane adsorption [13], polarizability [14], polarity [15,16], and molecular size and shape [6,7,14,17,18] have been noted. The complexities of CEC mass transfer have been scrutinized and reported on over the years and serve as the basis for additional investigations such as those presented herein.

The position of functional groups in structural isomers has also been shown to have significant effects on rejection by reverse osmosis membranes [19]. This suggests that the spatial arrangement of atoms plays a larger role in membrane process removal than currently understood. A solute property that has received little attention regarding behavior in a membrane process is chirality. Chiral molecules, or stereoisomers, are molecules with the same molecular formula and chemical bonding arrangement, but dissimilar spatial position of atoms. Enantiomers are pairs of stereoisomers that are non-superimposable mirror images. Although enantiomers have the same molecular formula and other chemical properties, some are known to behave differently.

A well-known example of a chiral molecule is ibuprofen (IBU). IBU is a weak propionic acid derivative and pharmaceutically active compound (PhAC) known for its non-steroidal anti-inflammatory (NSAID) properties. The molecule contains a chiral carbon, yielding two enantiomers, S-IBU and R-IBU. Although medically administered IBU is a racemic mixture of the two enantiomers, the S-form possesses most of the anti-inflammatory properties [20–22]. Physiochemical properties of IBU such as low Henry's law constant (1.5×10^{-7} atm-m^3/mol), moderately high log octanol-water partition constant (K_{ow}; 3.97), and soil adsorption constant (log K_{oc}; 2.60) suggest that the PhAC often persists in aquatic environments and can display adsorptive qualities to clay and other loamy solids [23]. Like other CECs, IBU has been detected in groundwater and surface waters from the ng/L to µg/L level [24,25]. The enantiomeric ratio (ER) of IBU in surface water has been recorded higher than 0.5, yielding disproportionate enantiomer concentrations in favor of the S-counterpart [26].

In this presented work, the capability of NF process to remove enantiomeric IBU (S-IBU, R-IBU) at acidic pH conditions is explored. The behavior of chirality with respect to removal by nanofiltration membranes has not been fully vetted. The results of this research provide insight into the differing behavior of chiral molecules, further elucidating the effect of solute properties on membrane rejection.

2. Materials and Methods

In this study, Dupont Filmtec NF270 (Edina, MN, USA) and Microdyn Nadir Trisep TS40 (Goleta, CA, USA) membranes were assessed. The polyamide thin-film composite NF270 and polypiperazine amide TS40 were acquired from Sterlitech Corp. (Kent, WA, USA). Membranes were received as flat sheets, cut to the appropriate size, and soaked in deionized water (DI) for at least 24 h prior to experimentation. Membrane operational parameters are listed in Table 1. Racemic IBU was purchased from Sigma-Aldrich (St. Louis, MO, USA) with reported purity of greater than 99%. A 400 mg/L standard solution was prepared in LC/MS grade methanol and sonicated for homogeneity. The standard was stored at 4 °C in a salinized amber bottle and used within one month.

Table 1. Membrane operational properties.

Membrane	MWCO (Da)	Water Flux Coefficient (L_p)	MgSO$_4$ Rejection (%)	Contact Angle (Virgin, °)	Contact Angle (Compacted, °)
NF270	200–400	0.460	>97	30.6	50.2
TS40	200–300	0.231	>98.5	28.7	43.3

2.1. Experimental Setup

A bench-scale, flat-sheet membrane testing apparatus was used in this research. The unit consisted of a Wanner Engineering M-03S Hydra-Cell 6.81 L/min pump (Minneapolis, MN, USA) with a Control Techniques variable frequency drive (VFD) (Eden Prairie, MN, USA), a 19 L Sterlitech stainless steel conical feed tank, two Sterlitech CF042 acrylic cells to house the membrane coupons (operated in parallel for duplicity) with 42 cm^2 effective membrane area, and accompanying appurtenances consisting of flowmeters, pressure gauges, check valves, and stainless steel braided hose. Two MyWeigh CTS-600 scales (Phoenix, AZ, USA) were utilized for permeate collection and flux measurements. Feed flow was controlled with the VFD and set at 1.0 L/min, corresponding to a crossflow velocity of 0.18 m/s. Feed pressure was controlled by the concentrate control valve. A chiller-coil system was utilized to sustain a feed water temperature of 20 ± 1 °C.

Prior to each experiment, membrane coupons were inserted into the bench-scale, flat-sheet unit and compacted at 6.9 bar (100 psi) with DI water for at least 24 h. After initial compaction, the mixture was replaced with a 10 L solution containing DI water spiked with racemic IBU and adjusted to acidic conditions (feed water pH of 4.0 to 6.0 units) with 1 M sodium hydroxide or 5.8 M hydrochloric acid. Then, the bench-scale, flat-sheet unit was repressured with the experimental feed water matrix and operated for 24 h. Experiments were conducted in recycle mode, where permeate and concentrate streams recycled back to the feed reservoir. Feed water samples were taken at 0 and 24 h, where permeate and concentrate aliquots were taken at 24 h. The unit was flushed at least twice with 5 L of DI water in between experiments. A new set of membrane coupons was used for each experiment.

2.2. Analytical Methods

Samples were collected in 150 mL salinized amber bottles, stored in a 4 °C refrigerator, and extracted and analyzed within 48 h and 7 d, respectively. A solid phase extraction (SPE) method was utilized to extract and preconcentrate R- and S-IBU enantiomers [27]. Extractions were performed utilizing a Waters vacuum manifold and Waters Oasis HLB 3 mL, 60 mg cartridges (Milford, MA, USA). The R- and S-IBU enantiomers were analyzed via a Perkin-Elmer Series 200 high performance liquid chromatography (HPLC) instrument (Santa Clara, CA, USA). Separations were carried out on a Chiral Technologies, Inc., (West Chester, PA, USA) Chiralcel OJ-H column (4.6 × 150 mm, i.d., 5 µm particle size). The column was operated in polar phase mode, with an isocratic mobile phase consisting of methanol/formic acid (100:0.1, v/v) at a flow rate of 1 mL/min.

A ramé-hart Model 100 goniometer (Succasunna, NJ, USA) was utilized to determine membrane hydrophobicity via contact angle. Contact angle measurements were attained utilizing the sessile drop technique. Membrane coupons were dried and inserted on the stage with the active layer facing up. A micrometer syringe delivered a droplet of DI water onto the membrane surface, and a contact angle was measured by the goniometer. To obtain a representative contact angle of the entire membrane surface, ten contact angle measurements were taken on various areas of the membrane coupon and averaged.

3. Results and Discussion

Adequate mass balance tests are recommended in bench-scale membrane filtration experiments to confirm that rejection is not affected by solute behavior such as volatilization, adsorption, or a reaction with the feed water matrix. Consequently, prior to the series of pressurized filtration tests, a mass balance confirmation experiment was conducted by circulating a feed solution containing 100 µg/L IBU at feed water pH of 4.0 units through the flat-sheet equipment without a membrane coupon for 24 h. After analysis, 23% loss of IBU was observed during the experiment.

3.1. Effect of Feed pH on Adsorption

The test was repeated at a feed water pH range from 3.0 to 7.0 units, resulting in an inverse relationship between loss of enantiomeric IBU and the water quality parameter, aligning with the acid dissociation constant (pKa) of IBU (4.4). Figure 1 illustrates the results of the flat-sheet equipment adsorption experiments, where the total IBU concentration adsorbed is presented as a function of the total initial IBU concentration.

Due to the low Henry's law constant (1.5×10^{-7} atm-m^3/mol), volatilization could not explain observed IBU losses. However, in acidic conditions, IBU is known to adsorb onto and protect metal from corrosion [28,29], and IBU has been documented to attach to chromium-based metal organic frameworks [30]. It is noted that the flat-sheet test equipment is comprised mainly of components that consist of stainless steel (16% chromium, 10% nickel, 2% molybdenum, and less than 0.02% carbon) [31]. Existing literature suggests that IBU adsorption onto the flat-sheet stainless steel components (comprised of the reservoir, tubing, and chiller coil) could occur, as similar results have been realized with 9-anthracenecarboxylic acid [32]. Minimal to no attachment to the acrylic holding cells and permeate polyethylene tubing was observed. The bonding mechanism of IBU adsorption onto the surface was postulated to be between the hydrogen on the carboxylate functional group of the solute and adsorption of oxygen on the metal from the hydroxide moiety [33,34].

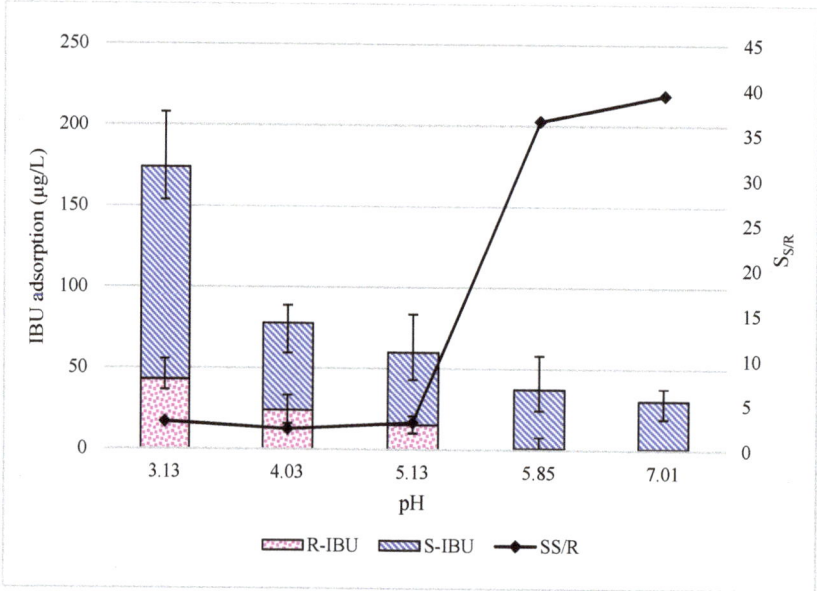

Figure 1. R- and S-enantiomers of ibuprofen (R-IBU and S-IBU) adsorption onto flat-sheet equipment as a function of feed water pH (temperature 20 ± 1 °C). Selectivity is a ratio of adsorption of S/R IBU.

3.2. Effect of Feed Concentration on Adsorption

Although significant adsorption onto the flat-sheet equipment was observed, IBU has also been known to attach onto membrane surfaces at pH values less than its pKa of 4.4 [35]. Hence, experiments deciphering the extent of IBU adsorption onto the flat-sheet equipment with and without a membrane coupon installed were conducted by repeating the experiment for a racemic solute concentration range from 100 µg/L to 1.50 mg/L. Feed samples were collected at 0 and 24 h. Triplicate feed concentration measurements were taken and averaged.

Results in Figure 2 indicate that the adsorption of IBU increases with initial feed concentration, which agree with prior PhAC-metal attachment studies [29]. It also appears that the sorbed IBU concentration may approach a saturated equilibrium in due course, and hence can be modeled by adsorption isotherms (see Supplementary Information).

An additional 19.6% to 39.2% IBU adsorption onto flat-sheet equipment with membrane coupon was recorded. These results suggest that additional IBU adsorption presumably occurred onto the membrane surface. IBU adsorption onto the membrane components was validated by an increase in hydrophobicity (measured by contact angle), illustrated in Figure 3. Hydrophobicity was found to have a positive direct relationship with the concentration of adsorbed IBU. Mechanisms of adsorption could include both hydrophobic interactions and the formation of hydrogen bonds between IBU and the membrane surface [36]. Previous studies have attributed IBU adsorption only to the membrane surface, neglecting to fully understand the behavior of IBU, hence, conceivably reporting inaccurate rejection values in flat-sheet studies [37–40].

At an initial racemic concentration of 1.08 mg/L, 83.7 µg/L IBU adsorbed onto the equipment and NF270 membrane. On the contrary, at an initial racemic content of 840 µg/L, 93.6 µg/L IBU adsorbed onto the equipment and TS40 membrane. Therefore, the TS40 membrane contained a slightly higher capacity to adsorb IBU. The difference of adsorption could not be explained by pore size or surface hydrophobicity. Although the NF270 membrane is more hydrophobic and has a larger MWCO, it did not adsorb as much IBU as the TS40 component. Others have postulated similar findings [35].

It should be noted that static batch experiments investigating IBU adsorptive capabilities on membrane coupons have been conducted elsewhere [35]. Batch adsorption experiments often do not represent actual attachment capacities of membrane while in pressurized operation [32], and thus were not included in the scope of this work.

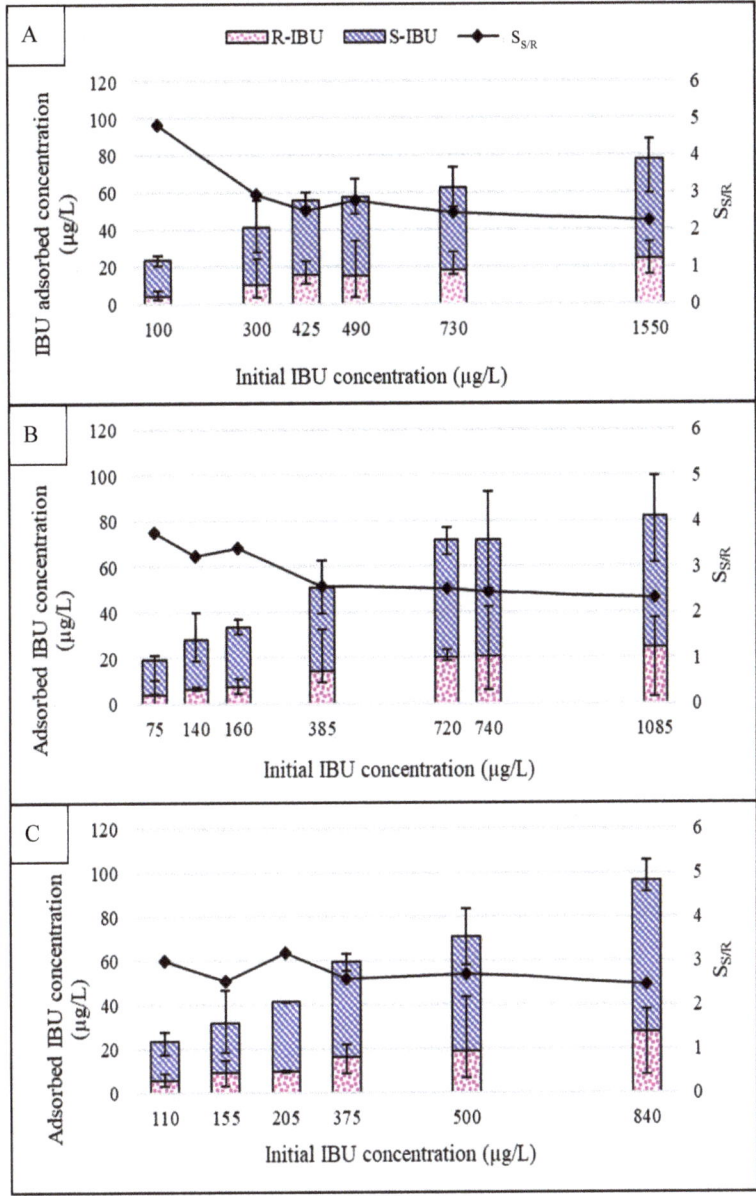

Figure 2. Adsorbed R- and S-IBU onto (**A**) Flat-sheet equipment; (**B**) Flat-sheet equipment with NF270 coupon; (**C**) Flat-sheet equipment with TS40 coupon (feed water pH of 4.0 units, temperature 20 ± 1 °C). Selectivity is a ratio of adsorption of S/R IBU. Error bars represent minimum and maximum values from triplicate analysis.

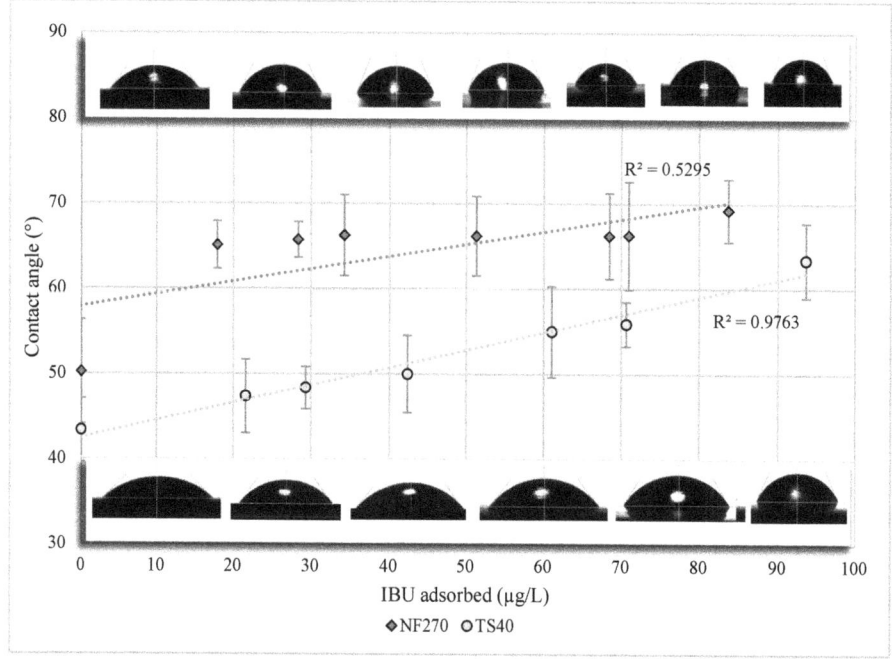

Figure 3. Contact angle of NF270 and TS40 membranes as a function of IBU adsorption. Error bars represent one standard deviation of uncertainty. Contact angle snapshots provide a visual image of the linear relationship between hydrophobicity and adsorbed IBU.

An apparent difference between the adsorption of R- and S-IBU onto the metal surface was noted. At an initial racemic concentration of 100 µg/L, S-IBU adsorbed 4.82 times more than its R-counterpart. The ratio fell to 2.25 at an initial racemic concentration of 1.50 mg/L. Although current literature on the adsorption behavior of enantiomers is scarce, some have claimed IBU can enantioselectively adsorb onto chromium- and vanadium-based metal organic frameworks [30]. Additionally, S-IBU has been reported to adsorb up to 10 times more than the R-enantiomer on a liposome membrane [41]. In liposomes, enantioselectivity was ascribed to hydrogen bonding or hydrophobic interactions between the asymmetric carbons of the chiral molecule and the spherical vesicle.

A possible explanation for the disparate enantiomer behavior could reside in optimized molecular geometry between R- and S-IBU performed by density functional theory (DFT) computations [42,43]. In conjunction with experimental studies, DFT computations can provide insight to the contrary behavior of chiral molecules, as previously illustrated by D-alanine's enantioselective adsorptive behavior [44]. The DFT framework in this study utilized the gradient correction non-local correlation functional of Lee, Yang, and Parr (B3LYP) with a basis set of 6-31G*, using the online GAMESS software [45,46]. Table 2 presents a comparison of the DFT-derived total energies and geometric properties of R- and S-IBU.

Table 2. Density functional theory (DFT) calculated energy and geometries of R- and S-IBU.

Compound	Method	Energy (Hartrees)	Surface Area ($Å^2$)	Molecular Volume ($Å^3$)	Dipole Moment (Debeye)
R-IBU	B3LYP/6-31G*	−656.3	179.5	199.3	2.018
S-IBU	B3LYP/6-31G*	−656.3	173.1	194.3	5.404

Results indicate approximately equal energies of R- and S-IBU, however, the surface area, volume, and dipole moment differ, which align with findings presented elsewhere [47]. The larger surface area, molecular volume, and smaller dipole moment of R-IBU suggest a bulkier, more hindered approach as compared with S-IBU. It should be noted that the ratio of S- to R-IBU dipole moments (2.68) compares well with the adsorption selectivity at initial racemic concentrations greater than 300 µg/L (2.46). Other conceivable explications for the dissonant chiral behavior include the Easson–Stedman three-point "lock and key" hypothesis between the chemical and binding site [48,49] or the slightly unequal opposite optical rotations of the enantiomers [50].

3.3. Adsorption Isotherm Modeling

Adsorption isotherms can be used to describe the relationship between the quantity of IBU attached on a solid surface in relation to its surrounding aqueous concentration at a constant temperature and pressure [51]. The concentration of IBU adsorbed to the solid surface at quasi-equilibrium (q_e) is calculated by Equation (1):

$$q_e = \frac{(C_o - C_e)V}{A} \tag{1}$$

where q_e is the concentration of IBU on solid surface (µg/cm^2), C_o is the initial concentration of IBU in aqueous solution (µg/L), C_e is the equilibrium concentration of IBU in aqueous solution (µg/L), V is the volume of aqueous solution (L), and A is the surface area of solid surface (cm^2).

In this work, Langmuir, Freundlich, and Temkin isotherms were utilized to model the adsorption behavior of IBU [51–57]. Manipulations of q_e and C_e for R- and S-IBU were plotted in accordance with the Langmuir, Freundlich, or Temkin isotherms to determine the best-fit model for the adsorption system. Isotherms were ascertained for error using coefficient of determination (r^2), relative percent difference (RPD), some of square errors (ERRSQ), and root mean square error (RMSE) [56]. Derived parameters and statistical error are shown in Table 3. The variables K_L and q_a represent Langmuir adsorption constants, K_F and n represents Freundlich adsorption constants, and the and K_T and b represent Temkin adsorption constants.

From Table 3, the Langmuir, Freundlich, and Temkin isotherms yielded r^2 values > 0.90, authenticating adsorption equilibrium tendencies for experimental data. Favorable adsorption was observed in the Freundlich isotherm as 1/n values were < 1 for R- and S-IBU. Adsorption intensities denoted by constants K_L, K_F, and K_T were higher for S-IBU. Furthermore, greater Langmuir maximum adsorption capacities (q_a) were also observed for S-IBU, aligning with the favorable adsorption presented herein. Larger concentrations of adsorbed IBU on the surface (q_e) were experienced in the equipment-IBU-membrane system, highlighting the additional adsorptive capacity of the membrane surface.

The Langmuir, Freundlich, and Temkin isotherms yielded similar predictability for IBU equilibrium concentrations up to 350 µg/L but diverged as solute content increased. Error analysis revealed that the Langmuir isotherm best modeled the equipment-IBU relationship. A Langmuir adsorption model fit insinuates an equal quantity of attachment free-energy changes and a monolater coating of IBU on the surface. Similar results have been realized in applications utilizing stainless steel as the adsorbent [57–59]. Therefore, the equipment-IBU relationship can be modeled via Langmuir > Temkin > Freundlich. On the contrary, the Freundlich isotherm revealed the closest representation to the range of equipment-IBU-membrane system experimental data based on error analysis. A best-fit Freundlich adsorption isotherm suggests heterogeneous adsorption free-energy changes and a multilayer of IBU chemisorption. These findings align with existing literature denoting Freundlich-type adsorption on a membrane surface due to its laminose structure [35,60]. However, it should be noted that the Langmuir and Freundlich isotherms produced analogous r^2 values for R-IBU (0.995 for NF270 and 0.996 for TS40), and contained similar error statistics. This suggests that as S-IBU has a stronger attachment affinity, weak interactions between R-IBU and the surface may yield a thinner adsorptive layer. A study of R- and S-IBU adsorption kinetics onto metal and membrane surfaces may elucidate the dissimilar attachment mechanisms. For the purposes of this work, the equipment-S-IBU-membrane system can be modeled via Freundlich > Temkin > Langmuir, whereas the equipment-R-IBU-membrane system can be modeled via Freundlich = Langmuir > Temkin.

Table 3. R- and S-IBU Langmuir, Freundlich, and Temkin isotherm parameters derived from bench-scale, flat-sheet experiments (feed water pH of 4.0 units, temperature of 20 ± 1 °C).

		\multicolumn{6}{c}{Isotherm}					
		\multicolumn{6}{c}{Langmuir}					
		K_L (L/μg)	q_a (μg/cm^2)	r^2	RPD	ERRSQ	RMSE
R-IBU	Equipment	3.93×10^{-3}	0.031	0.993	3.73	3.42×10^{-3}	2.40×10^{-2}
	NF270	4.08×10^{-3}	0.033	0.995	4.04	4.98×10^{-6}	8.43×10^{-4}
	TS40	3.88×10^{-3}	0.040	0.996	3.76	6.67×10^{-6}	1.05×10^{-3}
S-IBU	Equipment	1.74×10^{-2}	0.052	0.979	3.57	2.46×10^{-5}	2.02×10^{-3}
	NF270	1.44×10^{-2}	0.057	0.982	6.45	6.62×10^{-5}	3.08×10^{-3}
	TS40	8.95×10^{-2}	0.076	0.974	6.00	8.71×10^{-5}	3.81×10^{-3}
		\multicolumn{6}{c}{Freundlich}					
		K_F (L/cm^2)	1/n (-)	r^2	RPD	ERRSQ	RMSE
R-IBU	Equipment	6.35×10^{-4}	0.566	0.945	9.36	3.38×10^{-3}	2.40×10^{-2}
	NF270	4.05×10^{-4}	0.665	0.995	3.87	4.53×10^{-6}	8.05×10^{-4}
	TS40	4.74×10^{-4}	0.677	0.996	2.78	7.99×10^{-7}	3.65×10^{-4}
S-IBU	Equipment	5.89×10^{-3}	0.350	0.946	7.97	7.67×10^{-5}	3.58×10^{-3}
	NF270	4.00×10^{-3}	0.433	0.988	3.92	1.80×10^{-5}	1.60×10^{-3}
	TS40	2.43×10^{-3}	0.570	0.994	3.02	1.08×10^{-5}	1.34×10^{-3}
		\multicolumn{6}{c}{Temkin}					
		K_T (L/μg)	b (J/mol)	r^2	RPD	ERRSQ	RMSE
R-IBU	Equipment	0.040	3.58×10^5	0.981	5.42	3.39×10^{-3}	2.30×10^{-2}
	NF270	0.041	3.25×10^5	0.980	14.0	7.47×10^{-6}	1.03×10^{-3}
	TS40	0.038	2.65×10^5	0.963	10.4	1.11×10^{-5}	1.36×10^{-3}
S-IBU	Equipment	0.185	2.23×10^5	0.974	3.56	1.11×10^{-5}	1.36×10^{-3}
	NF270	0.110	1.78×10^5	0.988	5.19	1.86×10^{-5}	1.63×10^{-3}
	TS40	0.061	1.19×10^5	0.975	9.11	4.23×10^{-6}	2.66×10^{-3}

3.4. Rejection of Ibuprofen Enantiomers

The effect of pH on IBU rejection via NF was investigated by altering the feed pH to 4.0, 5.0, or 6.0 units with an initial IBU concentration of 1.5 mg/L (R- and S-enantiomer concentrations of 750 μg/L). Figure 4 displays the rejection of R- and S-IBU from the NF270 and TS40 membrane at a feed pH of 4.0 units. The total NF270 and TS40 IBU rejection was 34.5% and 49%, respectively. However, the adsorption of IBU affected the rejection value based on the time of collection. Adsorption accounted for 14.3% to 23.4% and 23.6% to 31.3% of R-IBU and S-IBU rejection, respectively.

The NF270 and TS40 exhibited poor (<50%) rejection at a feed water pH of 4.0 units. However, removal efficacy increased with feed water pH, as illustrated in Figure 4, aligning with findings from others [37,38]. The feed water pH affects the speciation of IBU and the magnitude of negative charge on the NF membrane. It should be noted that feed water pH has an opposing effect on IBU adsorption. At feed water pH values higher than 4.4 units, the membrane surface is negatively charged and IBU is dissociated, primarily existing in the anionic form. Anionic IBU is believed to be rejected by electrostatic repulsion and steric hindrance. Conversely, at a feed water pH less than 4.4 units, IBU principally exists as the neutral form, and the membrane is less negatively charged. Neutral IBU readily adsorbs onto stainless steel and the membrane surface. As available adsorptive sites become saturated, the NF membrane can partially reject neutral IBU due to size exclusion. Therefore, the mechanism of IBU removal at acidic conditions is postulated as initially adsorption and subsequently steric hindrance.

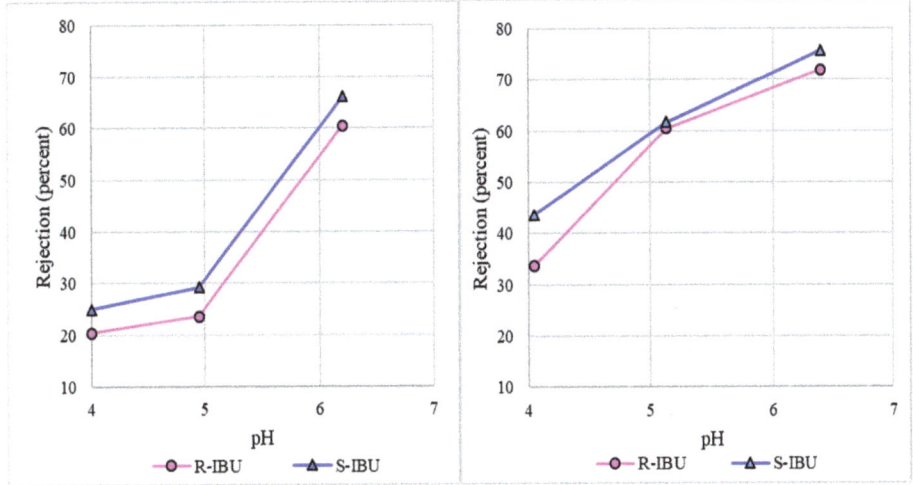

Figure 4. R- and S-IBU rejection as a function of feed water pH for NF270 membrane (**left**), TS40 membrane (**right**) (water flux 42.4 L/m^{2}h, temperature 20 ± 1 °C).

It should be noted that the rejection of S-IBU was consistently greater than the R-enantiomer for the feed pH conditions examined, due to the preferential attachment onto the metal flat-sheet equipment, shown in Figure 5. Although CEC adsorption impacts the overall rejection, it should not impact the mechanism of removal by the membrane. This indicates that the membrane may have a slight affinity for the rejection of S-IBU. A possible explanation for the increased rejection lies in DFT calculations, which revealed a dipole moment of 2.02 and 5.40 Debeye for R- and S-IBU, respectively. Existing literature suggests that a molecule's polarity influences the orientation of the solute relative to the membrane [15,16]. A molecule with a lower dipole moment is less polar, and hence contains an orientation more perpendicular to the membrane surface, increasing the probability of the solute to travel through the material without being rejected. Others have also found a direct relationship between CEC dipole moment and rejection [14–16,61].

The influence of sample time is important when recording removal of hydrophobic CECs like IBU from a NF process. In this work, adsorption was recorded over 24 h, and rejection was collected at 24 h. Others have agreed that 24 h of operation was adequate for equilibration of hydrophobic compounds [12,13,32]. However, additional time may be required to confirm complete adsorption of the compound. If rejection is collected shortly after start-up, the value may not account for the adsorption of the CEC onto the membrane or equipment. Therefore, system equilibration is important in obtaining accurate removal capacities.

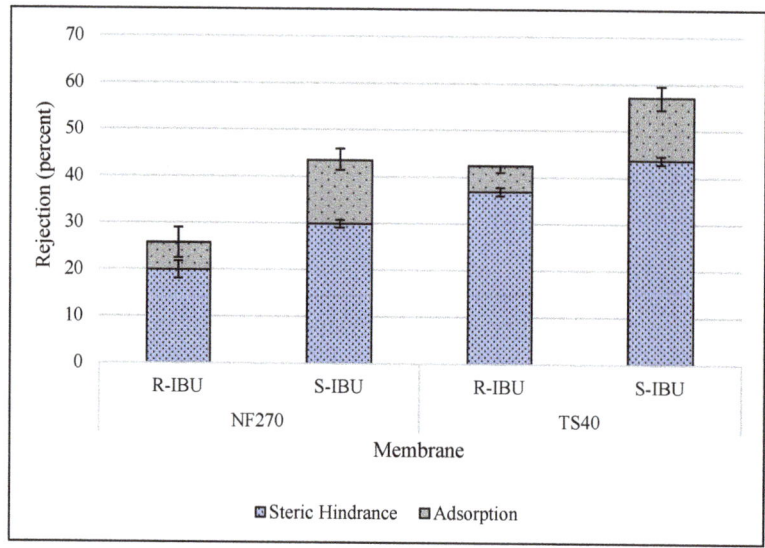

Figure 5. Rejection of R- and S-IBU at a feed water pH of 4.0 units (initial IBU concentration 400 µg/L, water flux 42.4 L/m^{2}h, temperature 20 ± 1 °C). Error bars represent minimum and maximum values from triplicate analysis.

4. Conclusions

The results of this study revealed the behavior of IBU enantiomers in aqueous solutions treated by NF membranes. Feed water characteristics (such as pH) have a substantial influence on the rejection and adsorption mechanisms of IBU. At low feed pH values, S-IBU adsorbed up to five times more than its R-counterpart onto stainless steel and showed preferential rejection in a NF process. DFT calculations could provide insight into the differing behavior of the enantiomers in terms of molecular volume and dipole moment. In bench-scale membrane removal studies, it is important to conduct initial mass balance experiments to determine possible losses of compound, which may impact overall removal. Furthermore, equilibration time proves vital in determination of the true removal capabilities of membrane processes.

Supplementary Materials: The following are available online at http://www.mdpi.com/2077-0375/10/12/383/s1, Figure S1: Enantiomers of IBU, Figure S2: IBU speciation based on pKa value (4.4), Figure S3: Flat-sheet bench-scale unit schematic operated in (A) recycle mode, (B), permeate collection mode, Figure S4: HPLC IBU enantiomer chromatogram, Figure S5: Dipole moment of (A) R-IBU and (B) S-IBU produced by DFT computations using GAMESS software. Figures S6–S13: Linearized adsorption isotherms for R- and S-IBU. Figure S14: Adsorption isotherm curves of R- and S-IBU, Figure S15: Adsorption isotherm curves of R- and S-IBU and (A) flat-sheet equipment, (B) flat-sheet equipment and NF270 coupon, (C) flat-sheet equipment and TS40 coupon (feed water pH of 4.0 units, temperature 20 ± 1°C). Error bars represent minimum and maximum values from triplicate analysis, Table S1: Chemical properties of IBU, Table S2: Operational parameters of NF270 and TS40 NF membrane coupons.

Author Contributions: C.J.H. conducted the investigation, methodology, software, formal analysis, data curation, validation, and writing—original draft preparation and visualization; S.J.D. provided and obtained the resources and funding acquisition for this study and conducted formal analysis, supported data curation, provided technical supervision, and project administration for the research. All authors have read and agreed to the published version of the manuscript.

Funding: This work was supported by Jupiter Water Utilities (17403 Central Blvd. Jupiter, FL 33458, USA) and Kimley-Horn & Associates, Inc. (1920 Wekiva Way Suite 200, West Palm Beach, FL 33411, USA) through UCF project 16208114.

Acknowledgments: The authors are in appreciation of funding provided by Jupiter Water Utilities (Jupiter, FL, USA) and Kimley-Horn & Associates, Inc. The authors would like to thank the support offered from Sterlitech, Inc. (Kent, WA, USA), especially Kensen Hirohata. Additional thanks are offered to Laura Gallindo and Steven Coker from Dupont Water Solutions (Edina, MN, USA) for providing FilmTec NF270 flat-sheet membrane coupons. Finally, the authors would like to thank the UCF Water Quality Engineering team for their assistance in this work. Any opinions, findings, and conclusions expressed in this material are those of the authors and do not necessarily reflect the view of UCF (Orlando, FL, USA) or its Research Foundation.

Conflicts of Interest: The authors declare no conflict of interest.

References

1. Bellona, C.; Drewes, J.E.; Xu, P.; Amy, G. Factors affecting the rejection of organic solutes during NF/RO treatment—A literature review. *Water Res.* **2004**, *38*, 2795–2809. [CrossRef]
2. Yangali-Quintanilla, V.; Maeng, S.K.; Fujioka, T.; Kennedy, M.; Amy, G. Proposing nanofiltration as acceptable barrier for organic contaminants in water reuse. *J. Membr. Sci.* **2010**, *362*, 334–345. [CrossRef]
3. Yoon, Y.; Westerhoff, P.; Snyder, S.A.; Wert, E.C. Nanofiltration and ultrafiltration of endocrine disrupting compounds, pharmaceuticals and personal care products. *J. Membr. Sci.* **2006**, *270*, 88–100. [CrossRef]
4. Radjenović, J.; Petrović, M.; Ventura, F.; Barceló, D. Rejection of pharmaceuticals in nanofiltration and reverse osmosis membrane drinking water treatment. *Water Res.* **2008**, *42*, 3601–3610. [CrossRef]
5. Kim, S.; Chu, K.H.; Al-Hamadani, Y.A.J.; Park, C.M.; Jang, M.; Kim, D.; Yu, M.; Heo, J.; Yoon, Y. Removal of contaminants of emerging concern by membranes in water and wastewater: A review. *Chem. Eng. J.* **2018**, *335*, 896–914. [CrossRef]
6. Yangali-Quintanilla, V.; Sadmani, A.; McConville, M.; Kennedy, M.; Amy, G. A QSAR model for predicting rejection of emerging contaminants (pharmaceuticals, endocrine disruptors) by nanofiltration membranes. *Water Res.* **2010**, *44*, 373–384. [CrossRef]
7. Yangali-Quintanilla, V.; Sadmani, A.; Kennedy, M.; Amy, G. A QSAR (quantitative structure-activity relationship) approach for modelling and prediction of rejection of emerging contaminants by NF membranes. *Des. Water Treat.* **2010**, *13*, 149–155. [CrossRef]
8. Van der Bruggen, B.; Verliefde, A.; Braeken, L.; Cornelissen, E.R.; Moons, K.; Verberk, J.Q.; van Dijk, H.J.; Amy, G. Assessment of a semi-quantitative method for estimation of the rejection of organic compounds in aqueous solution in nanofiltration. *J. Chem. Technol. Biotechnol.* **2006**, *81*, 1166–1176. [CrossRef]
9. Ozaki, H.; Li, H. Rejection of organic compounds by ultra-low pressure reverse osmosis membrane. *Water Res.* **2002**, *36*, 123–130. [CrossRef]
10. Van der Bruggen, B.; Vandecasteele, C. Removal of pollutants from surface water and groundwater by nanofiltration: Overview of possible applications in the drinking water industry. *Environ. Pollut.* **2003**, *122*, 435–445. [CrossRef]
11. Schutte, C.F. The rejection of specific organic compounds by reverse osmosis membranes. *Desalination* **2003**, *158*, 285–294. [CrossRef]
12. Comerton, A.M.; Andrews, R.C.; Bagley, D.M.; Hao, C. The rejection of endocrine disrupting and pharmaceutically active compounds by NF and RO membranes as a function of compound and water matrix properties. *J. Membr. Sci.* **2008**, *313*, 323–335. [CrossRef]
13. Comerton, A.M.; Andrews, R.C.; Bagley, D.M.; Yang, P. Membrane adsorption of endocrine disrupting compounds and pharmaceutically active compounds. *J. Membr. Sci.* **2007**, *303*, 267–277. [CrossRef]
14. Jeffery-Black, S.; Duranceau, S.J. The influence of solute polarizability and molecular volume on the rejection of trace organics in loose nanofiltration membrane processes. *Des. Water Treat.* **2016**, *57*, 29059–29069. [CrossRef]
15. Van der Bruggen, B.; Schaep, J.; Wilms, D.; Vandecasteele, C. Influence of molecular size, polarity and charge on the retention of organic molecules by nanofiltration. *J. Membr. Sci.* **1999**, *156*, 29–41. [CrossRef]
16. Darvishmanesh, S.; Vanneste, J.; Tocci, E.; Jansen, J.C.; Tasselli, F.; Degrevè, J.; Drioli, E.; Van der Bruggen, B. Physicochemical Characterization of Solute Retention in Solvent Resistant Nanofiltration: The Effect of Solute Size, Polarity, Dipole Moment, and Solubility Parameter. *J. Phys. Chem. B* **2011**, *115*, 14507–14517. [CrossRef]
17. Kiso, Y.; Mizuno, A.; Jung, Y.; Kumano, A.; Ariji, A. Rejection properties of pesticides with a hollow fiber NF membrane (HNF-1). *Desalination* **2002**, *143*, 147–157. [CrossRef]
18. Sadmani, A.H.M.; Andrews, R.C.; Bagley, D.M. Nanofiltration of pharmaceutically active and endocrine disrupting compounds as a function of compound interactions with DOM fractions and cations in natural water. *Sep. Purif. Technol.* **2014**, *122*, 462–471. [CrossRef]

19. Breitner, L.N.; Howe, K.J.; Minakata, D. Effect of Functional Chemistry on the Rejection of Low-Molecular Weight Neutral Organics through Reverse Osmosis Membranes for Potable Reuse. *Environ. Sci. Technol.* **2019**, *53*, 11401–11409. [CrossRef]
20. Davies, N.M. Clinical Pharmacokinetics of Ibuprofen: The First 30 Years. *Clin. Pharm.* **1998**, *34*, 101–154. [CrossRef]
21. Evans, A.M. Comparative Pharmacology of S(+)-Ibuprofen and (RS)-Ibuprofen. *Clin. Rheumatol.* **2001**, *20*, 9–14. [CrossRef] [PubMed]
22. Bonato, P.S.; Maria Perpetua, F.M.; de Carvalho, R. Enantioselective determination of ibuprofen in plasma by high-performance liquid chromatography–electrospray mass spectrometry. *J. Chromatogr. B.* **2003**, *796*, 413–420. [CrossRef] [PubMed]
23. Cho, H.; Huang, H.; Schwab, K. Effects of Solution Chemistry on the Adsorption of Ibuprofen and Triclosan onto Carbon Nanotubes. *Langmuir* **2011**, *27*, 12960–12967. [CrossRef] [PubMed]
24. Sui, Q.; Cao, X.; Lu, S.; Zhao, W.; Qiu, Z.; Yu, G. Occurrence, sources and fate of pharmaceuticals and personal care products in the groundwater: A review. *Emerg. Contam.* **2015**, *1*, 14–24. [CrossRef]
25. McEachran, A.D.; Shea, D.; Bodnar, W.; Nichols, E.G. Pharmaceutical occurrence in groundwater and surface waters in forests land-applied with municipal wastewater. *Environ. Toxicol. Chem.* **2016**, *35*, 898–905. [CrossRef]
26. Buser, H.; Poiger, T.; Müller, M.D. Occurrence and Environmental Behavior of the Chiral Pharmaceutical Drug Ibuprofen in Surface Waters and in Wastewater. *Environ. Sci. Technol.* **1999**, *33*, 2529–2535. [CrossRef]
27. Hashim, N.H.; Khan, S.J. Enantioselective analysis of ibuprofen, ketoprofen, and naproxen in wastewater and environmental water samples. *J. Chromatogr. A.* **2011**, *1218*, 4746–4754. [CrossRef]
28. Fajobi, M.; Fayomi, O.S.I.; Akande, G.; Odunlami, O. Inhibitive Performance of Ibuprofen Drug on Mild Steel in 0.5 M of H2SO4 Acid. *J. Bio. Tribo. Corros.* **2019**, *5*, 5. [CrossRef]
29. Tasić, Z.Z.; Mihajlović, M.B.P.; Simonović, A.T.; Radovanović, M.B.; Antonijević, M.M. Ibuprofen as a corrosion inhibitor for copper in synthetic acid rain solution. *Sci. Rep.* **2019**, *9*, 14710. [CrossRef]
30. Bueno-Perez, R.; Martin-Calvo, A.; Gómez-Álvarez, P.; Gutiérrez-Sevillano, J.J.; Merkling, P.J.; Vlugt, T.J.H.; van Erp, T.S.; Dubbeldam, D.; Calero, S. Enantioselective adsorption of ibuprofen and lysine in metal-organic frameworks. *Chem. Comm.* **2014**, *50*, 10849–10852. [CrossRef]
31. Davalos Monteiro, R.; van de Wetering, J.; Krawczyk, B.; Engelberg, D.L. Corrosion Behaviour of Type 316L Stainless Steel in Hot Caustic Aqueous Environments. *Met. Mater. Int.* **2019**, *26*, 630–640. [CrossRef]
32. Kimura, K.; Amy, G.; Drewes, J.; Watanabe, Y. Adsorption of hydrophobic compounds onto NF/RO membranes: An artifact leading to overestimation of rejection. *J. Membrane. Sci.* **2003**, *221*, 89–101. [CrossRef]
33. Tanaka, Y.; Saito, H.; Tsutsumi, Y.; Doi, H.; Imai, H.; Hanawa, T. Active Hydroxyl Groups on Surface Oxide Film of Titanium, 316L Stainless Steel, and Cobalt-Chromium-Molybdenum Alloy and Its Effect on the Immobilization of Poly(Ethylene Glycol). *Mater. Trans.* **2008**, *49*, 805–811. [CrossRef]
34. Seo, P.W.; Bhadra, B.N.; Ahmed, I.; Khan, N.A.; Jhung, S.H. Adsorptive Removal of Pharmaceuticals and Personal Care Products from Water with Functionalized Metal-organic Frameworks: Remarkable Adsorbents with Hydrogen-bonding Abilities. *Sci. Rep.* **2016**, *6*, 34462. [CrossRef] [PubMed]
35. Lin, Y.; Lee, C. Elucidating the Rejection Mechanisms of PPCPs by Nanofiltration and Reverse Osmosis Membranes. *Ind. Eng. Chem. Res.* **2014**, *53*, 6798–6806. [CrossRef]
36. Zhao, Y.; Kong, F.; Wang, Z.; Yang, H.; Wang, X.; Xie, Y.F.; Waite, T.D. Role of membrane and compound properties in affecting the rejection of pharmaceuticals by different RO/NF membranes. *Front. Environ. Sci. Eng.* **2017**, *11*, 20. [CrossRef]
37. Nghiem, L.D.; Schäfer, A.I.; Elimelech, M. Pharmaceutical Retention Mechanisms by Nanofiltration Membranes. *Environ. Sci. Technol.* **2005**, *39*, 7698–7705. [CrossRef]
38. Nghiem, L.; Hawkes, S. Effects of membrane fouling on the nanofiltration of pharmaceutically active compounds (PhACs): Mechanisms and role of membrane pore size. *Sep. Purif. Technol.* **2007**, *57*, 176–184. [CrossRef]
39. Simon, A.; Nghiem, L.D.; Le-Clech, P.; Khan, S.J.; Drewes, J.E. Effects of membrane degradation on the removal of pharmaceutically active compounds (PhACs) by NF/RO filtration processes. *J. Membr. Sci.* **2009**, *340*, 16–25. [CrossRef]
40. Ge, S.; Feng, L.; Zhang, L.; Xu, Q.; Yang, Y.; Wang, Z.; Kim, K. Rejection rate and mechanisms of drugs in drinking water by nanofiltration technology. *Environ. Eng. Res.* **2017**, *22*, 329–338. [CrossRef]
41. Okamoto, Y.; Kishi, Y.; Ishigami, T.; Suga, K.; Umakoshi, H. Chiral Selective Adsorption of Ibuprofen on a Liposome Membrane. *J. Phys. Chem. B.* **2016**, *120*, 2790–2795. [CrossRef]

42. Wu, X.; Kang, F.; Duan, W.; Li, L. Density functional theory calculations: A powerful tool to simulate and design high-performance energy storage and conversion materials. *Prog. Nat. Sci. Mater.* **2019**, *29*, 247–255. [CrossRef]
43. Tsuneda, T. *Density Functional Theory in Quantum Chemistry*; Springer: Tokyo, Japan, 2014.
44. Seshadri, H.; Venkatachalam, S.; Sangaranarayanan, M.; Malar, E.J.P. Adsorption of Enantiomers on Metal Surfaces: Application to D- and L-Alanine on Cu, Ni and Zn Electrodes. *J. Electrochem. Soc.* **2013**, *160*, G102–G110.
45. Lee, C.; Yang, W.; Parr, R.G. Development of the Colle-Salvetti correlation-energy formula into a functional of the electron density. *Phys. Rev. B* **1988**, *37*, 785–789. [CrossRef] [PubMed]
46. Perri, M.J.; Weber, S.H. Web-Based Job Submission Interface for the GAMESS Computational Chemistry Program. *J. Chem. Educ.* **2014**, *91*, 2206–2208. [CrossRef]
47. Raschi, A.; Romano, E.; Castillo, M.; Leyton, P.; Brandán, S. Structural and Vibrational Properties of the Two Enantiomers of Etodolac. A Complete Assignment of the Vibrational Spectra. In *Descriptors, Structural and Spectroscopic Properties of Heterocyclic Derivatives of Importance for Health and the Environment*; Brandán, S.A., Ed.; Nova Science Publishing, Inc.: Hauppauge, NY, USA, 2015; pp. 108–131.
48. McConalthy, J.; Owens, M.J. Sterochemistry in drug action. *J. Clin. Psychiatry Prim. Care Companion* **2003**, *5*, 70–73. [CrossRef] [PubMed]
49. Nguyen, L.A.; He, H.; Pham-Huy, C. Chiral drugs: An overview. *Int. J. Biomed. Sci.* **2006**, *2*, 85–100. [PubMed]
50. Lee, S.T.; Molyneux, R.J.; Panter, K.E.; Chang, C.T.; Gardner, D.R.; Pfister, J.A.; Garrossian, M.J. Ammodendrine and N-Methylammodendrine Enantiomers: Isolation, Optical Rotation, and Toxicity. *Nat. Prod.* **2005**, *68*, 681–685. [CrossRef] [PubMed]
51. Howe, K.J.; Hand, D.W.; Crittenden, J.C.; Trussell, R.R.; Tchobanoglous, G. *Principles of Water Treatment*; John Wiley & Sons, Inc.: Hoboken, NJ, USA, 2012.
52. Langmuir, I. The adsorption of gases on plane surfaces of glass, mica and platinum. *J. Am. Chem.* **1918**, *40*, 1361–1403. [CrossRef]
53. Freundlich, H.M. Over the adsorption in solution. *J. Phys. Chem.* **1906**, *57*, 385–470.
54. Temkin, M.J.; Pyzhev, V. Kinetics of the synthesis of ammonia on promoted iron catalysts. *Acta Physicochim.* **1940**, *12*, 217–222.
55. Alsehli, B.R.M. A simple approach for determining the maximum sorption capacity of chlorpropham from aqueous solution onto granular activated charcoal. *Processes* **2020**, *8*, 398. [CrossRef]
56. Hadi, M.; Samarghandi, R.; McKay, G. Equilibrium two-parameter isotherms of acid dyes sorption by activated carbons: Study of residual errors. *Chem. Eng. J.* **2010**, *160*, 408–416. [CrossRef]
57. Duduna, W.; Akeme, O.N.; Zinipere, T.M. Comparison of Various Adsorption Isotherm Models for Allium Cepa as Corrosion Inhibitor on Ausenitic Stainless Steel in Sea Water. *Int. J. Sci. Res.* **2019**, *8*, 961–964.
58. Omanovic, S.; Roscoe, S.G. Electrochemical Studies of the Adsorption Behavior of Bovine Serum Albumin on Stainless Steel. *Langmuir* **1999**, *15*, 8315–8321. [CrossRef]
59. Imamura, K.; Okamoto, T.M.; Sakiyama, T.; Nakanishi, K. Adsorption Behavior of Amino Acids on a Stainless Steel Surface. *J. Colloid Interf. Sci.* **2000**, *229*, 237–246. [CrossRef]
60. Liu, T.; Chang, E.; Chiang, P. Adsorption of CECs in the nanofiltration process. *Desal. Water Treat.* **2013**, *54*, 2658–2668. [CrossRef]
61. Shirley, J.; Mandale, S.; Kochkodan, V. Influence of solute concentration and dipole moment on the retention of uncharged molecules with nanofiltration. *Desalination* **2014**, *344*, 116–122. [CrossRef]

Publisher's Note: MDPI stays neutral with regard to jurisdictional claims in published maps and institutional affiliations.

© 2020 by the authors. Licensee MDPI, Basel, Switzerland. This article is an open access article distributed under the terms and conditions of the Creative Commons Attribution (CC BY) license (http://creativecommons.org/licenses/by/4.0/).

Review

Functions of Ionic Liquids in Preparing Membranes for Liquid Separations: A Review

Dayuan Zheng [1], Dan Hua [1,*], Yiping Hong [1], Abdul-Rauf Ibrahim [2], Ayan Yao [1], Junyang Pan [1] and Guowu Zhan [1,*]

1. Integrated Nanocatalysts Institute (INCI), College of Chemical Engineering, Huaqiao University, 668 Jimei Avenue, Xiamen 361021, Fujian, China; zhengdayuan@stu.hqu.edu.cn (D.Z.); 1626211007@stu.hqu.edu.cn (Y.H.); ayan8360@stu.hqu.edu.cn (A.Y.); 20014087024@stu.hqu.edu.cn (J.P.)
2. Department of Mechanical Engineering, Faculty of Engineering and Built Environment, Tamale Technical University, Education Ridge Avenue, Sagnarigu District, Tamale, Ghana; ghrauf@gmail.com
* Correspondence: huadan@hqu.edu.cn (D.H.); gwzhan@hqu.edu.cn (G.Z.)

Received: 12 November 2020; Accepted: 29 November 2020; Published: 5 December 2020

Abstract: Membranes are widely used for liquid separations such as removing solute components from solvents or liquid/liquid separations. Due to negligible vapor pressure, adjustable physical properties, and thermal stability, the application of ionic liquids (ILs) has been extended to fabricating a myriad of membranes for liquid separations. A comprehensive overview of the recent developments in ILs in fabricating membranes for liquid separations is highlighted in this review article. Four major functions of ILs are discussed in detail, including their usage as (i) raw membrane materials, (ii) physical additives, (iii) chemical modifiers, and (iv) solvents. Meanwhile, the applications of IL assisted membranes are discussed, highlighting the issues, challenges, and future perspectives of these IL assisted membranes in liquid separations.

Keywords: ionic liquids (ILs); membranes; liquid separation; modifier; solvent

---

1. Introduction

Membrane-based liquid separation technologies mainly include microfiltration (MF), ultrafiltration (UF), nanofiltration (NF), organic solvent nanofiltration (OSN), reverse osmosis (RO), forward osmosis (FO), and pervaporation (PV). These separation technologies play important roles in the industry and our daily life, because of their functions in the concentration, fractionation, and purification of liquid mixtures. Moreover, the separation technologies have experienced rapid growth in the past decades since they are believed to have fewer energy consumptions, smaller carbon footprints, and convenient operations compared to traditional separation technologies such as distillation, condensation, and crystallization [1].

However, most of the membranes are fabricated through the phase inversion method with the use of toxic organic solvents or strong acids, which often generate toxic volatile organic compounds and produce a large amount of wastewater containing toxic solvents [2]. To circumvent these environmental problems, one direct strategy is replacing the common toxic solvents with greener solvents that have lower volatility or less toxicity, such as TamiSolve® NxG [3], PolarClean [4], Cyrene™ [5], dimethyl isosorbide [6], dimethyl carbonate [7], etc. Ionic liquids (ILs) are also viewed as an alternative green solvent. Specifically, room-temperature ILs are molten organic salts that are in a liquid state at or near room temperatures. They have received tremendous attention due to their excellent properties such as strong polarity, negligible vapor pressure (above the liquid surface), low volatility, thermal and chemical stability, designable structure, and a good ability to dissolve many inorganic, organic, and polymeric materials. Consequently, they have been used in the preparation of various membranes [8–19]. The usage of ILs in fabricating membranes could be broadly categorized into four types as illustrated in Figure 1.

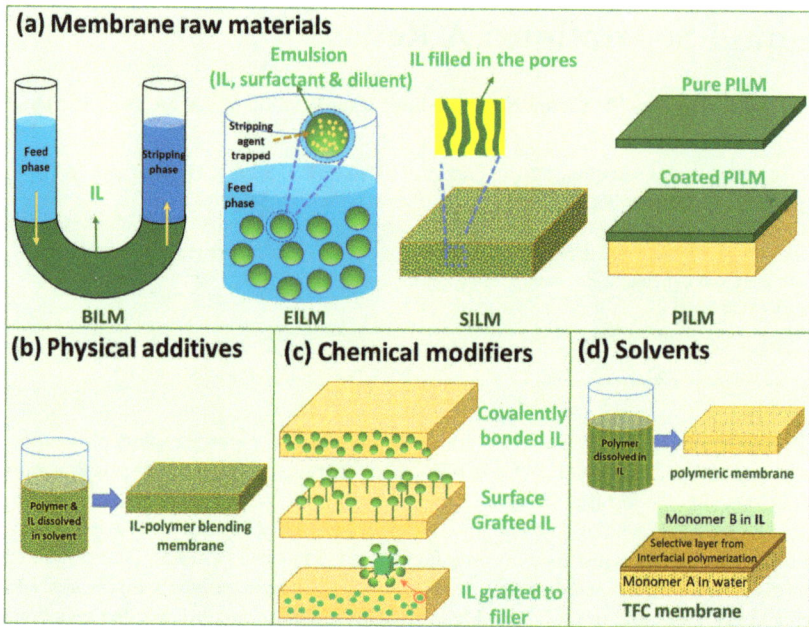

Figure 1. The four major functions of ionic liquids (ILs) in preparing membranes: (**a**) membranes raw materials (BILM: bulk IL membrane, EILM: emulsion IL membrane, SILM: supported IL membrane, PILM: poly (ionic liquid) membrane), (**b**) physical additives, (**c**) chemical modifiers, (**d**) solvents (TFC: thin-film composite).

ILs can be used directly as raw materials to fabricate membranes. As shown in Figure 1a, some of such membranes include bulk IL membranes (BILMs), emulsion IL membranes (EILMs), supported IL membranes (SILMs), and poly(ionic liquid) membranes (PILMs): (i) BILMs, the simplest non-supported IL membranes, usually consist of the aqueous feed phase and the stripping phase directly contacted with an IL membrane in a U-tube (refer to a setup in Figure 2a); (ii) EILMs are generally emulsions prepared by emulsifying an organic phase (i.e., the carrier containing IL, surfactant, and diluents) with an internal aqueous stripping agent. In the case of EILMs, the emulsion droplets are often stabilized by surfactant, enabling them to trap internal stripping agents inside them and form water-in-oil emulsion [20]; (iii) SILMs are a type of liquid membrane, wherein the IL is held by capillary forces in the pores of a polymeric/inorganic membrane support via direct immersion, vacuum infiltration, etc. The ILs play an important role in the operating performance of SILM due to the relatively high viscosity; (iv) PILMs are directly made of poly(ionic liquid)s (PILs), which are polyelectrolytes that feature IL species in each monomer repeating unit and connected through a polymeric backbone to form a macromolecular architecture. The synthesis methods for PILs could be referred to in a report developed by Yan et al. [21]. The PILMs could normally be made from PILs solutions via a phase inversion process and coating on a membrane support. As compared to the liquid membranes, PILMs made from high molecular weight PILs are much more stable, which form adjustable structures and morphologies [21].

The first three types of membranes (BILMs, EILMs, and SILMs) are types of solvent extraction based on liquid membranes. The IL membranes are much greener and more stable as compared with traditional liquid membranes made from organic solvents, meaning that the usage of ILs overcomes the evaporation loss of organic solvents. Among these types of membranes, BILMs have the lowest contact surface area for extraction while EILMs have the largest contact surface area. Hence, the permeation rate of BILMs is very low, making them technologically not very attractive. However, BILMs are the simplest membranes and are still widely used to study the transport properties of novel ILs as carriers [22]. EILMs have the advantages of a high surface area, non-dependence on equilibrium consideration, and relatively low cost, but their stability is a critical issue because the emulsions formed should be stable enough to avoid leakage during the separation process. On the other hand, the emulsions should not be too stable so that they can be destroyed and recycled after the separation process. Although the SILMs have lower flux as compared to EILMs due to having less contact surface area, high selectivity could be achieved based on small amounts of ILs. Thus, SILMs have gained much popularity in recent years. Similarly, ILs and polymerized ILs (PILs) have been used as physical additives into membranes to form IL/PIL-polymer blending membranes (Figure 1b). The IL-polymer blending membranes are also called polymer inclusion membranes (PIMs). As compared with pristine IL membranes, they possess improved stability due to the fact that ILs are trapped in the dense polymeric/inorganic membrane matrix. As compared with pristine membranes, they may have enhanced separation performance due to the special physical properties of ILs such as hydrophilicity, charge, special functional groups, etc. Moreover, ILs/PILs have been used to chemically modify the membranes or membrane components owing to the abundant active functional groups. As illustrated in Figure 1c, on one hand, ILs/PILs could be added into polymer bulk solutions and form a membrane layer possessing excellent stability due to covalent bonds; on the other hand, ILs could also be chemically grafted to the membrane surface or serve as membrane additives to improve the separation performance. Furthermore, ILs have been used as green solvents for dissolving polymers or as reaction media during membrane fabrication (Figure 1d). Due to the electrostatic nature of ILs, they are able to interact strongly with the polymers via pronounced hydrogen bonding, Coulombic forces, and van der Waals interactions [23–25]. Using IL as the solvent is greener than using organic solvents because ILs are non-volatile and can be recovered in some cases. Moreover, they can be used to dissolve some rigid polymers (e.g., cellulose, polybenzimidazole (PBI), and polyamides) which are not easily dissolved by organic solvents [25–27]. Besides, using ILs as the reaction media benefits not only from their non-volatility, but also from their different properties such as interfacial tension, viscosity, and the solubility of organic compounds, making the membrane synthesis less hazardous and more tunable to obtain membranes with better separation performance [28].

Therefore, ILs play an important role in membrane fabrication from the versatile aspects shown above. Currently, there have been many studies on developing these types of IL assisted membranes for gas separations (such as CO_2 and volatile organic compounds removal) because of their high solubilities for different gaseous species [10,11,15,17]. Besides this, the use of ILs has also been extended to fabricating membranes for other purposes such as electrochemical applications [29–33], osmotic power generation [16], liquid separations including desalination, the removal of organics from water or organic solvents [34,35], removal of heavy metal [36], organic solvent/water separations, and so forth. Although the ILs have been used extensively in developing membranes for liquid separations, few reviews have summarized the functions of ILs in the rational design of these membranes. Besides, the existing reviews are limited to IL membranes, in which ILs are used as fabricating materials [12,18,19].

Considering that liquid separations are in fact, more prominent than gas separations in industrial membrane separation processes [1] and that ILs have more versatile applications in membrane fabrications, it is thus important to systematically summarize the functions of ILs in developing membranes for liquid separations. In this regard, this review provides a comprehensive overview of four major functions of ILs in developing liquid separations membranes. Moreover, problems and challenges in IL related membrane processes for liquid separations are identified and discussed.

2. Functions of ILs in Developing Liquid Separation Membranes

2.1. ILs as Raw Membrane Materials

ILs have very low vapor pressure but they have good solubility for organic and inorganic compounds. Therefore, they can be used as materials directly to fabricate membranes such as bulk IL membrane, supported IL membrane, emulsion IL membrane, and poly (IL) membrane. Recent progress in the development of each of these membranes will be summarized in the following sections.

2.1.1. Bulk IL Membrane (BILM)

Unlike conventional bulk liquid membranes, BILMs use ILs as the hydrophobic liquid membrane phase instead of organic solvents, which has attracted the interest of researchers due to the 'green properties' of ILs such as low vapor pressure, low volatility, and good stability at high temperatures, making the BILM more stable and less hazardous because of the reduced evaporation of the membrane phase.

Several studies on the use of BILMs to remove various organic compounds in the liquid solutions have been reported, including phenols [37,38], organic acids [39–41], and others [39,40,42–44]. Lakshmi et al. utilized three different highly hydrophobic ILs to study the removal efficiency for chlorophenol [38]. Interestingly, high chlorophenol extraction and stripping efficiencies of 98.10% and 78.5%, respectively, were achieved by using 1-butyl-3-methylimidazolium tetrafluoroborate ([BMIM][BF_4]) with minimum membrane loss. Similarly, Mohammed and Hameed synthesized several hydrophobic ILs for extracting 4-nitrophenol from an aqueous solution [37]. They found that the distribution coefficients for the 4-nitrophenol in the ILs were higher than in conventional organic solvents. Furthermore, 1-butyl-3-methylimidazolium bis(trifluoromethylsulfonyl)imide ([BMIM][Tf_2N]) exhibited the greatest extraction and stripping efficiencies. Baylan and Çehreli used four hydrophobic imidazolium-based ILs as the membrane, tributyl phosphate (TBP) as a carrier for the membrane, and NaOH solutions as the stripping phase to remove levulinic acid [41] and acetic acid [40] from the aqueous solutions. Their results indicated that all the investigated ILs as a membrane phase had good transport selectivity. Moreover, TBP and NaOH concentrations had a major influence on both the extraction and stripping efficiencies. In addition, Branco et al. performed a systematical selective transport study using a 7-component mixture of representative organic compounds, and 10 different ILs based on five cation structures. Remarkably, they observed that the use of ILs with more polar cations (containing ether or hydroxyl functional groups) generally increased their affinity for all organic compounds but reduced the selective transport, especially for secondary and tertiary amines. Chakraborty and Bart successfully used 1-octyl-3-methylimidazolium chloride as a membrane solvent and Ag^+ as the carrier to remove toluene from n-heptane [42]. They revealed that the Ag^+ concentration, stirring speed, initial toluene concentration in the feed, and temperature greatly influenced the permeation rate and separation factor.

There have also been a few studies on the use of BILMs to remove metal ions from the aqueous solutions. For instance, Kogelnig et al. conducted a successful quantitative transport of Fe(III) ion from a hydrochloride (6 M) feed phase containing Ni(II) to a hydrochloride (0.5 M) receiving phase by using a commercialized IL trihexyl(tetradecyl)phosphonium chloride (Cyphos® IL101) as the membrane phase [45]. Both of the two metal ions have an ion association with the chloride anion to form a complex. The separation mechanism was based on the difference in the complex behavior depending on the concentration of HCl.

As can be seen from above, the studies on the use of BILM are very limited, especially the applications on metal ion removal. For instance, most researchers have screened several potential hydrophobic ILs to develop a suitable BILM for a certain organic solute. In addition, the selections of a suitable carrier and stripping agent have been found to be critical. Also, other operation factors (such as the feed phase, feed pH, carrier concentration, stripping agent concentration, contact time, stirring speed, temperature, etc.) greatly influence the permeation rate and separation factor. By tuning those parameters, the separation performance could be further enhanced, but BILMs are technologically unattractive due to their low contact surface area and slow mass transfer rate [22]. Despite this, BILMs are still meaningful because they are normally employed to study the transport properties of the novel carriers [22,43], which could give guidance for further developing other types of IL membranes.

2.1.2. Emulsion IL Membrane (EILM)

EILMs have a much higher surface area per unit of volume and lower thickness as compared to BILM because the membrane phase is made of numerous small emulsion droplets containing ILs, making the separation process and accumulation inside the emulsion vehicle fast. Similarly, the usage of ILs in the membrane emulsion makes the liquid membrane system more stable, which has been proved by the following studies.

Balasubramanian and Venkatesan built an EILM system by using a mixture of 1-butyl 3-methylimidazolium hexafluorophosphate and TBP as a mixed carrier, Span 80 as a surfactant, kerosene as a diluent agent, and NaOH as the internal stripping agent [46]. The EILM system was then applied for the removal of phenolic compounds; the scheme is illustrated in Figure 2b. They optimized the system parameters for achieving maximum removal of phenol with a higher treat ratio. The various parameters include the concentrations of IL, TBP, stripping reagent, surfactant, emulsification time, phase volume ratio, treat ratio, stirring speed, and external phase pH. Even though the IL was not purely the carrier in the system, the addition of the IL in the membrane phase increased the stability of the emulsion over 5 folds than that of the emulsion without the IL. Moreover, Kulkarni's group established an EILM system by using di-2-ethylhexyl phosphoric acid and 1-octyl-3-methylimidazolium hexafluorophosphate ([OMIM][PF_6]) as a carrier, Span 80 as a surfactant, hexane as a diluent agent, and sulphuric acid as an internal stripping agent [47], which was then used to remove Pb (II) ions from aqueous solution. Similarly, the various operating parameters were investigated and optimized. They found that the stability and the enrichment factor of the EILM were 2-3 folds greater than those for the system without the IL.

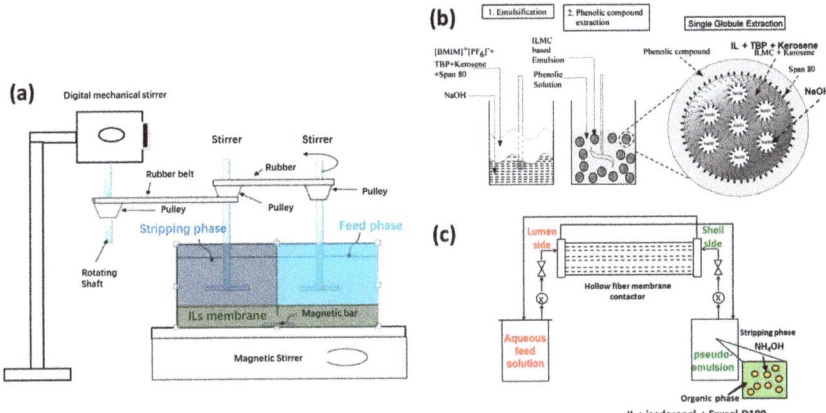

Figure 2. Schematic for (**a**) an extraction unit of 4-nitrophenol compounds using a BILM, adapted and modified from [37]; (**b**) the removal of phenolic compounds using an IL mixed carrier-based EILM, adapted from [46]; (**c**) the pseudo-emulsion based hollow fiber membrane with strip dispersion system, adapted from [48].

In addition, Alguacil et al. reported a smart IL membrane technology based on microporous polypropylene hollow fiber membrane contactor for the removal of Cd (II) from acidic chloride solutions. In this case, the Cd (II) feed solution and the pseudo-emulsion of the organic solution (tri-isooactylammonium chloride IL + isodecanol + Exxsol D100) plus the stripping agent (NH_4OH) were not directly in contact. Actually, they passed through the lumen side and shell side of a membrane contactor in a counter-current mode, as shown in Figure 2c [48]. After the optimization of several different parameters, efficient removal of Cd(II) with a mass transfer coefficient value of 2.68×10^{-4} cm/s was achieved. The authors believed their methodology was a promising alternative to conventional procedures because it combined the operational characteristics of liquid membranes and liquid-liquid extraction technologies. However, further studies on comparing this system with the traditional ones are needed to confirm the removal efficiency.

To sum up, the stability of EILMs is greatly enhanced as compared to the traditional emulsion liquid membranes with only organic solvents as the carrier, but the swelling and breakage of emulsion still exist during the separation process, which could be alleviated by optimizing the operational conditions (such as IL concentration, surfactant concentration, agitation speed, extractant concentration, etc.). Besides, other types of surfactants and ILs that have better chemical interactions could be investigated, since that they can avoid the coalescence of internal phase droplets and enhance stability [47].

2.1.3. Supported IL Membrane (SILM)

Similarly to EILMs, SILMs are more stable than a traditional supported liquid membrane owing to the negligible evaporation of IL. Compared with BILM and EILM, the SILM requires much less membrane solvent and yet, offers higher selectivity [22] making it more popular for membrane fabrication. There have been several review papers reporting the research progress of SILMs and their applications in the separation of organic compounds, gases, vapors, ions, and so on [12,22,49–52]. In this review, we focus more on SILM related research works for liquid separations reported in the last 5 years. PV is a membrane process where the liquid feed is in direct contact with one side of the membrane, while the permeate evaporates into sweeping gas or vacuum on the other side. Generally, the applications of PV include (i) dehydration of organic-water mixtures, (ii) removal of trace organic compounds from water, and (iii) organic-organic mixture separation. Currently, developed SILMs for PV are mainly applied in the latter two applications, with hydrophobic ILs as the membrane phase. According to the reports that were published previously, recovery of butanol from aqueous mixtures is the most investigated way to remove trace organic compounds from water [49,51]. In recent years, SILMs have been explored for separating other mixed systems, especially those organic-organic mixtures.

Mai et al. fabricated a SILM by depositing 1-octyl-3-methylimidazolium bis(trifluoromethylsulfonyl)imide ([OMIM][Tf$_2$N]) on a polydimethylsiloxane (PDMS) support in vacuum. The SILM was then used to recover acetone, butanol, and ethanol (ABE) from an aqueous solution by using PV [53]. However, they found that the SILM had a lower permeation flux and lower selectivity compared to the immobilized IL-PDMS membrane. Zhang et al. developed a SILM system for separating toluene and cyclohexane by impregnating porous polyvinylidene fluoride (PVDF) hollow fiber membrane with ILs [54]. They studied the interactions of several ILs with toluene and cyclohexane by using quantum chemical calculations and the liquid-liquid extraction process. The results showed that N-Butylpyridinium tetrafluoroborate ([BPy][BF$_4$]) has a stronger interaction with toluene than cyclohexane, and it also showed good long-term stability of over 100 h due to its good affinity for the hollow fiber support and the high viscosity. Luis's group prepared two SILMs based on [OMIM][Tf$_2$N] and N-octyl-N-methylpyrrolidinium bis(triuoromethanesulfonyl) imide by using a vacuum for PV separation of dimethyl carbonate (DMC)/methanol mixtures [55]. They found that the separation factor (methanol relative to DMC) of the SILMs was highly dependent on the feed concentration, which was high only at a high DMC concentration of 0.8 M.

Meanwhile, SILMs have also been used to remove organic compounds from aqueous solutions based on extraction. Fortunato et al. fabricated an [OMIM][PF$_6$] based SILM to extract amino acids or amino acid esters [39]. They found that the IL had a better selectivity for amino acid esters. However, the SILMs showed a significant loss of selectivity in a short testing period (2-4 h). This according to them could be ascribed to two reasons: (1) the loss of the organic phase from the membrane support to the adjacent aqueous solution caused by dissolution/emulsification; (2) the formation of water microenvironments inside the organic phase, which constitute new and non-selective environments for solute transport. Matsumoto et al. prepared several SILMs for separating lactic acid by impregnating 6 commercial ILs into the pores of the PVDF MF membrane support using the direct immersion method [56]. They found that CYPHOS IL-104 SILM showed a very low permeation rate, whereas the CYPHOS IL-109 and -111 SILMs were unstable due to the loss of IL from the membrane support. Aliquat 336, CYPHOS IL-101, and CYPHOS IL-102 were found to be suitable in terms of the membrane stability and permeation of lactic acid. The same ILs and hydrophilic PVDF were also used to fabricate SILMs to remove phenol from aqueous solutions by Pilli et al. [57]. For this case, CYPHOS IL-104 gave the highest permeation rate. However, it also showed a quick decline of permeation in the first 10 h although the decline was much less in the later evaluation time. Nevertheless, a longer testing time is still needed to further confirm the stability of the membranes in this study. Panigrahi et al. also fabricated several SILMs using PVDF as the membrane support by a direct immersion method, and the SILMs were then used to separate Bisphenol A (BPA) from the aqueous solution [58]. They got a BPA permeation rate order among different ILs and claimed that the IL weight loss was less than 2% after 24 h. However, their study was a preliminary one and the maximum permeation of BPA they reported needed to be improved upon. Abejón et al. studied five different membrane supports and nine ILs for removing or selective transport of two different technical lignins (i.e., Kraft lignin and lignosulphonate) and monosaccharides (xylose and glucose) in an aqueous solution [59]. However, only the SILM composed of 1-butyl-3-methylimidazolium dibutylphosphate and polytetrafluoroethylene membrane support allowed for the selective transport of the tested solutes. Some of their ILs dissolved some of the membrane supports, whereas others experienced precipitation. Moreover, the stability and separation efficiencies of their SILM need further studies.

Similarly, SILMs have also been applied for metal removal. Jean et al. reported the extraction of Hg(II), Cd(II), and Cr(III) ions from acidic media with a SILM using a novel synthesized IL (isooctylmethylimidazolium bis-2-ethylhexylphosphate) as the carrier [60]. The SILM was more suitable for the extraction of Cd (II) ions. During stability investigation, 11% of the IL was released after 4 cycles. Zante et al. evaluated the feasibility of selectively separating lithium cations from aqueous solutions containing sodium, cobalt, and nickel ions using a SILM fabricated by impregnating porous PVDF membrane support with a mixture of hydrophobic IL [BMIM] [Tf_2N] and TBP as the carrier [61]. Very importantly, their stability experiments indicated that the loss of IL into the aqueous phase could be reduced by the addition of salt in the feed or the stripping phase.

Although the SILMs have been widely studied for various liquid separations, stability issues a concern in some cases, which is due to the gradual solubilization or emulsification of the liquid phase of the membrane (carrier or organic solvent) in the surrounding aqueous phase. As indicated by the literature above, enhancing the interactions between the IL and membrane support is critical. Accordingly, various strategies to improve the stability are considered when developing SILMs, such as screening the strong affinity between the membrane support and IL, chemical modification, minimizing the pores of membrane supports to nano-size level, or directly mixing IL and another polymer prior to membrane casting or impregnation in the pores of membrane supports.

2.1.4. Polymerized IL Membrane (PILM)

As a type of polymer, PILs are more suitable to be directly used as membrane materials than ILs because they possess both the designability of ILs and the selectivity of polymer segments. The use of PILMs in separations offer undoubtedly engineering and economic advantages over other separation technologies for CO_2 capture from fossil fuels and flue gas streams, as well as in CH_4 separation and purification [21,62]. Moreover, PILs have already been studied as novel polyelectrolyte membranes and electrolytes for batteries, fuel cells, and dye-sensitized solar cells [62,63]. Considering that PILs could interact with other molecules through hydrophobic and hydrophilic interactions, hydrogen bonding, ion exchange, π-π stacking, or electrostatic interactions, PILMs have also been extended to several liquid separations, including the removal of metals, dyes, desalination, the concentration of proteins, oil/water separations, etc. To design these PILMs membranes with different structures for different applications, various fabrication methods have been developed.

Tang et al. reported a novel method to prepare positively charged NF membrane by using rapid counter-ion exchange of hydrophilic poly(1-vinyl-3-butylimidazolium) bromide (i.e., PIL/polysulfone (PSf) in aqueous KPF6 solution. The system was then transformed from the initial hydrophilic state to a hydrophobic state, and the scheme is shown in Figure 3 [64]. Interestingly, a thin film of hydrophobic PIL layer was formed in the interface of the hydrophilic PIL and KPF6 aqueous phase due to phase separation along with the self-inhibiting effect induced by the hydrophilic–hydrophobic transformation of the PIL chains. The designed PIL/PSf membrane showed pure water permeance (PWP) of 7.55 $Lm^{-2}h^{-1}bar^{-1}$ (LMH/bar), a rejection of 84% to $MgCl_2$, and a high rejection of about 90% to heavy metallic salts.

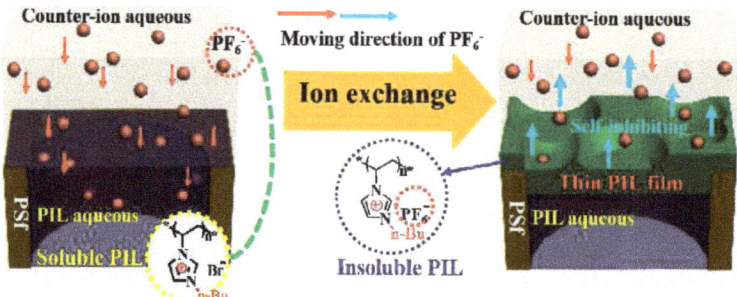

Figure 3. Schematic illustration of the formation mechanism of the poly(ionic liquid) (PIL)/polysulfone (PSf) preparation process. Adapted from [64].

Yuan's group fabricated porous polymeric membranes via simultaneous phase separation of a PIL and its ionic complexation with an acid, which occurred in a basic solution of a nonsolvent [65]. As shown in Figure 4, the membranes have stimuli-responsive porosity. This means they had open pores in isopropanol but close ones in the water, leading to higher isopropanol flux but lower water flux. This property made them potential prospective for stimuli-responsive filtration systems, smart sensors, or controlled loading and release systems. However, further studies are needed to explore their practical applications.

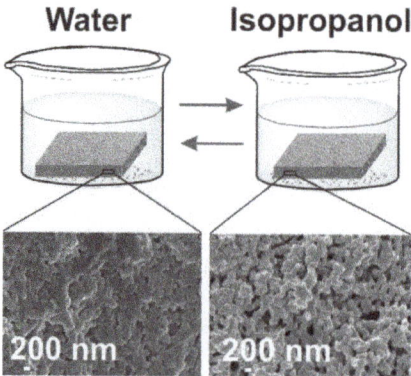

Figure 4. Schematic representation of the solvent experiment by reversibly immersing the membrane in water and isopropanol and the respective SEM images of the membrane structure. Adapted from [65].

Kohno et al. reported a novel type of thermo-responsive PILM that could control the partition of proteins via a lower critical solution temperature (LCST) behavior for protein concentration from aqueous media [66]. They studied the salt effects on the phase behavior of PIL materials, including PILMs in combination with different IL monomers and salt species. The results showed that the water content of a chemically cross-linked and sufficiently hydrated PILM 1, i.e., poly([P4444] [SS]0.3-co-[P4448] [SS]0.7)-type, exhibited reversible water uptake/release via LCST behavior, enabling selective concentration of proteins without significant loss of their higher-order structures.

Besides the direct use of PILs to fabricated liquid-separation membranes, PILs could also be made into porous carbon membranes. For instance, Shao et al. fabricated charged porous membranes (CPMs) with controllable pore architectures by using a rational choice of anions in PILs [67]. Afterwards, they also successfully synthesized hierarchically porous carbon membranes (HCMs) with micrometer-sized pores from CPMs by using one-step vacuum pyrolysis, which was uniform in the molecular distribution of nitrogen species. The HCMs as photothermal membranes exhibited high performance for seawater desalination as shown in Figure 5, revealing their great potential in portable water production technologies. Although there are only a few studies of PILMs for liquid separations at present, they show promising potential. Further investigations could unlock great potential applications and progress in this area.

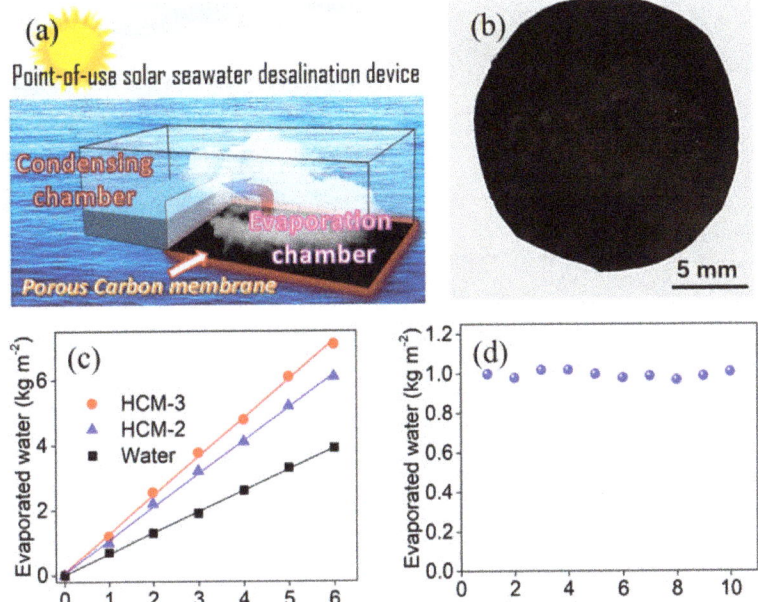

Figure 5. (a) Illustration of an air-water interface solar desalination device. (b) Digital photographs of hierarchically porous carbon membrane HCM-3 with a diameter of 0.8 cm and thickness of 160 μm. (c) Mass of the evaporated water as a function of the radiation time with and without HCMs. (d) Reusability of HCM-3 for solar-powered seawater desalination. Adapted from [67].

As indicated by the aforementioned research, PILs are able to form novel structures and morphologies that are not accessible using ILs due to the polymer segments. Accordingly, PILMs have more versatile applications than the IL membranes. Currently, the application of PILMs for liquid separations is still a relatively new research area. The future of this field lies in the development of new polymers and PIL-nanomaterial composites with improved properties.

2.2. ILs as Physical Additives

Recall that the various membranes synthesized with ILs as the fabricated materials which we have discussed so far have poor stability in liquid separations due to the loss of the ILs in the liquid phase. As a result, researchers have made efforts to enhance membrane stability and separation performance by utilizing ILs to physically modify polymeric/inorganic membranes. In other words, the ILs were used as physical additives; some representative research using ILs or PILs as such are stated below.

2.2.1. IL-Polymer Blending Membranes for PV

Blending ILs with hydrophobic polymers to fabricate PV membranes for solvent recovery has drawn much attention recently. To fabricate the IL/PIL-polymer membranes, the IL/PIL and polymer are normally mixed and dissolved by solvents to form a polymer solution, which is then cast into a membrane.

For example, Izak's group impregnated PDMS-1-ethenyl-3-ethyl-imidazolium hexafluorophosphate (IL1) and PDMS-tetrapropylammonium tetracyano-borate (IL2) blend into ceramic ultrafiltration membrane support to fabricate PV membranes for acetone and 1-butanol removal from water [68]. Compared with the pristine PDMS membrane, the PDMS-IL membranes greatly improved the enrichment factor. More importantly, they showed good stability under a low pressure of 20 Pa in an aqueous solution of acetone and 1-butanol for more than five months. In their follow-up work, the diffusion coefficients and sorption isotherms of 1-butanol in the pristine PDMS membrane and two PDMS-IL membranes at different pressures were determined [69]. The results indicated that the higher permeation flux and enrichment factors of the IL-PDMS membranes were probably caused by the higher diffusion coefficient.

Rdzanek et al. blended two ILs, namely trihexyl (tetradecyl) phosphonium tetracyanoborate ($P_{6,6,6,14}$ tcb) and 1-hexyl-3-methylimidazolium tetracyanoborate ($Im_{6,1}$ tcb), with polyether block amide (PEBA) and immobilized them into the pores of PSf or polypropylene membranes to fabricate PV membranes for ABE recovery from water [70]. The IL elution into the liquid feed solution could still pose some challenges, since the blending of IL and polymer is based on physical forces. To overcome this issue, the authors covered the PEBA+IL membrane with an additional thin silicone layer using two different arrangements, as shown in Figure 6. The experimental results showed that arrangement (2) was more suitable for the butanol separation because it had a smaller water flux. In their later work, the silicone layer was replaced by PEBA but they still used arrangement (2) and concluded that the addition of a hydrophobic IL in the PV membrane could decrease the water flux and enhance the enrichment factor. However, a long-term PV test of these membranes is still needed to confirm their stability.

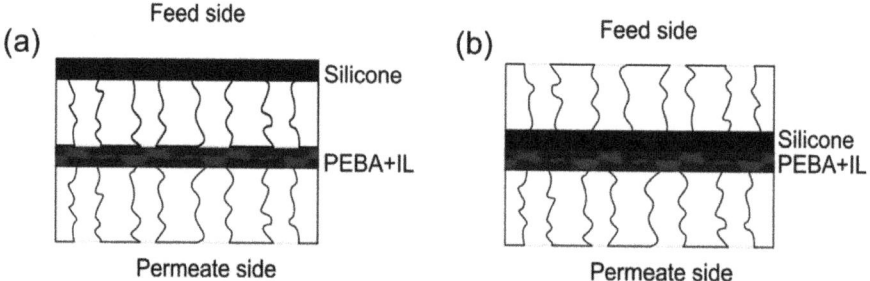

Figure 6. Arrangement of double-layer membranes. (**a**) consists of two separated layers, polyether block amide (PEBA) + IL and silicon layer oriented to the feed side; (**b**) PEBA + IL is located to feed side while the silicon layer is oriented to the permeate side. Adapted from [70].

Ong and Tan also blended [BMIM][BF_4] with polyvinylalcohol at a weight ratio of 70/30 and immobilized the [BMIM][BF_4]- polyvinylalcohol solution into a porous bucky paper made of carbon nanotubes by vacuum filtration to form a PV membrane, which was then cross-linked with glutaraldehyde [71]. The fabricated membrane successfully dehydrated water from a ternary azeotropic mixture of ethyl acetate/ethanol/water. The results suggested that the addition of IL could attain a good balance in the trade-off between the permeation flux and the separation factor.

Clearly, PILs can also be used as physical membrane additives. Tang et al. synthesized positively charged PIL (poly[1-cyanomethyl-3-vinylimidazolium bromide], PCMVImBr) and blended it with positively charged sodium carboxymethyl cellulose (CMCNa) to form a stable PV membrane based on electrostatic force, and the scheme is illustrated in Figure 7 [72]. The blended membranes with the PIL performed well by stably dehydrating 10 wt% acidic water-isopropanol mixtures. The separation factor was also much higher compared to the pristine CMCNa membrane.

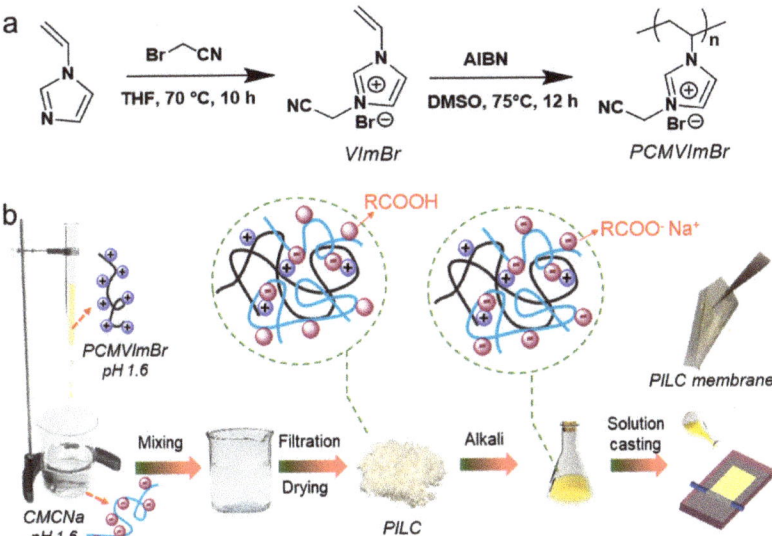

Figure 7. (a) Synthesis of IL monomer (VImBr) and its radical polymerization yielding PIL (PCMVImBr), (b) schematic preparation of the PCMVIm-CMCNa PILC membranes. Adapted and modified from [72].

2.2.2. IL-Polymer Blending Membranes for Separating Metal/Organic from Water

Meanwhile, IL-polymer blending membranes have been used for the separation of organics and various metal ions from aqueous solutions. In this context, He et al. fabricated a cellulose acetate (CA)-sulfonated polysulfone (SPS) blend imprinted membranes for selective adsorption of salicylic acid (SA) by using phase inversion [73]. To enhance the separation efficiency, polyethylene glycol-4000 (PEG 4000) and IL 1-butyl-3-methylimidazolium chloride ([BMIM]Cl) were mixed with the polymer dope as additives. The CA/SPS (90/10) + [BMIM]Cl membranes possessed higher membrane flux, stronger adsorption capacity, and higher selectivity for the SA relative to the competitive species. This performance was attributed to the fact that SPS improved the hydrophilicity of the membrane, whereas the IL promoted the formation of a dense and ordered porous structure.

Similarly, Chen's group designed an asymmetric PVDF membrane with addition of IL [tricaprylmethylammonium][di-(2-ethylhexyl)orthophosphinate] ([A336][P507]) for the preconcentration and separation of the heavy rare earth Lutetium [74]. Their study showed that the transport sequence of $LuCl_3$ and $YbCl_3$ in the membrane was different from that in liquid-liquid extractions, which benefited the separation between Yb and Lu. Moreover, the PVDF-[A336][P507] membrane showed good stability and reusability for $LuCl_3$ transport, albeit with weak physical interaction between them.

Elias et al. also fabricated a membrane for mercury preconcentration by incorporating two different ILs: trioctylmethylammonium thiosalicylate (TOMATS) and trioctylmethylammonium salicylate (TOMAS) into cellulose triacetate using nitrophenyl octyl ether as a plasticizer [75]. The membrane with TOMATS yielded effective transport of Hg, which was then made into a special device for global detection of low-concentration Hg in natural water. More interestingly, they investigated the growth of biofilm on the surface of the membrane for the first time and observed no significant differences in Hg transport between a fresh membrane and a membrane deployed for 7 days in a pond.

Yang's group, on the other hand, fabricated a polymer inclusion membrane (PIM) for the separation of low-concentration gold (I) from alkaline cyanide solutions using solvent evaporation with PVDF, [A336][SCN] as the carrier, and 2-nitrophenyl n-octyl ether as the plasticizer [76]. Their results indicated that the IL concentration greatly influenced the extraction because the mechanism involved an anion exchange reaction between the IL embedded in the PVDF and the gold cyanide complex in the feed. The membrane showed a high extraction efficiency of 98.6% for $Au(CN)^{2-}$ and a high gold recovery rate of 98.2% after 24 h using KSCN as the stripping phase. A re-usability investigation confirmed that the PIM maintained long-term stability and excellent durability. In their later work, the PIM system was integrated with an electroplating unit; the scheme is illustrated in Figure 8 [77]. The permeability coefficient of gold increased over two folds after the constant voltage was applied to the stripping solution. Thus, the membrane flux increased with high extraction and deposition percentages of gold. Furthermore, the metallic state gold was coated uniformly on the cathode.

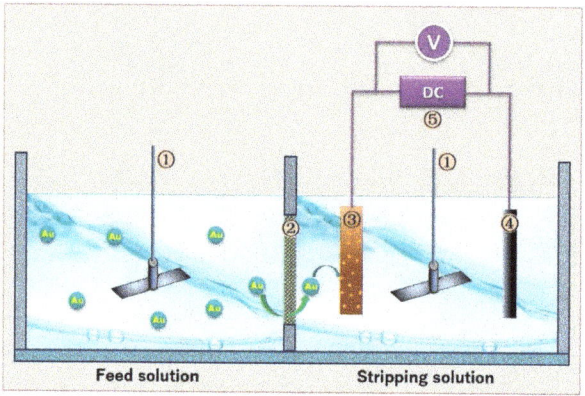

Figure 8. Schematic illustration of the permeation device with an electroplating module. 1: stirrer; 2: PIM; 3: copper cathode; 4: graphite anode; and 5: DC stabilized voltage source. Adapted from [77].

2.3. ILs as Chemical Modifiers

In fact, there have been concerns about IL leaching out to the feed phase for the IL-polymer blending membranes, since the interaction between the IL and polymer matrix is based on weak physical forces [70]. Considering that many ILs have functional groups that can covalently be bonded with other materials, researchers have also tried to use IL to chemically modify polymeric/inorganic membrane materials or additives.

2.3.1. Chemically Modify Membrane

On one hand, ILs could chemically bond to polymer chains before forming a membrane, which normally results in a membrane with different membrane structures and possesses improved membrane stability. For instance, Mai et al. fabricated an immobilized $[Tf_2N]^-$ based IL-PDMS membrane to recover acetone, n-butanol, and ethanol (ABE) from an aqueous solution by PV [53]. In order to covalently bond the $[Tf_2N]^-$ based IL to the PDMS backbone polymer, a $[Tf_2N]^-$ based

IL precursor which contains active groups which can react with the hydroxyl terminated PDMS was synthesized in advance. Compared with the conventional IL-PDMS supported membrane where the IL was filled in the void volume of the PDMS membrane, the immobilized IL-PDMS membrane exhibited much higher permeate flux, enhanced the recovery of accompanying products such as acetone and ethanol from ABE fermentation, and improved operational stability.

Xi et al. synthesized IL copolymerized waterborne polyurethane (IL-co-PU) membranes for the PV separation of benzene/cyclohexane mixtures based on the reaction mechanism shown in Figure 9 [78]. Both the permeation flux and separation factor (benzene/cyclohexane) of the IL-co-PU membranes increased when the IL content was increased, indicating that the IL might facilitate transportation in the membranes.

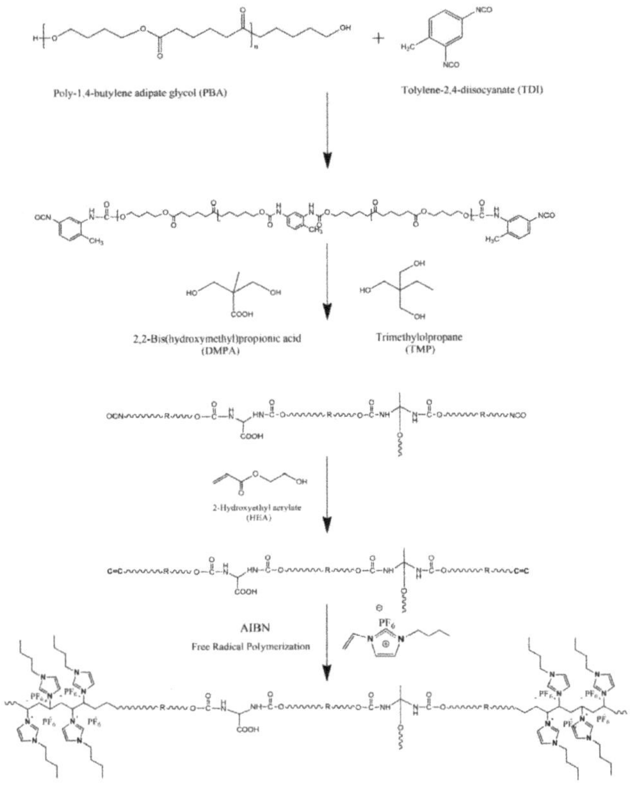

Figure 9. The schematic diagram for the preparation of IL copolymerized waterborne polyurethane (IL-co-PU). Adapted from [78].

On the other hand, ILs could be chemically bonded to the membrane surface after membrane formation, which tunes the surface properties of membranes and improves the separation performance, antifouling properties, stability, and so on. Most related studies focus on using ILs to surface modify polyamide membranes because the polyamide chains could be split by the hydrogen bonds with imidazolium ILs or react with amine-containing ILs via the "acyl chloride~amine" esterification. Zhang et al. also synthesized an IL (i.e., 1,3-dimethylimidazolium dimethyl phosphate ([MMIM][DMP])), which was adopted to modify the surface of the commercial RO membrane by an activation method [79]. It was revealed (Figure 10) that the IL modification mechanism was based on the effective breakage of inter- and intra-molecular hydrogen bonds in the polyamide chains accompanied by the breakage of polyamide chains dissolved in the IL. The results showed that the modification led to a thinner, smoother, and more hydrophilic PA layer of the RO membrane, resulting in a great improvement in the water permeability and anti-fouling property.

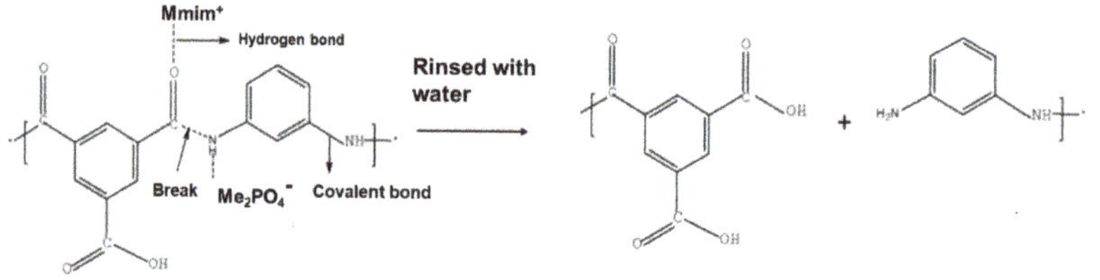

Figure 10. The modification mechanism of the PA active layer in the 1,3-dimethylimidazolium dimethyl phosphate ([MMIM][DMP]). Adapted from [79].

Sun's group used an amino acid IL (AAIL) to functionalize interfacial polymerized NF membranes, and the scheme is shown in Figure 11 [80]. The AAIL modification did not only improve the hydrophilicity and increase the pure water permeability by 63%, but it also caused the membrane surface to be more negatively charged, resulting in high $Na_2SO_4/NaCl$ selectivity. Furthermore, the amino acid end group of the AAIL could serve as a humectant, allowing the membrane to be heat-treated and stored in a dry state.

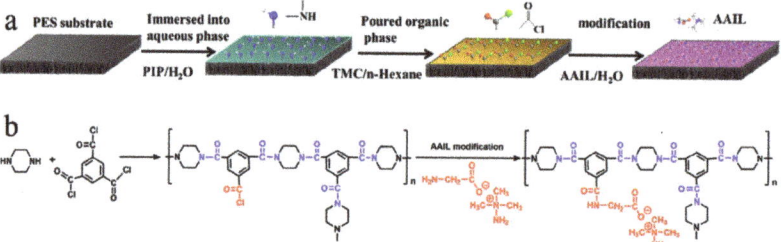

Figure 11. (a) Schematic diagram of an amino acid IL (AAIL) modified polyamide selective layer, and (b) the possible reaction formula. Adapted from [80].

In a similar fashion, He et al. utilized an imidazolium IL (1-aminoethyl-3-methylimidazolium bromide, AMIB) to surface modify the polyamide selective layer of thin-film composite NF membranes via "acyl chloride~amine" amidation, as shown in Figure 12 [81]. They also found a great improvement in the water flux with good salt rejection ($R_{Na_2SO_4}$ = 95%) after the IL modification. Moreover, the IL modified membrane showed good performance for antibiotic/salt separation as well as promising levels of stability and antibacterial ability.

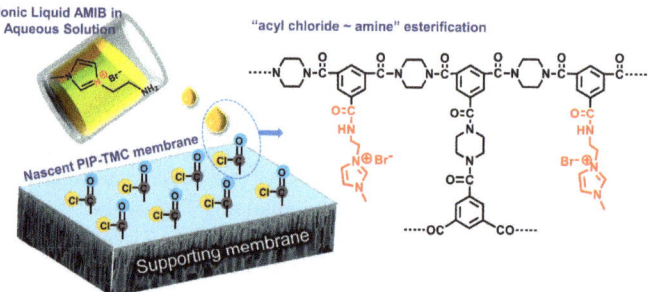

Figure 12. Surface modification of polyamide thin-film composite membranes by 1-aminoethyl-3-methylimidazolium bromide (AMIB). Adapted from [81].

Liu et al. used IL or polydopamine to chemically modified graphene oxide (that is, iGO or pGO), and then the modified GO nanosheets were assembled with polyelectrolytes on polyethersulfone (PES) membrane support to form composite GO membranes (PE-iGO or PE-pGO) for dye/salt fractionation [82]. The iGO nanosheets were formed by binding methylimidazolium IL with the -COOH groups on GO mediated by 1-ethyl-3-(3-dimethylaminopropyl) carbodiimide hydrochloride and N-hydroxy-succinimide. The PE-iGO membrane could be operated at a low operating pressure of 0.5 bar while achieving high permeability of ~38.4 LMH/bar (100 ppm direct red 80, 5 g/L NaCl). In fact, the salt rejection for 10 g/L NaCl was constantly lower than 5%. Moreover, the IL modification favored the elution of dye molecules from the IL moieties at higher pH, thus improving the efficiency of alkaline washing of the membrane.

2.3.2. Chemically Modify the Membrane Additives

For these products, Shi et al. synthesized IL-TiO_2 nanoparticles by chemically modifying TiO_2 with carboxyl-functional IL ([CH_2COOHmim]Cl) according to the reaction route shown in Figure 13. The modified product was then added into the PVDF/dimethyl phthalate solutions to fabricate PVDF/IL-TiO_2 hybrid microfiltration membranes via the thermally induced phase separation (TIPS) method [83]. The addition of IL-TiO_2 had a strong effect on the crystal formation in the TIPS process. Moreover, the increased amount of IL-TiO_2 initially increased pure water flux and porosity but then decreased the parameters eventually. However, both the stability and antifouling property were also enhanced, indicating that the PVDF/IL-TiO_2 hybrid membranes may have potential in catalytic water treatment.

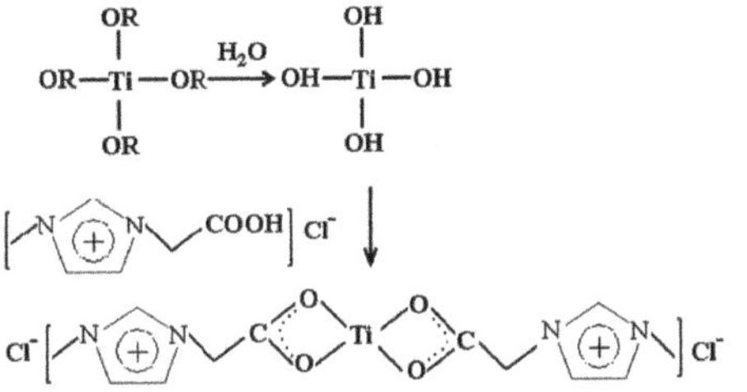

Figure 13. The schematic diagram for the formation of IL-TiO_2. Adapted from [83].

Abraham et al. then fabricated mixed matrix membranes (MMMs) for separating toluene/methanol azeotropic mixtures by incorporating IL (1-benzyl-3-methyl imidazolium chloride) functionalized multi-walled carbon nanotubes into styrene-butadiene rubber (SBR) [84]. The benzyl groups of the IL on the MWCNT surface possessed greater toluene affinity and higher repellency against methanol due to their aromatic π-π interactions with toluene molecules, leading to higher permeation flux and separation factor compared to pristine SBR membranes.

Tang et al. also fabricated pervaporative MMMs for butanol recovery from aqueous solutions by incorporating IL (N-octylpyridiniunm bis (trifluoromethyl) sulfonyl imide, [OPY][Tf_2N]) modified graphene oxide (IL-GO) nanosheets into PEBA matrix [85]. The reaction mechanism between [OPY][Tf_2N] and GO is shown in Figure 14. The author also found that the addition of IL-GO increased the separation factor and the permeation flux of the MMMs due to the good butanol affinity as well as the hydrophobicity of the IL. Furthermore, anchoring the IL to GO avoided the IL loss to the feed during the PV process, and thus enhanced the membrane stability. The long-term stability was conducted during a 180 h PV test of the IL-GO/PEBA MMM, in which the separation performance almost showed no change.

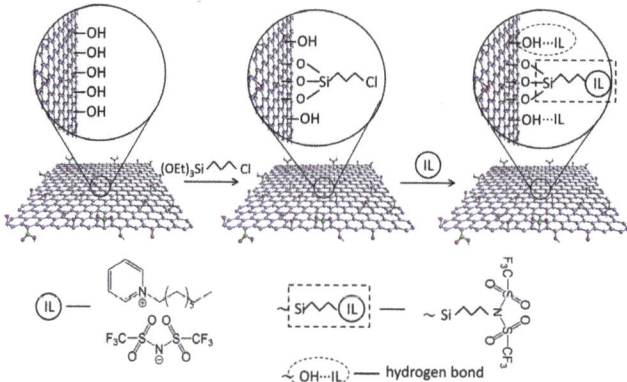

Figure 14. Preparation of IL modified graphene oxide and the structure of N-octylpyridiniunm bis (trifluoromethyl) sulfonyl imide ([OPY][Tf$_2$N]). Adapted from [85].

Likewise, PILs have been used for the chemical modifications of membranes. Zhang's group first synthesized positively charged nano-sized silica spheres modified by PIL brushes via atom transfer radical polymerization (ATRP), and then incorporated them into the PES solution to cast SiO$_2$-PIL/PES MMMs for NF (Figure 15) [86]. The designed positively charged MMMs showed higher water flux, low-molecular-weight organic rejection, and salt permeability. Furthermore, the salt concentration showed little effect on the separation property. The authors also used the same PIL to modify Mg-Al hydrotalcite (HT) nanosheets and fabricated HT-PIL/PES MMMs using a similar strategy [87]. The HT-PIL/PES MMM showed higher rejection of reactive dyes than the previously developed SiO$_2$-PIL/PES MMMs (90~95% vs 85~95%). This type of loose NF membrane may open opportunities for separating dyes from salts-containing textile wastewater.

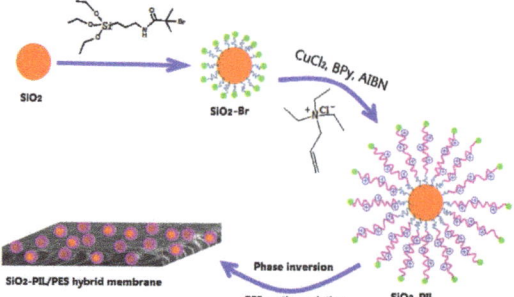

Figure 15. The preparation process for SiO$_2$-PIL/polyethersulfone (PES) hybrid membrane. Step 1: the anchoring of the initiator (BTPAm); step 2, the polymerization of IL monomers (ATEA-Cl) on the surface of SiO$_2$ particles by using the reverse ATRP method; step 3, the preparation of SiO$_2$-PIL/PES positively charged membrane by blending SiO$_2$-PIL particles with a PES casting solution. Adapted from [86].

2.4. ILs as Solvents

ILs have also been used to dissolve polymers; in this case, they serve as green solvents, especially for those polymers which have limited solubility in common organic solvents. This has extended the range of polymers that can be dissolved in order to fabricate membranes for phase separation. Meanwhile, ILs can also serve as reaction media for polymerization or other chemical reactions during membrane fabrication or modification.

2.4.1. Solvents for Polymer Dissolution

The liquid-separation membranes fabricated from the non-solvent induced phase inversion of polymer/ IL solutions, which have been reported in the recent 10 years are summarized in Table 1.

Table 1. Membranes fabricated from polymer/ IL solutions and their applications.

Polymer	IL	Applications	Year	Reference
Cellulose (surface modified)	[BMIM][Cl]	Human immunoglobulin G (IgG) purification by absorption	2010	[88]
Cellulose	[AMIM][Cl]	NF: dye rejection (<700 Da)	2011	[35]
Cellulose (TFC)	[EMIM][OAc]	UF: PEO rejection (3000 Da) NF: PEG rejection (<200 Da)	2015	[89]
Cellulose -TiO_2	[EMIM][OAc]	UF: humic acid (100 kDa) rejection	2015	[90]
Cellulose	[EMIM][OAc]	UF: oil/water separation	2018	[91]
Cellulose TFC	[EMIM][OAc]	OSN: dye rejection (500 Da)	2018	[92]
Cellulose-graphene quantum dot	[EMIM][OAc]	NF: dyes (300 < MWCO < 5000 Da) rejection	2018	[34]
Cellulose-iron/polyacrylic acid/lignin sulfonate	[EMIM][OAc]	NF: dye rejection (<300 Da)	2019	[93]
Cellulose HF	[EMIM][OAc] [EMIM][DEP] [DMIM][DMP]	UF: PEG rejection (~18 kDa) PS rejection (25 kDa) NF/OSN: Dye rejection (700–1500 Da)	2019	[94]
Cellulose-GO	[EMIM][OAc]	NF: heavy metal removal	2019	[36]
Cellulose from bamboo	[BMIM][Cl]	NF: dye rejections	2020	[95]
Cellulose acetate	[BMIM][SCN]	UF: PEG/PEO rejection	2010	[96]
Cellulose acetate HF	[BMIM][SCN]	UF: PEG/PEO rejection	2011	[24]
Cellulose acetate	[EMIM][OAc]	UF: BSA (66 kDa), γ-globulin (~140 kDa) rejection	2016	[97]
PBI	[EMIM][OAc]	OSN: dye rejection (600 Da)	2014	[26]
PBI/P84	[EMIM][OAc]	UF: PEG/PEO rejection (~100 kDa)	2013	[98]
Extem	[EMIM][SCN]	UF: BSA (66 kDa), γ-globulin (~140 kDa) rejection DNA (6.4 kDa)	2017	[99]
PVDF-HFP	[dema][TfO] [MIM] [Tf_2N] [MIM][Cl]	MF	2018	[100]
PMIA-TFC	[EMIM][OAc]	OSN: Dye rejection (470–730 Da)	2018	[27]
Polytriazole	[EMIM][DEP]	OSN: PEG rejection rom DMF (1~3 kDa)	2020	[101]

Note: PBI: polybenzimidazole; PVDF-HFP: poly(vinylidene fluoride-co-hexafluoropropylene); PMIA: Poly(m-phenylene isophthalamide); TFC: thin-film composite; PEO: poly(ethylene oxide); PEG: poly(ethylene glycol); [BMIM][Cl]: 1-butyl-3-methylimidazolium chloride, [BMIM][SCN]: 1-butyl-3-methylimidazolium thiocyanate; [AMIM][Cl]: 1-allyl-3-methylimidazolium chloride; [EMIM][OAc]: 1-ethyl-3-methylimidazolium acetate; [EMIM][DEP]: 1-ethyl-3-methyimidazolium diethyl phosphate; [DMIM][DEP]:1,3-dimethylimidazolium dimethyl phosphate; [dema][TfO]: trifluoromethanesulfonate; [MIM][Tf_2N]: 1-methylimidazolium bis(trifluoromethylsulfonyl); [EMIM][Cl]: 1-ethyl-3-methylimidazolium chloride; [MIM][Cl]: 1-methylimidazolium chloride; [DMIM][DMP]: 1,3-dimethylimidazolium dimethyl phosphate; BSA: Bovine serum albumin.

As shown in Table 1, cellulose is the most widely studied polymer to be dissolved by ILs and cast into liquid-separation membranes by non-solvent phase inversion. This may be because cellulose is the most abundant renewable biopolymer on the earth but it is very difficult to be dissolved and processed using traditional organic solvents. Besides, the most frequently studied ILs for dissolving these polymers are almost all hydrophilic imidazolium cations based ILs. They have a strong ability to disrupt hydrogen bonds in polymers and are also miscible with water to favor the membrane fabrication through non-solvent phase inversion. Related studies in different liquid separation areas are introduced based on the types of polymers used as follows.

Most of the cellulose membranes fabricated from cellulose-IL solutions by water-induced phase inversion have pores that are within the MF or UF range. Some of the MF/UF cellulose membranes have been directly used for the rejection of PEO [89], PEG [94], PS [94], humic acid [90], or oil [91]. There has also been some research work which has succeeded in fabricating cellulose NF membranes. For example,

Li et al. fabricated cellulose NF membrane by phase inversion from a cellulose - [AMIM][Cl] solution on PET non-woven fabric [35]. The rejection data for a series of dyes showed that the molecular weight cut-off was less than 700 Da. Falca et al. also fabricated cellulose HF membranes via spinning using three different ILs as solvents [94]. The HFs spun from solutions in [DMIM][DMP] and [EMIM][DEP] showed better results for dye rejection. However, it should be noted that the fabricated membranes showed good rejections for negatively charged dyes but poor rejection for positively charged dyes in water or ethanol, indicating that the charge effect rather than size exclusion was dominant during the separation. Therefore, they may not be suitable for OSN in non-polar solvents. Esfahani et al. fabricated loose NF membranes from cellulose-[BMIM][Cl] solutions; the cellulose was extracted from bamboo waste fiber, indicating the sustainability of the technique [95].

Meanwhile, other membranes fabricated from cellulose-IL solutions were not used directly but have to be further modified before they could be used for liquid separations. For example, Barroso et al. reported on the surface modification of a cellulose MF membrane which could also be surface modified with a synthetic ligand 2-(3-aminophenol)-6-(4-amino-1-naphthol)-4-chloro-s-triazine, and was used to absorb human immunoglobulin G (IgG) rather than BSA, achieving the separation goal [62]. Livazovic et al. designed a polyamide/cellulose TFC membrane by using interfacial polymerization between m-phenylenediamine (MPD) and trimesoyl chloride (TMC) on the cellulose surface. The MWCO was much smaller than f the pristine cellulose membrane (<200 vs. 3000 Da) [89]. Abdellah et al. fabricated a polyester/cellulose TFC membrane by using the interfacial reaction between catechin and terephthaloyl chloride [92]. The membrane showed dimethylformamide permeance of 1.2 LMH/bar with an MWCO around 500 Da. The membrane exhibited stable performance within 1 month, indicating great potential for application in the food and pharmaceutical industries.

Besides, other materials could be added to the cellulose-IL solution to fabricate the relevant cellulose composite membranes. In this case, Nevstrueva et al. reported that adding a small amount of TiO_2 can increase the PWF of the resultant membrane but also that doing this reduced the humic acid retention [90]. Colburn et al. incorporated graphene oxide quantum dots (GQDs) into cellulose. The abundant hydroxyl and carboxyl groups of the GQD made the cellulose membrane very stable due to the hydrogen bonding, made it negatively charge, and made it more hydrophilic. In their later work, iron oxide nanoparticles, polyacrylic acid, and lignin sulfonate were uniformly incorporated into their cellulose membranes [93]. The iron-cellulose membrane showed excellent dye rejections with MWCO less than 300 Da. Slusarczyk and Fryczkowska also reported that the introduction of nano-sized GO into the cellulose matrix influenced the membrane formation process, consequently, the physicochemical, transport, and separation properties of the composite membranes [36]. Yet, the addition of GO enhanced the PWF by up to 10 fold, with a rejection of heavy metals reaching higher than 90%.

The applications of ILs become very meaningful when they are used to dissolve polymers (e.g., cellulose, PBI, PMIA, polytriazole, etc.) which are not easily dissolved by traditional organic solvents. Chung's group used [EMIM][OAc] to dissolve the rigid polymers PBI [26] and PMIA [27], and fabricated membranes for OSN. The chemically crosslinked PBI membrane displayed good separation performance and impressive stability in many aggressive solvents. While the glutaraldehyde (GA) modified PMIA membranes with a selective layer synthesized by the water-based reaction between GA and hyper-branched polyethylenimine (HPEI) showed MWCO of 470-730 Da with acceptable ethanol permeance. Importantly, they also showed good performance for concentrating lecithin in hexane. Nunes's group utilized [EMIM][DEP] to dissolve polytriazole, followed by nonsolvent induced phase inversion and chemical crosslinking to fabricate effective membranes [101]. The crosslinked polytriazole membranes showed good performance for dipolar aprotic solvents (e.g., N,N'-dimethylformamide) with a MWCO within 1~3 kDa with an operational temperature varying from 25 °C to 105 °C, which have great potential for applications under high-temperature harsh organic solvent environments.

The ILs have also been studied to dissolve other polymers which are easier to dissolve by using traditional organic solvents. This is because the nonvolatility of ILs avoids the emission of volatile organic compounds that are toxic. Cellulose acetate, a derivative polymer of cellulose, was also dissolved in imidazolium cations based ILs to fabricate flat sheet [96,97] or HF [24] membranes with a pore size in the UF range. However, these membranes were only suitable for the separation of aqueous solutions because cellulose acetate is not as stable as cellulose in organic solvents. Besides, other polymers such as Extem, PVDF-HFP, and P84/PBI blend were also dissolved by using ILs and they can be fabricated into UF or MF membranes [98–100]. In addition, researchers are using IL-organic solvent mixtures to dissolve polymers in order to reduce the viscosity of the polymer solution for easier processing and/or adjusting the membrane structure [102–104]. In these cases, the selection of less toxic organic solvents is critical due to environmental factors.

Overall, ILs are good solvents for dissolving polymers for fabricating membranes. They have a strong ability to dissolve rigid polymers. More importantly, the negligible vapor pressure of ILs makes the membrane fabrication processes more environmentally benign. However, the types of ILs that could be used for the dissolution of polymers are limited currently. Moreover, the mechanism for the dissolution of polymers in ILs and the phase inversion of IL-polymer solutions to form membranes are not widely studied, making the selection of suitable ILs for desired polymers time-consuming.

2.4.2. Solvents as Reaction Media during Membrane Fabrication

ILs could also serve as solvents (reaction media) for the interfacial polymerization reaction for fabricating TFC membranes. For instance, Vankelecom's group conducted several studies on this research area in recent years. A water-immiscible IL, [BMIM][Tf$_2$N] was used as the organic reaction phase for the interfacial polymerization between MPD and TMC to form the TFC RO membrane. The membrane showed 350% higher water permeance with comparable selectivity (96.8% NaCl retention) to the TFC membranes synthesized using traditional hexane as the organic phase [28]. In their subsequent work, the synthesis conditions were further optimized in terms of reaction time, cost-efficiency, and environmental impact [105]. The time for interfacial polymerization was reduced to 10 seconds. More importantly, the IL and TMC could be recycled, proving the eco-friendliness and sustainability of their technique. In the meantime, a polyamide/crosslinked Matrimid TFC membrane for FO was fabricated by using [BMIM][Tf$_2$N] or [BMIM][Tf$_2$N]/hexyl acetate mixture as the organic phase for the interfacial polymerization [106]. The use of the IL instead of hexane made the fabrication process less hazardous but with similar FO separation performance. These studies have shown the great potential of ILs as the reaction media for interfacial polymerization in fabricating liquid-separation TFC membranes. However, the ILs which could be used for such purposes are still very limited, and the reaction mechanism and reaction kinetics need to be studied further to precisely control such reaction processes.

3. The Sustainability of ILs in Developing Liquid Separation Membranes

More and more ILs have been explored in assisting the development of liquid separation membranes. The information of the commercial ILs used in membrane fabrication is summarized in Table 2. As shown, the prices of most ILs are much higher than common organic solvents. Besides this economic aspect, the environmental aspects of ILs such as stability, biodegradability, toxicity should be considered as well [107–109]. These aspects are major hurdles for sustainable development and commercialization of the ILs related technologies. Therefore, the recovery of ILs is of great importance. Various separation methods have been developed for the regeneration, recovery, or removal of ILs, including phase separation methods (i.e., evaporation, vacuum distillation, and crystallization), phase addition methods (i.e., liquid-liquid extraction and supercritical fluid extraction), adsorption by solid agents or solid extraction, separation by a barrier (i.e., membrane filtration methods and PV), separation by an external force-field (e.g., decantation, magnetic-field separation), which have been comprehensively summarized by several review papers [108,110–112].

Table 2. The information of the commercial ILs used in membrane fabrication.

Chemical Name	Abbreviation	Formula	CAS registry Number	Molecular Weight	Density [a] (kg/m^3)	Viscosity [a] (Pa×s)	Price (RMB, 100 g Weight Basis) [b]
1-butyl-3-methylimidazolium tetrafluoroborate	[BMIM][BF$_4$]	C$_8$H$_{15}$BF$_4$N$_2$	174501-65-6	226.03	1201	0.108	400
1-butyl-3-methylimidazolium bis(trifluoromethylsulfonyl)imide	[BMIM][Tf$_2$N]	C$_{10}$H$_{15}$F$_6$N$_3$O$_4$S$_2$	174899-83-3	419.36	1436	0.0508	1500
1-octyl-3-methylimidazolium chloride	[OMIM][Cl]	C$_{12}$H$_{25}$ClN$_2$	64697-40-1	230.78	1013	13.3	300
1-butyl-3-methylimidazolium hexafluorophosphate	[BMIM][PF$_6$]	C$_8$H$_{15}$F$_6$N$_2$P	174501-64-5	284.19	1367	0.274	400
1-octyl-3-methylimidazolium hexafluorophosphate	[OMIM][PF$_6$]	C$_{12}$H$_{23}$F$_6$N$_2$P	304680-36-2	340.29	1236	0.691	500
1-octyl-3-methylimidazolium bis(trifluoromethylsulfonyl)imide	[OMIM][Tf$_2$N]	C$_{14}$H$_{23}$F$_6$N$_3$O$_4$S$_2$	862731-66-6	475.47	1320	0.0931	1500
N-butylpyridinium tetrafluoroborate	[BPy][BF$_4$]	C$_9$H$_{14}$BF$_4$N	203389-28-0	223.02	1214	0.1603	500
1-butyl-3-methylimidazolium chloride	[BMIM][Cl]	C$_8$H$_{15}$ClN$_2$	79917-90-1	174.67	1082	0.00604	200
N-octylpyridinium bis(trifluoromethyl) sulfonyl imide	[OPy][Tf$_2$N]	C$_{15}$H$_{22}$F$_6$N$_2$O$_4$S$_2$	384347-06-2	472.47	1327	0.1143	1600
1-allyl-3-methylimidazolium chloride	[AMIM][Cl]	C$_7$H$_{11}$ClN$_2$	65039-10-3	158.63	1166	0.82	300
1-ethyl-3-methylimidazolium acetate	[EMIM][OAc]	C$_8$H$_{14}$N$_2$O$_2$	143314-17-4	170.21	1100	0.1436	900
1-ethyl-3-methylimidazolium diethyl phosphate	[EMIM][DEP]	C$_{10}$H$_{21}$N$_2$O$_4$P	848641-69-0	264.26	1144	0.41	400
1-butyl-3-methylimidazolium thiocyanate	[BMIM][SCN]	C$_9$H$_{15}$N$_3$S	344790-87-0	197.30	1070	0.05652	1600

[a] Data was measured at a temperature of 298.15 K and pressure of 100 kPa. Database: https://ilthermo.boulder.nist.gov/. [b] Data was provided by Lanzhou Greenchem ILs, CAS, China.

Even though versatile methods for recovering ILs have been studied, suitable ones should be chosen by analyzing the physicochemical properties of ILs and other components. Among cases of using ILs for fabricating liquid separation membranes, the ILs recovery is most urgent in the case where ILs are used as solvents to dissolve the polymers or as the reaction media during membrane fabrication because a large amount of ILs is used.

In the former situation, the mixture to be separated is normally composed of ILs and water. Xing et al. used a simple evaporation method to remove the water in a coagulant bath, and thus recovered [BMIM][SCN] after the membrane fabrication [96]. Interestingly, the recovered [BMIM][SCN] was reused for casting cellulose acetate flat sheet membranes, which showed morphological and performance characteristics similar to those prepared by using fresh [BMIM][SCN]. However, the coagulant bath contains much water, thus making the evaporation method energy-consuming. Therefore, more research focuses on using NF for recovering ILs from aqueous solutions [110,112]. In the latter situation, the recovery of ILs is a bit difficult because the ILs contain monomers (precursors of the thin-film composites) that are also not volatile. As aforementioned, Vankelecom's group fabricated TFC RO membrane via the interfacial polymerization between MPD and TMC using [BMIM][Tf$_2$N] as the organic reaction phase. Though TMC and [BMIM][Tf$_2$N] were not separated, the mixture could be recycled several times for membrane synthesis, without loss of performance [105].

Nevertheless, the studies on the regeneration, recovery, and removal of ILs after the fabrication of liquid-separation membranes are limited. Future work should provide more efforts towards comparing these methods and finding suitable and efficient methods for different special cases in developing liquid-separation membranes, which would boost the usage of ILs to be more sustainable and beneficial.

4. Conclusions and Future Directions

There have been increasing studies on the use of ILs to fabricate liquid-separation membranes in the last decade. Therefore, four major functions of ILs in developing these membranes have been highlighted and discussed. Consequently, some conclusions and future directions are listed as follows:

(1) Membranes that are fabricated directly with ILs (i.e., BILM, EILM, and SILM) are mainly used for removing organics and metal ions based on extraction. Such membranes often suffer from poor stability due to either the leaching of ILs into the liquid phase or the emulsion swelling and breakage (specifically involving EILM).

(2) PILMs are more stable than IL membranes due to their larger molecular structures. They exhibited good performance in applications such as the removal of metal ions and organic dyes, desalination, the concentration of proteins, and oil/water separation. The PILMs have shown great potential but further studies on them are required.

(3) The stability of the membranes can be improved tremendously if ILs/PILs are blended with polymers due to physical interactions such as hydrogen bonds, π-π stacking, or electrostatic interactions. These types of membranes have been widely studied for organophilic PV and separation of metal/organic from water. However, the issue of IL leaching may still exist due to weak physical interactions.

(4) ILs/PILs can be used to chemically modify polymeric membranes or membrane components (like fillers) to improve the separation performance and membrane stability due to their strong covalent bonds. These membranes showed promising performance and excellent stability in PV, RO, MF, and RO, etc. Some large-scale demonstrations are needed to promote industrial applications.

(5) The use of ILs as a solvent to dissolve polymers (especially those which are too rigid to be dissolved by traditional solvents) and as reaction media for the interfacial polymerization to fabricate liquid membranes are important because this option is less hazardous and more sustainable. Currently, the use of ILs to dissolve polymers and fabricate membranes have been studied a great deal. Therefore, further efforts must be directed to the study of the mechanisms for the dissolution and phase inversion of IL-polymer solutions. Since the use of ILs as reaction media for the thin-film

membranes just started recently, more ILs must be investigated for applications in this area, and their reaction mechanism must be explored to fully understand how to precisely control their reaction processes.

(6) To make the usage of ILs in developing liquid separation membranes more sustainable and economic, more efforts should be paid to looking for efficient methods for the regeneration, recovery, and removal of ILs in these special cases.

Author Contributions: D.Z. did literature retrieval, collected data, and drafted the manuscript. D.H. designed the manuscript framework, made charts, analyzed data, wrote and revised the manuscript. Y.H. participated in the literature search, extraction, analysis of the data. A.-R.I., A.Y., and J.P. made suggestions on data analysis and the revision of the manuscript. G.Z. proposed and revised the manuscript. All authors have read and agreed to the published version of the manuscript.

Funding: This research was funded by the National Natural Science Foundation of China (No. 21808072), Natural Science Foundation of Fujian Province (No. 2019J01075), Quanzhou City Science & Technology Program of China (No. 2018C129R), and Postgraduates' Innovative Fund in Scientific Research of Huaqiao University.

Conflicts of Interest: The authors declare no competing financial interest.

References

1. Li, X.; Liu, Y.; Wang, J.; Gascon, J.; Li, J.; van der Bruggen, B. Metal-organic frameworks based membranes for liquid separation. *Chem. Soc. Rev.* **2017**, *46*, 7124–7144. [CrossRef] [PubMed]
2. Razali, M.; Kim, J.F.; Attfield, M.; Budd, P.M.; Drioli, E.; Lee, Y.M.; Szekely, G. Sustainable wastewater treatment and recycling in membrane manufacturing. *Green Chem.* **2015**, *17*, 5196–5205. [CrossRef]
3. Marino, T.; Russo, F.; Criscuoli, A.; Figoli, A. TamiSolve® NxG as novel solvent for polymeric membrane preparation. *J. Membr. Sci.* **2017**, *542*, 418–429. [CrossRef]
4. Cseri, L.; Szekely, G. Towards cleaner PolarClean: Efficient synthesis and extended applications of the polar aprotic solvent methyl 5-(dimethylamino)-2-methyl-5-oxopentanoate. *Green Chem.* **2019**, *21*, 4178–4188. [CrossRef]
5. Marino, T.; Galiano, F.; Molino, A.; Figoli, A. New frontiers in sustainable membrane preparation: Cyrene™ as green bioderived solvent. *J. Membr. Sci.* **2019**, *580*, 224–234. [CrossRef]
6. Russo, F.; Galiano, F.; Pedace, F.; Aricò, F.; Figoli, A. Dimethyl isosorbide as a green solvent for sustainable ultrafiltration and microfiltration membrane preparation. *ACS Sustain. Chem. Eng.* **2019**, *8*, 659–668. [CrossRef]
7. Le Phuong, H.A.; Ayob, N.A.I.; Blanford, C.F.; Rawi, N.F.M.; Szekely, G. Nonwoven membrane supports from renewable resources: Bamboo fiber reinforced poly(lactic acid) composites. *ACS Sustain. Chem. Eng.* **2019**, *7*, 11885–11893. [CrossRef]
8. Lu, J.; Yan, F.; Texter, J. Advanced applications of ionic liquids in polymer science. *Prog. Polym. Sci.* **2009**, *34*, 431–448. [CrossRef]
9. Han, D.; Row, K.H. Recent applications of ionic liquids in separation technology. *Molecules* **2010**, *15*, 2405–2426. [CrossRef]
10. Noble, R.D.; Gin, D.L. Perspective on ionic liquids and ionic liquid membranes. *J. Membr. Sci.* **2011**, *369*, 1–4. [CrossRef]
11. Karkhanechi, H.; Salmani, S.; Asghari, M. A review on gas separation applications of supported ionic liquid membranes. *ChemBioEng Rev.* **2015**, *2*, 290–302. [CrossRef]
12. Wang, J.; Luo, J.; Feng, S.; Li, H.; Wan, Y.; Zhang, X. Recent development of ionic liquid membranes. *Green Energy Environ.* **2016**, *1*, 43–61. [CrossRef]
13. Salar-Garcia, M.J.; Ortiz-Martinez, V.M.; Hernandez-Fernandez, F.J.; Rios, A.P.d.L.; Quesada-Medina, J. Ionic liquid technology to recover volatile organic compounds (VOCs). *J. Hazard. Mater.* **2017**, *321*, 484–499. [CrossRef] [PubMed]
14. Rynkowska, E.; Fatyeyeva, K.; Kujawski, W. Application of polymer-based membranes containing ionic liquids in membrane separation processes: A critical review. *Rev. Chem. Eng.* **2018**, *34*, 341–363. [CrossRef]
15. Yan, X.; Anguille, S.; Bendahan, M.; Moulin, P. Ionic liquids combined with membrane separation processes: A review. *Sep. Purif. Technol.* **2019**, *222*, 230–253. [CrossRef]

16. Li, J.L.; Zhang, Y.; Zhang, S.; Liu, M.; Li, X.; Cai, T. Hyperbranched poly(ionic liquid) functionalized poly(ether sulfone) membranes as healable antifouling coatings for osmotic power generation. *J. Mater. Chem. A* **2019**, *7*, 8167–8176. [CrossRef]
17. Mustafa, M.Z.u.; bin Mukhtar, H.; Md Nordin, N.A.H.; Mannan, H.A.; Nasir, R.; Fazil, N. Recent developments and applications of ionic liquids in gas separation membranes. *Chem. Eng. Technol.* **2019**, *42*, 2580–2593. [CrossRef]
18. Isosaari, P.; Srivastava, V.; Sillanpää, M. Ionic liquid-based water treatment technologies for organic pollutants: Current status and future prospects of ionic liquid mediated technologies. *Sci. Total. Environ.* **2019**, *690*, 604–619. [CrossRef]
19. Foong, C.Y.; Wirzal, M.D.H.; Bustam, M.A. A review on nanofibers membrane with amino-based ionic liquid for heavy metal removal. *J. Mol. Liq.* **2020**, *297*, 111793. [CrossRef]
20. Aguilar, M.; Luis, C.J. *Solvent Extraction and Liquid Membranes: Fundamentals and Applications Innew Materials*; CRC Press Taylor & Francis Group: Boca Raton, FL, USA, 2008.
21. Qian, W.; Texter, J.; Yan, F. Frontiers in poly(ionic liquid)s: Syntheses and applications. *Chem. Soc. Rev.* **2017**, *46*, 1124–1159. [CrossRef]
22. Malik, M.A.; Hashim, M.A.; Nabi, F. Ionic liquids in supported liquid membrane technology. *Chem. Eng. J.* **2011**, *171*, 242–254. [CrossRef]
23. Winterton, N. Solubilization of polymers by ionic liquids. *J. Mater. Chem.* **2006**, *16*, 4281–4293. [CrossRef]
24. Xing, D.Y.; Peng, N.; Chung, T.-S. Investigation of unique interactions between cellulose acetate and ionic liquid [EMIM]SCN, and their influences on hollow fiber ultrafiltration membranes. *J. Membr. Sci.* **2011**, *380*, 87–97. [CrossRef]
25. Zhang, S.-J.; Liu, Y.-R.; Nie, Y. Research review of dissolving natural polymer materials with ionic liquids and green spinning technology. *J. Light Ind.* **2016**, *31*, 1–14.
26. Xing, D.Y.; Chan, S.Y.; Chung, T.S. The ionic liquid [EMIM]OAc as a solvent to fabricate stable polybenzimidazole membranes for organic solvent nanofiltration. *Green Chem.* **2014**, *16*, 1383–1392. [CrossRef]
27. Hua, D.; Japip, S.; Wang, K.Y.; Chung, T.-S. Green design of poly(m-phenylene isophthalamide)-based thin-film composite membranes for organic solvent nanofiltration and concentrating lecithin in hexane. *ACS Sustain. Chem. Eng.* **2018**, *6*, 10696–10705. [CrossRef]
28. Mariën, H.; Bellings, L.; Hermans, S.; Vankelecom, I.F. Sustainable process for the preparation of high-performance thin-film composite membranes using ionic liquids as the reaction medium. *ChemSusChem* **2016**, *9*, 1101–1111. [CrossRef] [PubMed]
29. Díaz, M.; Ortiz, A.; Ortiz, I. Progress in the use of ionic liquids as electrolyte membranes in fuel cells. *J. Membr. Sci.* **2014**, *469*, 379–396. [CrossRef]
30. Shen, M.; Han, Y.; Lin, X.; Ding, B.; Zhang, L.; Zhang, X. Preparation and electrochemical performances of porous polypyrrole film by interfacial polymerization. *J. Appl. Polym. Sci.* **2013**, *127*, 2938–2944. [CrossRef]
31. Yasuda, T.; Watanabe, M.J.M.B. Protic ionic liquids: Fuel cell applications. *Mrs. Bull.* **2013**, *38*, 560–566. [CrossRef]
32. Leones, R.; Reis, P.M.; Sabadini, R.C.; Esperança, J.M.S.S.; Pawlicka, A.; Silva, M.M. Chitosan polymer electrolytes doped with a dysprosium ionic liquid. *J. Polym. Res.* **2020**, *27*, 45. [CrossRef]
33. Bakonyi, P.; Koók, L.; Rozsenberszki, T.; Tóth, G.; Bélafi-Bakó, K.; Nemestóthy, N. Development and application of supported ionic liquid membranes in microbial fuel cell technology: A concise overview. *Membranes* **2020**, *10*, 16. [CrossRef] [PubMed]
34. Colburn, A.; Wanninayake, N.; Kim, D.Y.; Bhattacharyya, D. Cellulose-graphene quantum dot composite membranes using ionic liquid. *J. Membr. Sci.* **2018**, *556*, 293–302. [CrossRef] [PubMed]
35. Li, X.-L.; Zhu, L.-P.; Zhu, B.-K.; Xu, Y.-Y. High-flux and anti-fouling cellulose nanofiltration membranes prepared via phase inversion with ionic liquid as solvent. *Sep. Purif. Technol.* **2011**, *83*, 66–73. [CrossRef]
36. Ślusarczyk, C.; Fryczkowska, B. Structure-property relationships of pure cellulose and GO/CEL membranes regenerated from ionic liquid solutions. *Polymers* **2019**, *11*, 1178. [CrossRef] [PubMed]
37. Mohammed, S.; Hameed, M.S. Extraction of 4-nitrophenol from aqueous solutions using bulk ionic liquid membranes. *Int. J. Curr. Eng. Technol.* **2016**, *6*, 542–550.
38. Lakshmi, A.B.; Sindhu, S.; Venkatesan, S. Performance of ionic liquid as bulk liquid membrane for chlorophenol removal. *Int. J. ChemTech Res.* **2013**, *5*, 1129–1137.

39. Fortunato, R.; González-Muñoz, M.J.; Kubasiewicz, M.; Luque, S.; Alvarez, J.R.; Afonso, C.A.M.; Coelhoso, I.M.; Crespo, J.G. Liquid membranes using ionic liquids: The influence of water on solute transport. *J. Membr. Sci.* **2005**, *249*, 153–162. [CrossRef]
40. Baylan, N.; Çehreli, S. Removal of acetic acid from aqueous solutions using bulk ionic liquid membranes: A transport and experimental design study. *Sep. Purif. Technol.* **2019**, *224*, 51–61. [CrossRef]
41. Baylan, N.; Çehreli, S. Ionic liquids as bulk liquid membranes on levulinic acid removal: A design study. *J. Mol. Liq.* **2018**, *266*, 299–308. [CrossRef]
42. Chakraborty, M.; Bart, H.-J. Highly selective and efficient transport of toluene in bulk ionic liquid membranes containing Ag^+ as carrier. *Fuel Process. Technol.* **2007**, *88*, 43–49. [CrossRef]
43. Branco, L.C.; Crespo, J.G.; Afonso, C.A.M. Studies on the selective transport of organic compounds by using ionic liquids as novel supported liquid membranes. *Chem.-A Eur. J.* **2002**, *8*, 3865–3871. [CrossRef]
44. Branco, L.C.; Crespo, J.G.; Afonso, C.A.M. Ionic liquids as an efficient bulk membrane for the selective transport of organic compounds. *J. Phys. Org. Chem.* **2008**, *21*, 718–723. [CrossRef]
45. Kogelnig, D.; Stojanovic, A.; Jirsa, F.; Körner, W.; Krachler, R.; Keppler, B.K. Transport and separation of iron(III) from nickel(II) with the ionic liquid trihexyl(tetradecyl)phosphonium chloride. *Sep. Purif. Technol.* **2010**, *72*, 56–60. [CrossRef]
46. Balasubramanian, A.; Venkatesan, S. Removal of phenolic compounds from aqueous solutions by emulsion liquid membrane containing Ionic Liquid $[BMIM]^+[PF_6]^-$ in Tributyl phosphate. *Desalination* **2012**, *289*, 27–34. [CrossRef]
47. Lende, A.B.; Dinker, M.K.; Bhosale, V.K.; Kamble, S.P.; Meshram, P.D.; Kulkarni, P.S. Emulsion ionic liquid membranes (EILMs) for removal of Pb(ii) from aqueous solutions. *RSC Adv.* **2014**, *4*, 52316–52323. [CrossRef]
48. Alguacil, F.J.; López, F.A.; García-Díaz, I.; Rodriguez, O. Cadmium(II) transfer using (TiOAC) ionic liquid as carrier in a smart liquid membrane technology. *Chem. Eng. Process. Process Intensif.* **2016**, *99*, 192–196. [CrossRef]
49. Fadeev, A.G.; Meagher, M.M. Opportunities for ionic liquids in recovery of biofuels. *Chem. Commun.* **2001**, 295–296. [CrossRef]
50. Lozano, L.J.; Godínez, C.; Ríos, A.P.d.L.; Hernández-Fernández, F.J.; Sánchez-Segado, S.; Alguacil, F.J. Recent advances in supported ionic liquid membrane technology. *J. Membr. Sci.* **2011**, *376*, 1–14. [CrossRef]
51. Kárászová, M.; Kacirková, M.; Friess, K.; Izák, P. Progress in separation of gases by permeation and liquids by pervaporation using ionic liquids: A review. *Sep. Purif. Technol.* **2014**, *132*, 93–101. [CrossRef]
52. Riisagera, A.; Fehrmanna, R.; Haumannb, M.; Wasserscheidb, P. Supported ionic liquids: Versatile reaction and separation media. *Top. Catal.* **2006**, *40*, 91–102. [CrossRef]
53. Mai, N.L.; Kim, S.H.; Ha, S.H.; Shin, H.S.; Koo, Y.-M. Selective recovery of acetone-butanol-ethanol from aqueous mixture by pervaporation using immobilized ionic liquid polydimethylsiloxane membrane. *Korean J. Chem. Eng.* **2013**, *30*, 1804–1809. [CrossRef]
54. Zhang, F.; Sun, W.; Liu, J.; Zhang, W.; Ren, Z. Extraction separation of toluene/cyclohexane with hollow fiber supported ionic liquid membrane. *Korean J. Chem. Eng.* **2014**, *31*, 1049–1056. [CrossRef]
55. Li, W.; Molina-Fernández, C.; Estager, J.; Monbaliu, J.-C.M.; Debecker, D.P.; Luis, P. Supported ionic liquid membranes for the separation of methanol/dimethyl carbonate mixtures by pervaporation. *J. Membr. Sci.* **2020**, *598*, 117790. [CrossRef]
56. Matsumoto, M.; Panigrahi, A.; Murakami, Y.; Kondo, K. Effect of ammonium- and phosphonium-based ionic liquids on the separation of lactic acid by supported ionic liquid membranes (SILMs). *Membranes* **2011**, *1*, 98–108. [CrossRef] [PubMed]
57. Pilli, S.R.; Banerjee, T.; Mohanty, K. Performance of different ionic liquids to remove phenol from aqueous solutions using supported liquid membrane. *Desalin. Water Treat.* **2014**, *54*, 3062–3072. [CrossRef]
58. Panigrahi, A.; Pilli, S.R.; Mohanty, K. Selective separation of Bisphenol A from aqueous solution using supported ionic liquid membrane. *Sep. Purif. Technol.* **2013**, *107*, 70–78. [CrossRef]
59. Abejón, R.; Rabadán, J.; Lanza, S.; Abejón, A.; Garea, A.; Irabien, A. Supported ionic liquid membranes for separation of lignin aqueous solutions. *Processes* **2018**, *6*, 143. [CrossRef]
60. Jean, E.; Villemin, D.; Hlaibi, M.; Lebrun, L. Heavy metal ions extraction using new supported liquid membranes containing ionic liquid as carrier. *Sep. Purif. Technol.* **2018**, *201*, 1–9. [CrossRef]
61. Zante, G.; Boltoeva, M.; Masmoudi, A.; Barillon, R.; Trébouet, D. Lithium extraction from complex aqueous solutions using supported ionic liquid membranes. *J. Membr. Sci.* **2019**, *580*, 62–76. [CrossRef]

62. Yuan, J.; Mecerreyes, D.; Antonietti, M. Poly(ionic liquid)s: An update. *Prog. Polym. Sci.* **2013**, *38*, 1009–1036. [CrossRef]
63. Kausar, A. Research progress in frontiers of poly(ionic liquid)s: A review. *Polym. Technol.* **2017**, *56*, 1823–1838. [CrossRef]
64. Tang, Y.; Tang, B.; Wu, P. Preparation of a positively charged nanofiltration membrane based on hydrophilic–hydrophobic transformation of a poly(ionic liquid). *J. Mater. Chem. A* **2015**, *3*, 12367–12376. [CrossRef]
65. Täuber, K.; Zimathies, A.; Yuan, J. Porous membranes built up from hydrophilic poly(ionic liquid)s. *Macromol. Rapid Commun.* **2015**, *36*, 2176–2180. [CrossRef] [PubMed]
66. Kohno, Y.; Gin, D.L.; Noble, R.D.; Ohno, H. A thermoresponsive poly(ionic liquid) membrane enables concentration of proteins from aqueous media. *Chem. Commun.* **2016**, *52*, 7497–7500. [CrossRef] [PubMed]
67. Shao, Y.; Jiang, Z.; Zhang, Y.; Wang, T.; Zhao, P.; Zhang, Z.; Yuan, J.; Wang, H. All-poly(ionic liquid) membrane-derived porous carbon membranes: Scalable synthesis and application for photothermal conversion in seawater desalination. *ACS Nano* **2018**, *12*, 11704–11710. [CrossRef] [PubMed]
68. Izák, P.; Ruth, W.; Fei, Z.; Dyson, P.J.; Kragl, U. Selective removal of acetone and butan-1-ol from water with supported ionic liquid–polydimethylsiloxane membrane by pervaporation. *Chem. Eng. J.* **2008**, *139*, 318–321. [CrossRef]
69. Izák, P.; Friess, K.; Hynek, V.; Ruth, W.; Fei, Z.; Dyson, J.P.; Kragl, U. Separation properties of supported ionic liquid–polydimethylsiloxane membrane in pervaporation process. *Desalination* **2009**, *241*, 182–187. [CrossRef]
70. Rdzanek, P.; Heitmann, S.; Górak, A.; Kamiński, W. Application of supported ionic liquid membranes (SILMs) for biobutanol pervaporation. *Sep. Purif. Technol.* **2015**, *155*, 83–88. [CrossRef]
71. Ong, Y.T.; Tan, S.H. Pervaporation separation of a ternary azeotrope containing ethyl acetate, ethanol and water using a buckypaper supported ionic liquid membrane. *Chem. Eng. Res. Des.* **2016**, *109*, 116–126. [CrossRef]
72. Tang, S.; Dong, Z.; Zhu, X.; Zhao, Q. A poly(ionic liquid) complex membrane for pervaporation dehydration of acidic water-isopropanol mixtures. *J. Membr. Sci.* **2019**, *576*, 59–65. [CrossRef]
73. He, Z.; Meng, M.; Yan, L.; Zhu, W.; Sun, F.; Yan, Y.; Liu, Y.; Liu, S. Fabrication of new cellulose acetate blend imprinted membrane assisted with ionic liquid ([BMIM]Cl) for selective adsorption of salicylic acid from industrial wastewater. *Sep. Purif. Technol.* **2015**, *145*, 63–74. [CrossRef]
74. Chen, L.; Chen, J. Asymmetric membrane containing ionic liquid [A336][P507] for the preconcentration and separation of heavy rare earth lutetium. *ACS Sustain. Chem. Eng.* **2016**, *4*, 2644–2650. [CrossRef]
75. Elias, G.; Díez, S.; Fontàs, C. System for mercury preconcentration in natural waters based on a polymer inclusion membrane incorporating an ionic liquid. *J. Hazard. Mater.* **2019**, *371*, 316–322. [CrossRef] [PubMed]
76. Wang, Z.; Sun, Y.; Tang, N.; Miao, C.; Wang, Y.; Tang, L.; Wang, S.; Yang, X. Simultaneous extraction and recovery of gold(I) from alkaline solutions using an environmentally benign polymer inclusion membrane with ionic liquid as the carrier. *Sep. Purif. Technol.* **2019**, *222*, 136–144. [CrossRef]
77. Sun, Y.; Wang, Z.; Wang, Y.; Liu, M.; Li, S.; Tang, L.; Wang, S.; Yang, X.; Ji, S. Improved transport of gold(I) from aurocyanide solution using a green ionic liquid-based polymer inclusion membrane with in-situ electrodeposition. *Chem. Eng. Res. Des.* **2020**, *153*, 136–145. [CrossRef]
78. Xi, T.; Lu, Y.; Ai, X.; Tang, L.; Yao, L.; Hao, W.; Cui, P. Ionic liquid copolymerized polyurethane membranes for pervaporation separation of benzene/cyclohexane mixtures. *Polymer* **2019**, *185*, 121948. [CrossRef]
79. Zhang, J.; Qin, Z.; Yang, L.; Guo, H.; Han, S. Activation promoted ionic liquid modification of reverse osmosis membrane towards enhanced permeability for desalination. *J. Taiwan Inst. Chem. Eng.* **2017**, *80*, 25–33. [CrossRef]
80. Xiao, H.-F.; Chu, C.-H.; Xu, W.-T.; Chen, B.-Z.; Ju, X.-H.; Xing, W.; Sun, S.-P. Amphibian-inspired amino acid ionic liquid functionalized nanofiltration membranes with high water permeability and ion selectivity for pigment wastewater treatment. *J. Membr. Sci.* **2019**, *586*, 44–52. [CrossRef]
81. He, B.; Peng, H.; Chen, Y.; Zhao, Q. High performance polyamide nanofiltration membranes enabled by surface modification of imidazolium ionic liquid. *J. Membr. Sci.* **2020**, *608*, 118202. [CrossRef]
82. Liu, L.; Xie, X.; Zambare, R.S.; Selvaraj, A.P.J.; Sowrirajalu, B.N.; Song, X.; Tang, C.Y.; Gao, C. Functionalized graphene oxide modified polyethersulfone membranes for low-pressure anionic dye/salt Fractionation. *Polymer* **2018**, *10*, 795. [CrossRef] [PubMed]

83. Shi, F.; Ma, Y.; Ma, J.; Wang, P.; Sun, W. Preparation and characterization of PVDF/TiO$_2$ hybrid membranes with ionic liquid modified nano-TiO$_2$ particles. *J. Membr. Sci.* **2013**, *427*, 259–269. [CrossRef]
84. Abraham, J.; Jose, T.; Moni, G.; George, S.C.; Kalarikkal, N.; Thomas, S. Ionic liquid modified multiwalled carbon nanotube embedded styrene butadiene rubber membranes for the selective removal of toluene from toluene/methanol mixture via pervaporation. *J. Taiwan Inst. Chem. Eng.* **2019**, *95*, 594–601. [CrossRef]
85. Tang, W.; Lou, H.; Li, Y.; Kong, X.; Wu, Y.; Gu, X. Ionic liquid modified graphene oxide-PEBA mixed matrix membrane for pervaporation of butanol aqueous solutions. *J. Membr. Sci.* **2019**, *581*, 93–104. [CrossRef]
86. Yu, L.; Zhang, Y.; Wang, Y.; Zhang, H.; Liu, J. High flux, positively charged loose nanofiltration membrane by blending with poly (ionic liquid) brushes grafted silica spheres. *J. Hazard. Mater.* **2015**, *287*, 373–383. [CrossRef] [PubMed]
87. Yu, L.; Deng, J.; Wang, H.; Liu, J.; Zhang, Y. Improved salts transportation of a positively charged loose nanofiltration membrane by introduction of poly(ionic liquid) functionalized hydrotalcite nanosheets. *ACS Sustain. Chem. Eng.* **2016**, *4*, 3292–3304. [CrossRef]
88. Barroso, T.; Temtem, M.; Hussain, A.; Aguiar-Ricardo, A.; Roque, A.C.A. Preparation and characterization of a cellulose affinity membrane for human immunoglobulin G (IgG) purification. *J. Membr. Sci.* **2010**, *348*, 224–230. [CrossRef]
89. Livazovic, S.; Li, Z.; Behzad, A.R.; Peinemann, K.V.; Nunes, S.P. Cellulose multilayer membranes manufacture with ionic liquid. *J. Membr. Sci.* **2015**, *490*, 282–293. [CrossRef]
90. Nevstrueva, D.; Pihlajamäki, A.; Mänttäri, M. Effect of a TiO$_2$ additive on the morphology and permeability of cellulose ultrafiltration membranes prepared via immersion precipitation with ionic liquid as a solvent. *Cellulose* **2015**, *22*, 3865–3876. [CrossRef]
91. Kim, D.; Livazovic, S.; Falca, G.; Nunes, S.P. Oil–water separation using membranes manufactured from cellulose/Ionic liquid solutions. *ACS Sustain. Chem. Eng.* **2018**, *7*, 5649–5659. [CrossRef]
92. Abdellah, M.H.; Pérez-Manríquez, L.; Puspasari, T.; Scholes, C.A.; Kentish, S.E.; Peinemann, K.-V. A catechin/cellulose composite membrane for organic solvent nanofiltration. *J. Membr. Sci.* **2018**, *567*, 139–145. [CrossRef]
93. Colburn, A.; Vogler, R.J.; Patel, A.; Bezold, M.; Craven, J.; Liu, C.; Bhattacharyya, D. Composite membranes derived from cellulose and lignin sulfonate for selective separations and antifouling aspects. *Nanomaterials* **2019**, *9*, 867. [CrossRef] [PubMed]
94. Falca, G.; Musteata, V.-E.; Behzad, A.R.; Chisca, S.; Nunes, S.P. Cellulose hollow fibers for organic resistant nanofiltration. *J. Membr. Sci.* **2019**, *586*, 151–161. [CrossRef]
95. Esfahani, M.R.; Taylor, A.; Serwinowski, N.; Parkerson, Z.J.; Confer, M.P.; Kammakakam, I.; Bara, J.E.; Esfahani, A.R.; Mahmoodi, S.N.; Koutahzadeh, N.; et al. Sustainable novel bamboo-based membranes for water treatment fabricated by regeneration of bamboo waste fibers. *ACS Sustain. Chem. Eng.* **2020**, *8*, 4225–4235. [CrossRef]
96. Xing, D.Y.; Peng, N.; Chung, T.-S. Formation of cellulose acetate membranes via phase inversion using ionic liquid, [BMIM]SCN, as the solvent. *Ind. Eng. Chem. Res.* **2010**, *49*, 8761–8769. [CrossRef]
97. Kim, D.; Le, N.L.; Nunes, S.P. The effects of a co-solvent on fabrication of cellulose acetate membranes from solutions in 1-ethyl-3-methylimidazolium acetate. *J. Membr. Sci.* **2016**, *520*, 540–549. [CrossRef]
98. Xing, D.Y.; Chan, S.Y.; Chung, T.-S. Fabrication of porous and interconnected PBI/P84 ultrafiltration membranes using [EMIM]OAc as the green solvent. *Chem. Eng. Sci.* **2013**, *87*, 194–203. [CrossRef]
99. Kim, D.; Nunes, S.P. Poly(ether imide sulfone) membranes from solutions in ionic liquids. *Ind. Eng. Chem. Res.* **2017**, *56*, 14914–14922. [CrossRef]
100. Gsaiz, P.; Lopes, A.C.; Barker, S.E.; de Luis, R.F.; Arriortua, M.I. Ionic liquids for the control of the morphology in poly(vinylidene fluoride-co-hexafluoropropylene) membranes. *Mater. Des.* **2018**, *155*, 325–333. [CrossRef]
101. Chisca, S.; Marchesi, T.; Falca, G.; Musteata, V.-E.; Huang, T.; Abou-Hamad, E.; Nunes, S.P. Organic solvent and thermal resistant polytriazole membranes with enhanced mechanical properties cast from solutions in non-toxic solvents. *J. Membr. Sci.* **2020**, *597*, 117634. [CrossRef]
102. Sukma, F.M.; Çulfaz-Emecen, P.Z. Cellulose membranes for organic solvent nanofiltration. *J. Membr. Sci.* **2018**, *545*, 329–336. [CrossRef]
103. Li, C.; Zhu, Y.; Lv, R.; Na, B.; Chen, B. Poly(vinylidene fluoride) membrane with piezoelectric β-form prepared by immersion precipitation from mixed solvents containing an ionic liquid. *J. Appl. Polym. Sci.* **2014**, *131*, 40505. [CrossRef]

104. Kim, D.; Moreno, N.; Nunes, S.P. Fabrication of polyacrylonitrile hollow fiber membranes from ionic liquid solutions. *Polym. Chem.* **2016**, *7*, 113–124. [CrossRef]
105. Mariën, H.; Vankelecom, I.F.J. Optimization of the ionic liquid-based interfacial polymerization system for the preparation of high-performance, low-fouling RO membranes. *J. Membr. Sci.* **2018**, *556*, 342–351. [CrossRef]
106. Hartanto, Y.; Corvilain, M.; Mariën, H.; Janssen, J.; Vankelecom, I.F.J. Interfacial polymerization of thin-film composite forward osmosis membranes using ionic liquids as organic reagent phase. *J. Membr. Sci.* **2020**, *601*, 117869. [CrossRef]
107. Feng, R.; Zhao, D.; Guo, Y. Revisiting characteristics of ionic liquids: A review for further application development. *J. Environ. Prot.* **2010**, *01*, 95–104. [CrossRef]
108. Fernández, J.F.; Neumann, J.; Thöming, J. Regeneration, recovery and removal of ionic Liquids. *Curr. Org. Chem.* **2011**, *15*, 1992–2014. [CrossRef]
109. Kudlak, B.; Owczarek, K.; Namiesnik, J. Selected issues related to the toxicity of ionic liquids and deep eutectic solvents -a review. *Environ. Sci. Pollut. Res.* **2015**, *22*, 11975–11992. [CrossRef]
110. Liu, X.; Wang, W. The application of nanofiltration technology in recovery of ionic liquids from spinning wastewater. *Appl. Mech. Mater.* **2012**, *178–181*, 499–502. [CrossRef]
111. Mai, N.L.; Ahn, K.; Koo, Y.-M. Methods for recovery of ionic liquids-a review. *Process. Biochem.* **2014**, *49*, 872–881. [CrossRef]
112. Avram, A.M.; Ahmadiannamini, P.; Qian, X.; Wickramasinghe, S.R. Nanofiltration membranes for ionic liquid recovery. *Sep. Sci. Technol.* **2017**, *52*, 2098–2107. [CrossRef]

Publisher's Note: MDPI stays neutral with regard to jurisdictional claims in published maps and institutional affiliations.

© 2020 by the authors. Licensee MDPI, Basel, Switzerland. This article is an open access article distributed under the terms and conditions of the Creative Commons Attribution (CC BY) license (http://creativecommons.org/licenses/by/4.0/).

Article

Understanding the Role of Pattern Geometry on Nanofiltration Threshold Flux

Anna Malakian [1], Zuo Zhou [2], Lucas Messick [1], Tara N. Spitzer [1], David A. Ladner [2] and Scott M. Husson [1,*]

[1] Department of Chemical and Biomolecular Engineering, Clemson University, Clemson, SC 29634, USA; amalaki@clemson.edu (A.M.); lmessic@clemson.edu (L.M.); tspitze@clemson.edu (T.N.S.)

[2] Department of Environmental Engineering and Earth Sciences, Clemson University, Clemson, SC 29634, USA; zuoz@clemson.edu (Z.Z.); ladner@clemson.edu (D.A.L.)

* Correspondence: shusson@clemson.edu

Received: 30 November 2020; Accepted: 18 December 2020; Published: 21 December 2020

Abstract: Colloidal fouling can be mitigated by membrane surface patterning. This contribution identifies the effect of different pattern geometries on fouling behavior. Nanoscale line-and-groove patterns with different feature sizes were applied by thermal embossing on commercial nanofiltration membranes. Threshold flux values of as-received, pressed, and patterned membranes were determined using constant flux, cross-flow filtration experiments. A previously derived combined intermediate pore blocking and cake filtration model was applied to the experimental data to determine threshold flux values. The threshold fluxes of all patterned membranes were higher than the as-received and pressed membranes. The pattern fraction ratio (PFR), defined as the quotient of line width and groove width, was used to analyze the relationship between threshold flux and pattern geometry quantitatively. Experimental work combined with computational fluid dynamics simulations showed that increasing the PFR leads to higher threshold flux. As the PFR increases, the percentage of vortex-forming area within the pattern grooves increases, and vortex-induced shielding increases. This study suggests that the PFR should be higher than 1 to produce patterned membranes with maximal threshold flux values. Knowledge generated in this study can be applied to other feature types to design patterned membranes for improved control over colloidal fouling.

Keywords: colloidal fouling; membrane patterning; membrane surface modification; threshold flux; thin-film composite membranes

1. Introduction

Colloidal fouling is an impediment to pressure-driven membrane operations. For nanofiltration (NF) membranes, fouling can occur by accumulation of particles such as colloidal silica, iron, aluminum, manganese oxides, and calcium carbonate precipitates on the membrane surface [1]. This foulant layer introduces an additional hydraulic resistance that reduces water permeability [2]. There have been systematic studies on reverse osmosis (RO) and NF colloidal fouling that examined the physical and chemical aspects of particle-membrane interactions and the application of antifouling functional groups on membranes surfaces [3–7]. Chemical modification is the most common method to decrease fouling [8–11]. While somewhat effective for this purpose, chemical modification can negatively impact other membrane performance metrics such as flux and selectivity [12].

Physical modification of membrane surfaces attempts to mitigate the fouling by decreasing the random roughness [3,13]. Introduction of ordered geometric or biomimetic patterns on membrane surfaces decreases fouling by influencing the hydrodynamics at the solid–liquid interface [14–18]. Different research groups have investigated different pattern shapes and sizes. Lee et al. [19] and

Won et al. [20] both designed microscale prism patterns. Jang et al. [21] studied both nanometer-and micrometer-scale patterns achieved through nanoimprint lithography. Zhou et al. [22] investigated several line-and-groove patterns, rectangular and circular pillars, and pyramids within the nanometer- to micrometer-scale range. Ling et al. [23] studied microscale pillars on RO membranes. Irregular shapes such as sharkskin mimetic patterns also have been investigated for improving biofouling resistance [24].

Recently, we demonstrated that the introduction of a nanoscale line-and-groove pattern on NF membranes could increase their threshold flux by 20–30% during filtration of colloidal suspensions [25]. Using atomic force microscopy of particle-fouled membranes, we observed below threshold flux that particles aligned along the pattern grooves in a way that appears to be consistent with the mechanism of selective deposition proposed by ElSherbiny et al. [26]. According to this mechanism, particles accumulate selectively in the pattern valleys due to lower shear stress. In our case, these low-shear regions occur within the pattern grooves, which delays cake layer formation and leads to higher values of threshold flux. Based on that work, we theorized that the pattern geometry would affect the threshold flux by altering the shear stress profiles at the interface, as we have seen in computational simulation work [22].

Computational simulations contribute to the analysis of transport phenomena adjacent to membrane surfaces, as the details of fluid flow within a membrane device can be difficult to measure experimentally. In the simulations, the Navier–Stokes, continuity, and convection–diffusion equations are fully coupled to describe fluid flow and mass transfer within a membrane channel [27,28]. Several research groups, ours included, have conducted computational fluid dynamics (CFD) simulations on patterned membranes, trying to understand their mechanisms for fouling control. One hypothesis is that increased turbulence at the apex of the pattern surface could lead to reduced deposition of foulants [15,20]. Others have posited that higher shear stress at the apex region efficiently reduces the attachment of the particles, keeping them away from the fouling region [19,23,29].

In this study, we investigated the influence of pattern geometry on threshold flux. We formed nanoscale line-and-groove patterns with different spacing dimensions on NF membranes through thermal embossing. Threshold flux was determined by applying a combined intermediate pore blocking and cake filtration model to the experimental data, as described previously [25,30]. CFD simulations were performed to analyze the velocity streamlines and shear stress profiles adjacent to the patterned membrane surfaces.

2. Materials and Methods

2.1. Materials

Ammonium hydroxide solution (ACS grade, 28.0–30.0% NH_3 basis), ethyl alcohol (anhydrous, 200 proof), N,N-dimethylformamide (DMF, 99.8%), succinic anhydride (99%), and tetraethyl orthosilicate (TEOS, 99.99%) were purchased from MilliporeSigma (St. Louis, MO, USA) and used as received. Polyamide thin-film NF membrane sheets (GE HL) were purchased from Sterlitech Corp. (Kent, WA, USA). The silicon line-and-groove stamps were purchased from Digi-Key Electronics (Thief River Falls, MN, USA). Deionized (DI) water was used to prepare the solutions.

2.2. Membrane Patterning

Membranes were patterned using thermal embossing. Membrane coupons (1.50 cm × 4.25 cm) were cut from membrane sheets. Silicon stamps with different feature sizes (Table 1) were used to pattern the membranes in a Carver press (model 3851-0, Wabash, IN, USA). The temperature of the hot press plates was set to 65 °C. A stamp was placed on top of the membrane with its polyamide selective layer facing upward, and the sample was sandwiched between two pieces of thermally conductive silicone rubber to distribute the force evenly across the membrane. The press plates were closed until the set pressure was reached. Embossing was conducted for 15 min with an applied pressure of

3.55 MPa. For "pressed" membranes, the same procedure was applied using a flat silicon wafer in place of a pattern stamp.

Table 1. Cross-sectional image and dimensions of the line-and-groove pattern for each stamp.

Pattern ID	Pattern Geometry
P27	150 nm / 60 nm / 550 nm
P31	200 nm / 60 nm / 640 nm
P50	300 nm / 60 nm / 600 nm
P60	300 nm / 60 nm / 500 nm

2.3. Membrane Surface Geometry Characterization

Non-contact tapping-mode atomic force microscopy (AFM) was employed to observe the membrane surface morphology. Images were taken over 25 μm × 25 μm areas at a scan rate of 0.5 Hz using a Bioscope AFM (Bruker, Inc., Billerica, MA, USA) with silicon cantilever probe tips (MikroMasch, Inc., HQ:NSC16/AL BS, Watsonville, CA, USA).

2.4. Filtration Experiments

Feed solutions for all experiments were prepared by dispersing 200 mg/L silica nanoparticles (SiNPs) in DI water. Monodisperse SiNPs with a particle size of 70 ± 30 nm were prepared using the Stöber–Fink–Bohn method described elsewhere [25]. Briefly, a solution comprising ethanol (25 mL), ammonium hydroxide solution (5 mL), and DI water (1 mL) was stirred vigorously while 10 mL of TEOS was added dropwise to the solution. The mixture solution was stirred overnight at room temperature, sonicated and centrifuged to separate nanoparticles from solution. SiNPs were dried under vacuum (0.06 bar) at 100 °C for 1 h.

Data collection for threshold flux determination was performed using a custom filtration system designed to operate with constant permeate flux in the recycle mode to avoid changes in concentration of the feed solution. The filtration system details were described elsewhere [25]. Cross-flow velocity was held constant at 0.25 m s^{-1}, and membrane samples were tested with the flow perpendicular to the patterns to reduce the rate of fouling relative to other orientations [31]. All experiments were conducted at 23 ± 1 °C. Prior to the filtration experiments, the feed solution was sonicated for 1 h, and membranes were preconditioned by operating the system with DI water for 15 min. Threshold flux values were measured for all six membrane types using the flux-stepping method. A starting permeate flux of 55 ± 5 L m^{-2} h^{-1} (LMH) was established, and transmembrane pressure (TMP) was adjusted to maintain constant flux for 10 min. Flux was increased stepwise in 20 LMH increments by adjusting the TMP. For each flux step, TMP again was adjusted to maintain constant flux for 10 min. TMP values ranged from 2.2 to 17.2 bar over the course of operation.

Constant-pressure flux decline experiments were conducted for all membrane types using an initial permeate flux of 140 ± 2 LMH, which was below the threshold flux for all samples. TMP was adjusted to have the same starting permeate flux for all samples. Cross-flow velocity was held constant at 0.25 m s^{-1}, and the flow direction was perpendicular to the surface patterns. Flux data were collected every minute for patterned, pressed, and as-received membranes. Each filtration measurement was repeated three times.

2.5. Computational Fluid Dynamics Simulations

Multiple membrane models with line-and-groove surface patterns were built for analysis. Details of the developed model were explained elsewhere [22]. Table 2 summarizes the parameters of the four geometries (four stamps) that were studied. The velocity and pressure values are consistent with the experimental settings. The models were run on COMSOL Multiphysics 5.3.

Table 2. Parameters of the geometries.

Pattern ID	Pattern Depth, h (nm)	Groove Size, d (nm)	Line Size, l (nm)	Cycle Length, $W = l + d$ (nm)
P27	60	400	150	550
P31	60	440	200	640
P50	60	300	300	600
P60	60	200	300	500

Equations (1) and (2) are governing equations. Equation (1) is the Navier–Stokes equation that is used to describe the motion of the fluid, and Equation (2) is the continuity equation. Details of the modeling conditions were described elsewhere [22].

$$\rho(\nabla \cdot \mathbf{u})\mathbf{u} = -\nabla P + \mu \nabla \cdot (\nabla \mathbf{u} + \nabla \mathbf{u}^T), \tag{1}$$

$$\nabla \cdot \mathbf{u} = 0, \tag{2}$$

3. Results and Discussion

3.1. Membrane Patterning

Table 1 shows the cross-sectional image and dimensions of each silicon stamp. Patterns consist of line-and-groove features with different spacing, from 150 to 300 nm for lines and 200 to 400 nm for grooves. The sample codes denote the percentage of the projected line surface area to total projected surface area of stamps. Pattern height was not considered as an independent parameter in this study due to the low thickness of the polyamide active layer. Precautions were made to avoid fracturing of the NF membrane active layer during patterning. The embossing pressure was set at 3.55 MPa to keep the local strain, defined by the height-to-pitch ratio of the pattern, below the cracking strain of nanoscale cross-linked polyamide films (14.04 ± 4.12% [32]), as Weinman and Husson [33] have suggested.

Figure 1 presents representative AFM images of all six membrane types. The groove depths and peak-to-peak distance for patterned membranes were determined by NanoScope Analysis 1.5 software and summarized along with local strain values in Table 3. While the average groove depth was limited to about 60 nm to avoid cracking of the active layer, the measured peak-to-peak distances are close to the stamp feature sizes, demonstrating successful replication of the stamp patterns on the membrane surfaces.

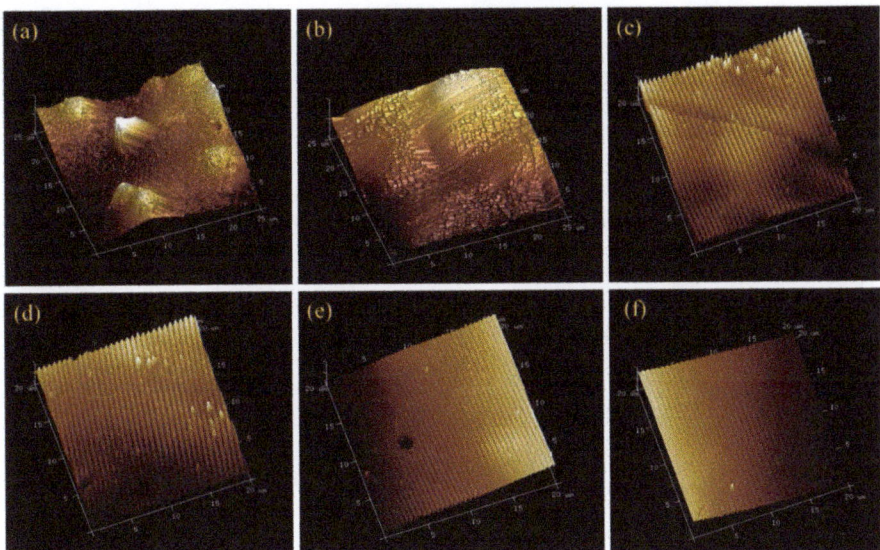

Figure 1. Representative AFM images of the six membrane types. (**a**) As-received membrane, (**b**) pressed membrane, (**c**) P27, (**d**) P31, (**e**) P50, and (**f**) P60. The scales are 25 µm × 25 µm × 600 nm for images (**a**,**b**) and 20 µm × 20 µm × 600 nm for images (**c**–**f**).

Table 3. Feature sizes on patterned membranes measured by AFM.

Pattern ID	Averaged Pattern Depth (nm)	Averaged Peak-to-Peak Distance (nm)	Local Strain (%)
P27	58 ± 6	530 ± 20	10.1 ± 3.2
P31	60 ± 8	600 ± 30	10.3 ± 5.6
P50	66 ± 7	680 ± 70	10.9 ± 5.4
P60	65 ± 6	510 ± 20	10.8 ± 6.2

3.2. Threshold Flux Measurements

Colloidal fouling is a surface phenomenon that is affected by physical and chemical properties of the membrane surface [12] and hydrodynamics at the membrane–solution interface [25]. All membranes in this study had the same chemical properties, and cross-flow velocity and SiNP type and concentration were held constant. Thus, differences in fouling behavior can be attributed to differences in membrane surface morphologies due to patterning.

To assess the effect of different pattern geometries on fouling behavior, threshold flux values of as-received, pressed, and four types of patterned membranes were determined using constant flux cross-flow filtration experiments. We selected threshold flux as the test metric because of its importance to industrial practice. Operating just below threshold flux offers the possibility of achieving a continuous high flux while consuming less energy and decreasing the frequency of membrane cleaning [34]. A previously derived combined intermediate pore blocking and cake filtration model [30] was applied to the experimental data to determine threshold flux values quantitatively and in a systematized way. The time-dependent transmembrane pressure (TMP_t) for the combined model is defined by Equation (3) [30]:

$$TMP_t = \frac{TMP_0(1 + K_c Jt)}{\left(\frac{1}{K_i} + \left(1 - \frac{1}{K_i}\right)\exp(-K_i Bt)\right)}, \qquad (3)$$

TMP_0 is initial transmembrane pressure [N/m^2], K_c is the cake filtration constant for cross-flow filtration [m^{-1}], K_i is the intermediate pore blocking constant for cross-flow filtration, and B is the particle resuspension rate [s^{-1}]. At the early stage of fouling, where the cake formation mechanism is absent, data can be fitted to the combined model with K_i equal to 1, and K_c values can be calculated. Details on this procedure are described elsewhere [25].

Figure 2 shows fitted values of K_c versus flux. The sharp rise in K_c shows the transition from intermediate pore blocking to cake filtration, which we use to define threshold flux. Figure 3 summarizes the threshold flux results, which depend strongly on the membrane surface structure. The threshold flux of all patterned membranes is higher than the as-received and pressed membranes. For patterned membranes, more fouling is observed when there is a low fraction of projected line surface area. Pressed membranes had higher threshold flux than as-received membranes. This result agrees with previous studies that showed how decreasing the nanoscale surface roughness of a membrane by pressing the membrane can decrease fouling [1,35]. The effectiveness of the "ordered roughness" on patterned membranes for increasing threshold flux could be related to increased membrane surface area and different hydrodynamic properties at the membrane–solution interface [26,33]. Patterning increases the overall surface area of the membrane. Since flux is calculated based on projected membrane area, increasing the overall area leads to higher flux values. It is known that the groove region of patterned membranes is more disposed to fouling as a result of unequal feed flow distribution [26], lower shear stress in groove regions compared to line regions [36], and lower hydrodynamic drag force compared to attractive interactions between foulant and membrane [37]. Therefore, more severe fouling may be expected for membranes with higher fractions of groove surface area, which agrees with the experimental results of this study; threshold flux increased in the following order: as-received < pressed < P27 < P31 < P50 < P60. To analyze the relationship between threshold flux and pattern geometry quantitatively, we defined the pattern fraction ratio (PFR) as the quotient of line width (b) and groove width (a).

$$PFR = \frac{b}{a}, \qquad (4)$$

To better understand the basis for this result, CFD simulations were carried out to analyze shear stresses and local flow behavior close to the membrane surfaces. Figure 5 shows the shear stress profiles of patterned membranes. Figures on the left show the 2D profiles on and near the membrane surface including part of the channel, and those on the right show the 3D profiles of the shear stress distribution on the membrane surface for all geometries. Higher shear stresses develop on the peaks and lower shear stresses are found in the valleys. Values decrease along the length of the channel. Simulations indicate that the P60 pattern has the highest average shear stress (0.46 Pa), while the P27 pattern has the lowest average shear stress (0.28 Pa). This finding agrees with the experimental results that show threshold flux is highest for P60 and lowest for P27 among patterned membranes; however, the differences in shear stress are not large. Therefore, streamline profiles for all pattern geometries were studied. Figure 6 shows streamline profiles of flat and patterned membranes. Vortices developed in all valley regions near the patterned membrane surfaces. Vortex formation separates bulk flow from flow inside the grooves, and the chance for particle transfer from grooves to bulk flow decreases, leading to increased fouling. Choi et al. [12] referred to this phenomenon as vortex-induced shielding. In addition, the velocity is lower inside the grooves than in the rest of the channel, which contributes to the enhanced fouling inside of grooves. Vortex formation was more complicated for the P27, P31, P50 patterns, as shown in Figure 5. Two small vortices developed, which may explain the higher rate of colloidal fouling. The area percentage of the vortex-forming region was calculated as Choi et al. [24] suggested. By increasing the groove width, the area percentage increases and therefore, the effect of vortex-induced shielding is higher.

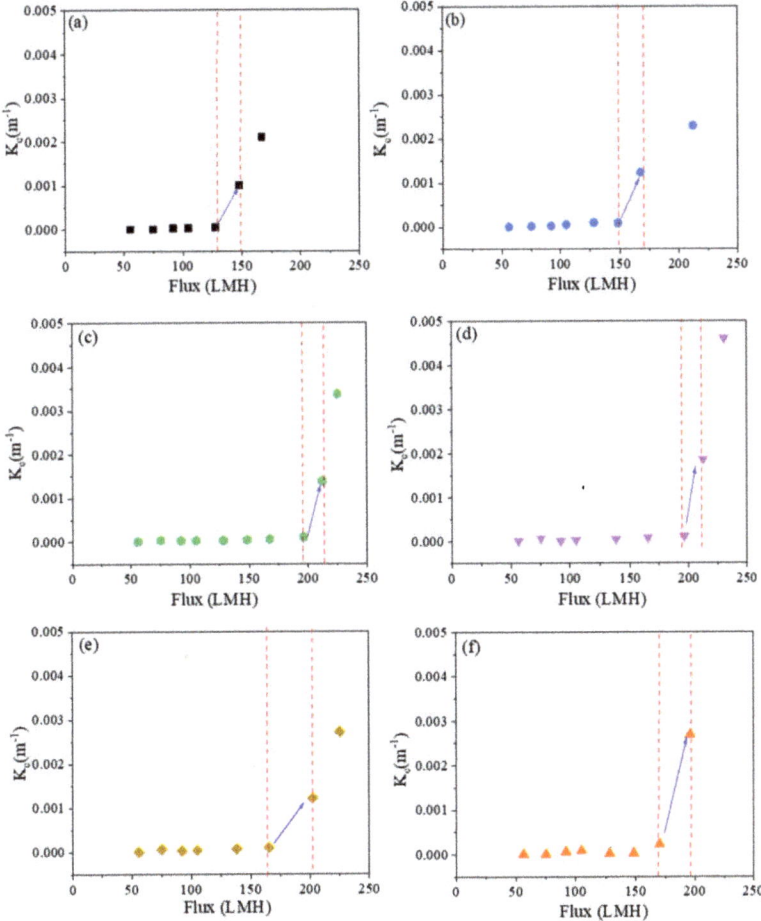

Figure 2. Dependence of the cake filtration constant on flux for (**a**) as-received, (**b**) pressed, (**c**) P27, (**d**) P31, (**e**) P50, and (**f**) P60 membranes.

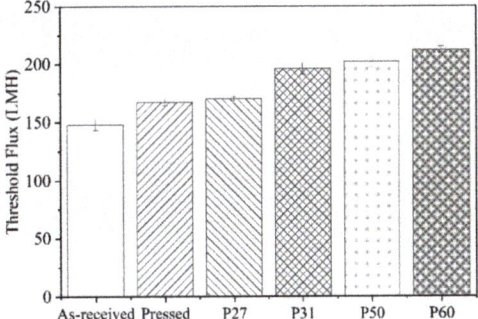

Figure 3. Threshold flux of as-received, pressed and patterned membranes. Error bars represent 95% confidence.

Figure 4 shows that there is a linear correlation between threshold flux and the PFR, with the exception of P60 at the highest PFR value.

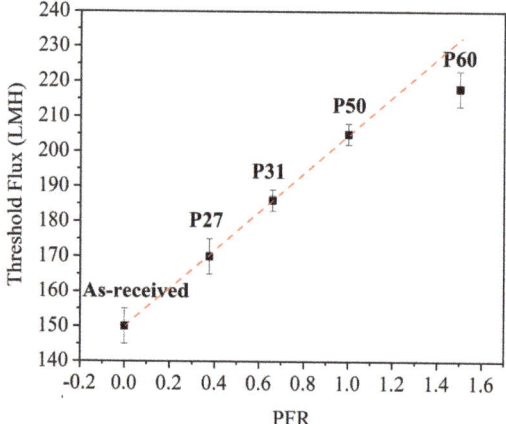

Figure 4. Relationship between threshold flux and pattern fraction ratio. Error bars represent standard deviation among three threshold flux measurements.

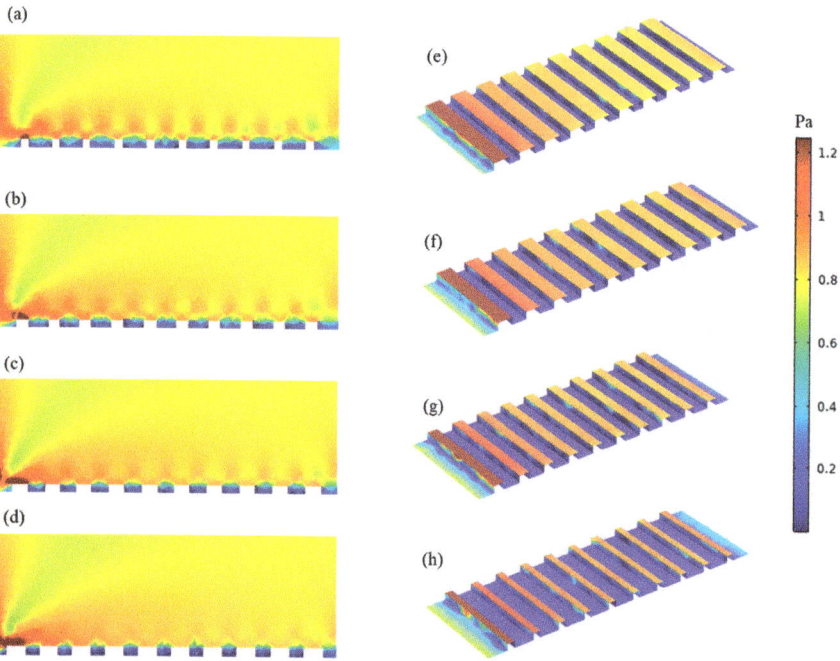

Figure 5. Shear stress profile along the membrane: (**a–d**) 2D ((**a**) P27, (**b**) P31, (**c**) P50 and (**d**) P60); (**e–h**) 3D ((**e**) P27, (**f**) P31, (**g**) P50 and (**h**) P60). Red color indicates a higher shear stress and blue indicates a lower shear stress.

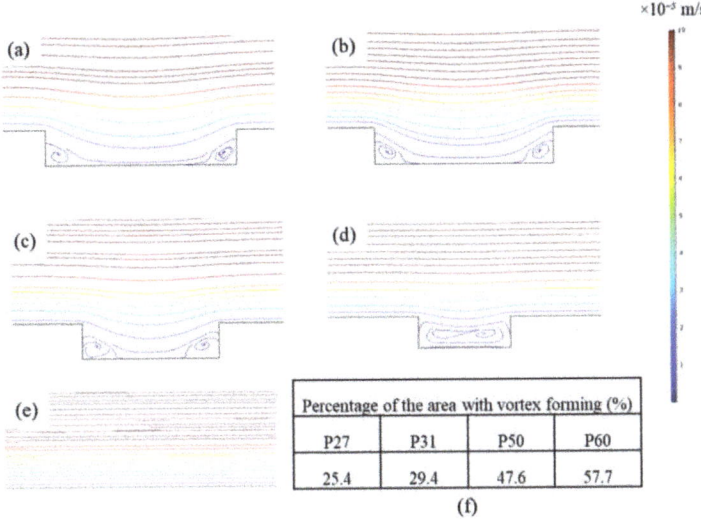

Figure 6. Streamline profiles for (**a**) P27, (**b**) P31, (**c**) P50, (**d**) P60, and (**e**) flat membrane; and (**f**) percentage of the area with vortex forming for patterned membranes. Red color indicates a higher velocity and blue indicates a lower velocity.

CFD simulation results explained the cause for the increasing trend of threshold flux with the PFR. As the PFR increases, the percentage of vortex-forming area increases, and vortex-induced shielding increases. In Figure 5, bulk flow streamlines, which have the highest velocity, are separated from vortices that have a lower velocity by a transitional region. A high thickness of transitional region between bulk and vortex flow can decrease the chance of particle skipping from vortex flow to bulk flow [20]. Therefore, fouling rate increases and threshold flux decreases. Our study suggests that the PFR should be higher than 1 to maximize threshold flux values of patterned membranes.

Noteworthy is that the CFD simulations were carried out without the introduction of particles. Thus, given the correlation between the PFR and the percentage of vortex-forming area determined by the particle-free CFD simulations, we expect the findings to be generalized to other particle sizes, provided that the particles are smaller than the groove width. As discussed by Maruf et al. [38], the highest critical (threshold) flux is found for colloidal particles closest in size to the groove width. Particles with diameters below and above this width can be expected to yield lower threshold flux.

As discussed earlier, patterning increases overall membrane surface area. The surface area increased in the following order: P60 < P27 < P50 < P31. Our results show that there is no correlation between the increase in threshold flux and increase in surface area. This finding further suggests that the hydrodynamic properties at the membrane–solution interface have greater influence on fouling than membrane surface area.

3.3. Flux Decline Measurements

Figure 7a shows flux versus time data collected for filtration of 200 mg/L SiNPs with all membrane types. To allow direct comparison of results, the initial permeate flux for all membranes was set at 140 ± 2 LMH, which is 10 LMH lower than lowest threshold flux value of 150 LMH for as-received membrane. Figure 7b shows that as-received membranes experienced the largest decrease in the flux (35%) over the 2 h test run. Pressing the membrane improved the fouling resistance due to the decrease in intrinsic membrane surface roughness, as Weinman and Husson [33] showed. Lower surface roughness decreases the rate of colloidal particles attachment to the membrane by providing less contact surface area between the membrane surface and particles [13]. In all cases, patterning lessened the flux decline more than pressing alone. Patterning changes the hydrodynamic properties at the membrane–solution interface, as discussed above. Overall, flux decline results aligned with threshold flux measurements. By measuring and operating below the threshold flux, less fouling occurs, and the frequency of membrane cleaning can be decreased.

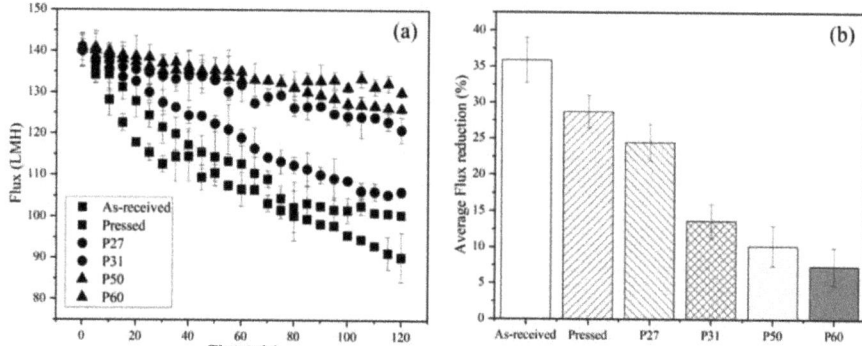

Figure 7. (a) Flux decline data for all six membrane types. (b) Average flux reduction over the 2 h test runs for all membranes. Temperature was 22–23 °C, cross-flow velocity was 0.25 m s^{-1}, and tested membrane area was 6.37 cm^2. The initial flux was 140 ± 2 LMH for all samples. The error bars represent standard deviation among three filtration tests.

4. Conclusions

This study revealed the effect of line-and-groove pattern geometry on threshold flux for filtration of colloidal nanoparticle suspensions through patterned nanofiltration membranes. Experimental work combined with CFD simulations showed that increasing the pattern ratio fraction leads to higher threshold flux, which is important for increasing the volumetric productivity of water treatment systems while maintaining low rates of fouling. Modeling provided insights into the fouling mechanism of colloidal particles on these membranes, and the transition from intermediate pore blocking to cake filtration was used to calculate threshold flux quantitatively in a systematic way. The results of this study can be extended to investigate the effect of pattern geometry for other feature types such as herringbone, pyramid, and biomimetic patterns.

Author Contributions: Conceptualization, A.M., Z.Z., D.A.L. and S.M.H.; methodology, A.M., Z.Z., D.A.L. and S.M.H.; validation, A.M., Z.Z.; formal analysis, A.M., Z.Z., D.A.L. and S.M.H.; investigation, A.M., Z.Z., L.M., T.N.S., D.A.L. and S.M.H.; resources, D.A.L. and S.M.H.; data curation, A.M., Z.Z., L.M., T.N.S., D.A.L. and S.M.H.; writing—original draft preparation, A.M., Z.Z., L.M. and T.N.S.; writing—review and editing, D.A.L. and S.M.H.; visualization, A.M. and Z.Z.; supervision, D.A.L. and S.M.H.; project administration, D.A.L. and S.M.H.; funding acquisition, D.A.L. and S.M.H. All authors have read and agreed to the published version of the manuscript.

Funding: This work was supported by funding from the National Science Foundation (NSF) under Award CBET-1534304 and the Stimulus Research Program of the South Carolina EPSCoR Program under Award 18-SR02. AM has been supported by a GAANN Fellowship (P200A180076) from the Department of Education. Any opinions expressed are those of the authors and do not necessarily reflect the views of the NSF, SC EPSCoR, or DoED.

Acknowledgments: We thank Clemson Machining and Technical Services for fabrication of the custom cross-flow cell. We thank Stephen Creager and Mansour Saberi for their assistance with membrane patterning work.

Conflicts of Interest: The authors declare no conflict of interest.

References

1. Zhu, X.; Elimelech, M. Colloidal Fouling of Reverse Osmosis Membranes: Measurements and Fouling Mechanisms. *Environ. Sci. Technol.* **1997**, *31*, 3654–3662. [CrossRef]

2. Tang, C.Y.; Chong, T.; Fane, A.G. Colloidal interactions and fouling of NF and RO membranes: A review. *Adv. Colloid Interface Sci.* **2011**, *164*, 126–143. [CrossRef] [PubMed]
3. Hoek, E.M.V.; Bhattacharjee, S.; Elimelech, M. Effect of Membrane Surface Roughness on Colloid–Membrane DLVO Interactions. *Langmuir* **2003**, *19*, 4836–4847. [CrossRef]
4. Li, Q.; Elimelech, M. Synergistic effects in combined fouling of a loose nanofiltration membrane by colloidal materials and natural organic matter. *J. Membr. Sci.* **2006**, *278*, 72–82. [CrossRef]
5. Cohen, R.; Probstein, R. Colloidal fouling of reverse osmosis membranes. *J. Colloid Interface Sci.* **1986**, *114*, 194–207. [CrossRef]
6. Boussu, K.; Belpaire, A.; Volodin, A.; Van Haesendonck, C.; Van Der Meeren, P.; Vandecasteele, C.; Van Der Bruggen, B. Influence of membrane and colloid characteristics on fouling of nanofiltration membranes. *J. Membr. Sci.* **2007**, *289*, 220–230. [CrossRef]
7. Boussu, K.; Van Der Bruggen, B.; Volodin, A.; Snauwaert, J.; Van Haesendonck, C.; Vandecasteele, C. Roughness and hydrophobicity studies of nanofiltration membranes using different modes of AFM. *J. Colloid Interface Sci.* **2005**, *286*, 632–638. [CrossRef]
8. Shafi, H.Z.; Khan, Z.; Yang, R.; Gleason, K.K. Surface modification of reverse osmosis membranes with zwitterionic coating for improved resistance to fouling. *Desalination* **2015**, *362*, 93–103. [CrossRef]
9. Choi, H.; Jung, Y.; Han, S.; Tak, T.; Kwon, Y.-N. Surface modification of SWRO membranes using hydroxyl poly(oxyethylene) methacrylate and zwitterionic carboxylated polyethyleneimine. *J. Membr. Sci.* **2015**, *486*, 97–105. [CrossRef]
10. Ma, W.; Yang, P.; Li, J.; Li, S.; Li, P.; Zhao, Y.; Huang, N. Immobilization of poly(MPC) brushes onto titanium surface by combining dopamine self-polymerization and ATRP: Preparation, characterization and evaluation of hemocompatibility in vitro. *Appl. Surf. Sci.* **2015**, *349*, 445–451. [CrossRef]
11. Hong, S.H.; Hong, S.; Ryou, M.-H.; Choi, J.W.; Kang, S.M.; Lee, H. Sprayable Ultrafast Polydopamine Surface Modifications. *Adv. Mater. Interfaces* **2016**, *3*, 1500857. [CrossRef]
12. Choi, W.; Lee, C.; Lee, D.; Won, Y.J.; Lee, G.W.; Shin, M.G.; Chun, B.; Kim, T.-S.; Park, H.-D.; Jung, H.W.; et al. Sharkskin-mimetic desalination membranes with ultralow biofouling. *J. Mater. Chem. A* **2018**, *6*, 23034–23045. [CrossRef]
13. Elimelech, M.; Zhu, X.; Childress, A.E.; Hong, S. Rapid communication Role of membrane surface morphology in colloidal fouling of cellulose acetate and composite aromatic polyamide reverse osmosis membranes. *J. Membr. Sci.* **1997**, *127*, 101–109. [CrossRef]
14. Vasudevan, R.; Kennedy, A.J.; Merritt, M.; Crocker, F.H.; Baney, R.H. Microscale patterned surfaces reduce bacterial fouling-microscopic and theoretical analysis. *Colloids Surf. B Biointerfaces* **2014**, *117*, 225–232. [CrossRef] [PubMed]
15. Won, Y.-J.; Lee, J.; Choi, D.-C.; Chae, H.R.; Kim, I.; Lee, C.-H.; Kim, I.-C. Preparation and Application of Patterned Membranes for Wastewater Treatment. *Environ. Sci. Technol.* **2012**, *46*, 11021–11027. [CrossRef] [PubMed]
16. Won, Y.-J.; Choi, D.-C.; Jang, J.H.; Lee, J.; Chae, H.R.; Kim, I.; Ahn, K.H.; Lee, C.-H.; Kim, I.-C. Factors affecting pattern fidelity and performance of a patterned membrane. *J. Membr. Sci.* **2014**, *462*, 1–8. [CrossRef]
17. Maruf, S.H.; Rickman, M.; Wang, L.; Iv, J.M.; Greenberg, A.R.; Pellegrino, J.; Ding, Y. Influence of sub-micron surface patterns on the deposition of model proteins during active filtration. *J. Membr. Sci.* **2013**, *444*, 420–428. [CrossRef]
18. Weinman, S.T.; Fierce, E.M.; Husson, S.M. Nanopatterning commercial nanofiltration and reverse osmosis membranes. *Sep. Purif. Technol.* **2019**, *209*, 646–6578. [CrossRef]
19. Lee, Y.K.; Won, Y.-J.; Yoo, J.H.; Ahn, K.H.; Lee, C.-H. Flow analysis and fouling on the patterned membrane surface. *J. Membr. Sci.* **2013**, *427*, 320–325. [CrossRef]
20. Won, Y.-J.; Jung, S.-Y.; Jang, J.-H.; Lee, J.-W.; Chae, H.-R.; Choi, D.-C.; Ahn, K.H.; Lee, C.-H.; Park, P.-K. Correlation of membrane fouling with topography of patterned membranes for water treatment. *J. Membr. Sci.* **2016**, *498*, 14–19. [CrossRef]
21. Jang, J.H.; Lee, J.; Jung, S.-Y.; Choi, D.-C.; Won, Y.-J.; Ahn, K.H.; Park, P.-K.; Lee, C.-H. Correlation between particle deposition and the size ratio of particles to patterns in nano- and micro-patterned membrane filtration systems. *Sep. Purif. Technol.* **2015**, *156*, 608–616. [CrossRef]
22. Zhou, Z.; Ling, B.; Battiato, I.; Husson, S.M.; Ladner, D.A. Concentration polarization over reverse osmosis membranes with engineered surface features. *J. Membr. Sci.* **2021**, *617*, 118199. [CrossRef]

23. Ling, B.; Battiato, I. Rough or wiggly? Membrane topology and morphology for fouling control. *J. Fluid Mech.* **2019**, *862*, 753–780. [CrossRef]
24. Choi, W.; Lee, C.; Yoo, C.H.; Shin, M.G.; Lee, G.W.; Kim, T.-S.; Jung, H.W.; Lee, J.S.; Lee, J.-H. Structural tailoring of sharkskin-mimetic patterned reverse osmosis membranes for optimizing biofouling resistance. *J. Membr. Sci.* **2020**, *595*, 117602. [CrossRef]
25. Malakian, A.; Husson, S.M. Understanding the roles of patterning and foulant chemistry on nanofiltration threshold flux. *J. Membr. Sci.* **2020**, *597*, 117746. [CrossRef]
26. Elsherbiny, I.M.; Khalil, A.S.; Ulbricht, M. Influence of Surface Micro-Patterning and Hydrogel Coating on Colloidal Silica Fouling of Polyamide Thin-Film Composite Membranes. *Membr.* **2019**, *9*, 67. [CrossRef]
27. Kim, S.; Hoek, E.M. Modeling concentration polarization in reverse osmosis processes. *Desalination* **2005**, *186*, 111–128. [CrossRef]
28. Lyster, E.; Cohen, Y. Numerical study of concentration polarization in a rectangular reverse osmosis membrane channel: Permeate flux variation and hydrodynamic end effects. *J. Membr. Sci.* **2007**, *303*, 140–153. [CrossRef]
29. Choi, D.-C.; Jung, S.-Y.; Won, Y.-J.; Jang, J.H.; Lee, J.; Chae, H.-R.; Ahn, K.H.; Lee, S.; Park, P.-K.; Lee, C.-H. Three-dimensional hydraulic modeling of particle deposition on the patterned isopore membrane in crossflow microfiltration. *J. Membr. Sci.* **2015**, *492*, 156–163. [CrossRef]
30. Kirschner, A.Y.; Cheng, Y.-H.; Paul, D.R.; Field, R.W.; Freeman, B.D. Fouling mechanisms in constant flux crossflow ultrafiltration. *J. Membr. Sci.* **2019**, *574*, 65–75. [CrossRef]
31. Gohari, R.J.; Lau, W.; Matsuura, T.; Ismail, A. Effect of surface pattern formation on membrane fouling and its control in phase inversion process. *J. Membr. Sci.* **2013**, *446*, 326–331. [CrossRef]
32. Chung, J.Y.; Lee, J.-H.; Beers, K.L.; Stafford, C.M. Stiffness, Strength, and Ductility of Nanoscale Thin Films and Membranes: A Combined Wrinkling–Cracking Methodology. *Nano Lett.* **2011**, *11*, 3361–3365. [CrossRef] [PubMed]
33. Weinman, S.T.; Husson, S.M. Influence of chemical coating combined with nanopatterning on alginate fouling during nanofiltration. *J. Membr. Sci.* **2016**, *513*, 146–154. [CrossRef]
34. Zhou, J.; Wandera, D.; Husson, S.M. Mechanisms and control of fouling during ultrafiltration of high strength wastewater without pretreatment. *J. Membr. Sci.* **2015**, *488*, 103–110. [CrossRef]
35. Hobbs, C.; Taylor, J.; Hong, S. Effect of surface roughness on fouling of RO and NF membranes during filtration of a high organic surficial groundwater. *J. Water Supply Res. Technol.* **2006**, *55*, 559–570. [CrossRef]
36. Yoo, C.H.; Lee, G.W.; Choi, W.; Shin, M.G.; Lee, C.; Shin, J.H.; Son, Y.; Chun, B.; Lee, J.-H.; Jung, H.W.; et al. Identifying the colloidal fouling behavior on the sharkskin-mimetic surface: In-situ monitoring and lattice Boltzmann simulation. *Chem. Eng. J.* **2021**, *405*, 126617. [CrossRef]
37. Lee, C.; Lee, G.W.; Choi, W.; Yoo, C.H.; Chun, B.; Lee, J.S.; Lee, J.-H.; Jung, H.W. Pattern flow dynamics over rectangular Sharklet patterned membrane surfaces. *Appl. Surf. Sci.* **2020**, *514*, 145961. [CrossRef]
38. Maruf, S.H.; Wang, L.; Greenberg, A.R.; Pellegrino, J.; Ding, Y. Use of nanoimprinted surface patterns to mitigate colloidal deposition on ultrafiltration membranes. *J. Membr. Sci.* **2013**, *428*, 598–607. [CrossRef]

Publisher's Note: MDPI stays neutral with regard to jurisdictional claims in published maps and institutional affiliations.

© 2020 by the authors. Licensee MDPI, Basel, Switzerland. This article is an open access article distributed under the terms and conditions of the Creative Commons Attribution (CC BY) license (http://creativecommons.org/licenses/by/4.0/).

Article

Groundwater Remediation of Volatile Organic Compounds Using Nanofiltration and Reverse Osmosis Membranes—A Field Study

Thomas J. Ainscough [1,2], Darren L. Oatley-Radcliffe [1,2,*] and Andrew R. Barron [1,3,4]

1. Energy Safety Research Institute (ESRI), Bay Campus, Swansea University, Fabian Way, Swansea SA1 8EN, UK; T.J.Ainscough@Swansea.ac.uk (T.J.A.); A.R.Barron@Swansea.ac.uk (A.R.B.)
2. Centre for Water Advanced Technologies and Environmental Research (CWATER), College of Engineering, Bay Campus, Swansea University, Fabian Way, Swansea SA1 8EN, UK
3. Department of Chemistry and Department of Materials Science and Nanoengineering, Rice University, Houston, TX 77005, USA
4. Faculty of Engineering, Universiti Teknologi Brunei, Jalan Tungku Link, Gadong BE1410, Brunei
* Correspondence: d.l.oatley@swansea.ac.uk; Tel.: +44-(0)1792-606668

Citation: Ainscough, T.J.; Oatley-Radcliffe, D.L.; Barron, A.R. Groundwater Remediation of Volatile Organic Compounds Using Nanofiltration and Reverse Osmosis Membranes—A Field Study. *Membranes* **2021**, *11*, 61. https://doi.org/10.3390/membranes11010061

Received: 9 December 2020
Accepted: 11 January 2021
Published: 16 January 2021

Publisher's Note: MDPI stays neutral with regard to jurisdictional claims in published maps and institutional affiliations.

Copyright: © 2021 by the authors. Licensee MDPI, Basel, Switzerland. This article is an open access article distributed under the terms and conditions of the Creative Commons Attribution (CC BY) license (https://creativecommons.org/licenses/by/4.0/).

Abstract: Groundwater contamination by chlorinated hydrocarbons represents a particularly difficult separation to achieve and very little is published on the subject. In this paper, we explore the potential for the removal of chlorinated volatile and non-volatile organics from a site in Bedfordshire UK. The compounds of interest include trichloroethylene (TCE), tetrachloroethylene (PCE), cis-1,2-dichloroethylene (DCE), 2,2-dichloropropane (DCP) and vinyl chloride (VC). The separations were first tested in the laboratory. Microfiltration membranes were of no use in this separation. Nanofiltration membranes performed well and rejections of 70–93% were observed for synthetic solutions and up to 100% for real groundwater samples. Site trials were limited by space and power availability, which resulted in a maximum operating pressure of only 3 bar. Under these conditions, the nanofiltration membrane removed organic materials, but failed to remove VOCs to any significant extent. Initial results with a reverse osmosis membrane were positive, with 93% removal of the VOCs. However, subsequent samples taken demonstrated little removal. Several hypotheses were presented to explain this behavior and the most likely cause of the issue was fouling leading to adsorption of the VOCs onto the membrane and allowing passage through the membrane matrix.

Keywords: groundwater; reclamation; nanofiltration; VOC

1. Introduction

The remediation of contaminated groundwater is a costly and complex process typically involving multiple stages and systems [1]. Water treatment plants vary greatly depending on the compounds and contaminants in the water source that need to be reduced or removed to below trace levels to meet local environmental standards [2]. A treatment plant can use a combination of physical, chemical and biological steps. A standard setup for hydrocarbon (including volatiles) remediation could include an oil–water separator (OWS) as an initial physical separation stage followed by an activated biological sludge and chemical dosing stage to remove oil emulsions below 150 microns and dissolved organics [3,4]. Often an adsorption stage is included and the selection of the adsorbent material is critical to success [5]. Other combinations of technology for VOC removal are also available and these may include the use of plasma, adsorption and catalysis [6] to name but a few.

Integrated membrane systems for desalination used in drinking water production typically combine a microfiltration (MF) or ultrafiltration (UF) membrane as a prefilter to preserve the integrity and avoid particulate damage of the reverse osmosis (RO) membrane by removing suspended solids and microorganisms [7,8]. Nanofiltration (NF) also plays a predominant role in the drinking water industry [9,10]—NF membranes were initially deployed as water softeners due to their unique capabilities of screening divalent and multivalent ions such as calcium (Ca^{2+}) and magnesium (Mg^{2+}) found in hard water areas [11]. Advances in the development of NF membranes has given NF membranes additional usages in water purification, specifically the removal of naturally occurring organic material including viruses and pesticides [12,13]. NFs are not capable of removing all organic material [14]. However, their inclusion is warranted to reduce fouling of the RO membranes by the organic material they can filter [15,16]. RO membranes will exclude all remaining divalent and multivalent ions, RO will also remove monovalent ions such as sodium (Na^+) salts. One of the major costs in the remediation of a contaminated water source is the use of activated carbon adsorption removal of volatile organic compounds (VOCs) [17]. The occurrence of VOCs within local water sources is highly regulated due to the associated health concerns [18–20]. Due to the prohibitively expensive nature of activated carbon, the used cartridges can be regenerated using solvent regeneration [21,22], steam regeneration [23,24] or more commonly used thermal regeneration [25–27]. Thermal regeneration is conducted by pyrolysis, burning off adsorbed VOCs along with a carrier gas, mainly nitrogen, to remove the VOCs and regenerate the activated carbon. There are two major drawbacks with thermal regeneration—it requires considerable investment in a suitable furnace which will only become financially viable with a high enough quantity of activated carbon [28], and it causes carbon losses of 5–15 wt% [29] or a reduction in adsorptive capacity [30], the addition of new material to make up the absorptive losses is required. Thus, small to medium wastewater treatment plants will dispose of the spent activated carbon through an external company where it will be recycled or regenerated for a cost and a discount on fresh material that would be required is applied.

In this work, our research group was approached by a large multinational company, which will be referred to as Company A for confidentiality purposes, to investigate the use of membrane technology for the removal of VOCs to reduce the expenditure of activated carbon on Company A's water treatment plants. Previous work [31,32] suggested that a novel super-hydrophilic ceramic microfiltration membrane could be used for this purpose. The study also investigated the use of nanofiltration and reverse osmosis membranes for the application which was conducted using laboratory trials and on site technology deployment.

2. Experimental

2.1. Background Information and Preliminary Testing

The remediation process at the Bedfordshire site involves the clean up of contaminated groundwater and the treatment process involves both liquid and vapor VOC abatement processes; both of which contain granular activated carbon as a treatment process. The groundwater from specifically dug wells in high-contamination areas is pumped to an OWS. The decanted water is passed through an air stripper to remove pollutants with high Henry's Law coefficients such as trichloroethylene (TCE) and tetrachloroethylene (PCE). The air-stripped VOCs are directed to the vapor treatment process, and the groundwater with any remaining dissolved VOCs is passed through a biological treatment step followed by the granular activated carbon adsorption stage. There is a chemical additive stage prior to being reinjected back into the ground wells.

The contaminated water source is an historically polluted source from a redundant production process, and the water contained over 10 VOCs above trace GC–MS detection. The EPA environmental limit for the compounds found in the water source is 5 ppb, and five compounds were regularly detected above this level—they were TCE, PCE cis-1,2-dichloroethylene (DCE), 2,2-dichloropropane (DCP) and vinyl chloride (VC). The volatile organic compounds had molecular weights between 62.50 g mol^{-1} (VC) and 165.83 g mol^{-1} (PCE). The intrinsic hydrophobicity of these compounds varied significantly, as was reflected by the logarithm of their octanol–water partitioning coefficient (XLogP3). As can be seen in Table 1, the properties of the selected volatile organic compounds demonstrated that some compounds are hydrophilic (XLogP3 > 2.5) while others are hydrophobic (XLogP3 < 2.5) and ranged between 1.5 and 3.4. The groundwater at the Bedfordshire site has very little organic content compared to Company A's other global sites. However, the global scope of the project covers any contaminated groundwater source Company A may have. The project initiated between Swansea University and Company A was to develop a system of removing organic matter contamination from groundwaters on various sites with an initial test site located in the UK. The methodology used in this study was to be based upon the development of a novel super-hydrophilic ceramic membrane system as previously reported [31,32]. Preliminary investigations from this project increased operational understanding of the separation mechanisms of the modified ceramic membrane and have shown that the membrane is very good for separations where the contamination levels are very high. In heavily contaminated systems, there is in effect a biphasic mixture of organic and aqueous components (with some low-level organics dissolved in the aqueous phase). The modified ceramic membrane has been shown to be extremely effective at separating these two phases. However, the low-level organics dissolved within the aqueous phase are not separated to a high degree in this process. Thus, to produce an ultra-clean aqueous stream, an additional NF or RO process is proposed as a polishing step. For this reason, the site strategy for deployment of a pilot system at the site was modified to use a polymeric NF and/or RO membrane.

Table 1. Summary of physio-chemical properties of selected VOCs.

Compound	CAS #	Formula	MW (g/mol)	XLogP3
Vinyl Chloride	75-01-4	C_2H_3Cl	62.50	1.5
Cis-1,2-Dichloroethylene	156-59-2	$C_2H_2CL_2$	96.94	1.9
2,2-Dichloropropane	594-20-7	$C_3H_6Cl_2$	112.981	2.1
Trichloroethylene	79-01-6	C_2HCl_3	131.39	2.6
Toluene	108-88-3	C_7H_8	92.141	2.7
Tetrachloroethylene	127-18-4	C_2Cl_4	165.83	3.4

2.1.1. Pilot-Scale Testing of Ceramic Microfiltration Membranes

All preliminary ceramic membrane filtrations were conducted on a Swansea University MF/UF pilot-scale membrane rig, as pictured in Figure 1. The system used is overengineered for the application required. However, the use of two pumps allows the pressure and cross-flow velocity to be independently controlled, which is beneficial in a research environment. The system was originally designed to be operated as both a micro and ultrafiltration system, capable of operating at pressures of up to 6 bar. However, the system can operate at the much lower pressures required for microfiltration using inverters for the pumps. The system consists of a 100 L stainless steel feed tank, which feeds into the feed pump (Fristam FPE 742, Fristam Pumpen KG, Hamburg, Germany) which can deliver up to 5.5 bar pressure at a flow rate of up to 10 m^3/h. The flow then enters the second pump (P2: Fristam FPE 722, Fristam Pumpen KG, Hamburg, Germany), which is capable of delivering up to 3.8 bar at a flow rate of up to 10 m^3/h. The flow from P2 then passes through the membrane module. The retentate flows out of the membrane module and enters the optional heat exchanger system. The heat exchanger consists of two shell and tube heat exchangers connected to a cold water and/or steam supply as required. Cold water supply is controlled by a solenoid (Burkert 6213 A, Burkert, Ingelfingen, Germany) connected to a thermostat on the control board. The retentate re-circulates around the loop and returns to the feed tank via a diaphragm valve which sets the loop pressure (large black valve at the top of Figure 1). The system was operated at several pressures controlled via the pumps (P1 and P2) and the return valve. A temperature of 25 °C was maintained throughout the experimental work controlled by the solenoid valves attached to the heat exchangers controlling the steam and water flow rates.

Figure 1. Swansea University ceramic pilot microfiltration membrane rig.

An industrial-scale Pall Membralox 0.2 μm ceramic microfiltration membrane was used (Pall, Portsmouth, UK). The membrane active layer and support are made from alpha-alumina, α-Al$_2$O$_3$, and with an active area of 0.24 m^2, the membrane is capable of handling temperatures of up to 95 °C, with a full pH range of 0–14 and a maximum pressure of 8 bar.

2.1.2. Small-Scale Laboratory Testing Equipment for Nanofiltration and Reverse Osmosis

All preliminary NF experiments were conducted at room temperature (22 ± 1 °C) and pH 6.5 ± 0.2, which is the pH of the deionized water (DI) water used throughout the study (Millipore Elix 5, Watford, UK). The filtrations were carried out using a commercially available stirred frontal filtration system (Membranology HP350 Filtration Cell, Membranology Ltd., Swansea, UK), previously described by Oatley-Radcliffe et al. [33] and illustrated in Figure 2. The cell has an operating capacity of 350 mL feed solution and an effective membrane surface area of 41.8 cm^2. The filtration solutions were stirred magnetically at 300 rpm, the maximum practical stirrer speed previously determined [34].

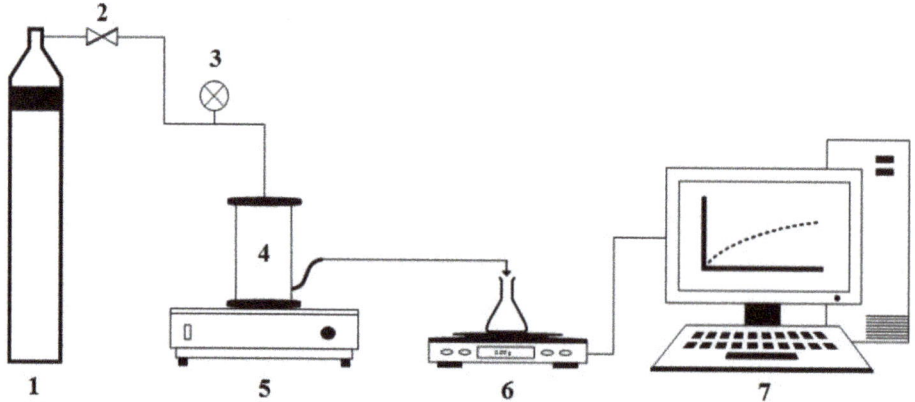

Figure 2. Frontal filtration experimental setup (1—nitrogen gas bottle, 2—pressure regulator, 3—pressure indicator, 4—Membranology HP350 stirred cell, 5—magnetic stirrer plate, 6—weight balance, 7—computer data logger).

Prior to first use, all membranes were soaked in DI water for 24 h. At the start of each series of experiments, the membrane was flushed with DI water at 30 bar for 1 h or 350 mL of filtrate, whichever was achieved first, to reach a steady compressed permeate flux. Following the compression experiment, a membrane clean water flux was recorded for 5, 10, 15, 20, 25 and 30 bar. Rejection experiments were then conducted at the previously stated pressures using toluene at a concentration of 1 g L^{-1} and a sample of groundwater prefiltered through a 0.2 μm microfilter to remove sediment. The concentrations of toluene and groundwater for the feed solution and permeate samples were analyzed using a total organic carbon analyzer (Shimadzu TOC-LCPH, Shimadzu Corporation, Milton Keynes, UK). Rejection measurements were based on 20 mL of permeate once the initial 5 mL of permeate was discarded, with 25 mL removed in total. After each rejection experiment the membrane was rinsed with DI water to remove any residual materials. Following a period of testing, the membrane pure water flux was retested to assess any deterioration in the membrane performance; this is a simple check for any potential fouling.

2.2. Nanofiltration/Reverse Osmosis Pilot for Deployment at the Bedfordshire Site

The Bedfordshire site trial was conducted over a 4 month period, with an initial setup and commissioning exercise lasting two days. Due to the limited capacity of the electrical supply identified at the water processing unit, the deployment of a full-scale pilot system was not possible. A simplified pilot rig was constructed to avoid overloading the electrical capacity at the site; however, this simplified rig was not capable of the full operational range normally expected for a membrane process. Most notably, the applied operating feed pressure of this simple rig was limited to only 3.0 bar, with a maximum deliverable pressure of 3.5 bar. The rig was capable of operating in several modes using both polymeric NF and RO 2.5" membrane modules separately. The membrane system, illustrated in Figure 3, consisted of a custom fabricated stainless steel feed tank (Axium Process Ltd., Hendy, UK) and in house-built Unistrut frame designed to be fully adjustable to allow the tank to be gravity fed from the OWS whilst also preventing the OWS from being fully drained. From the feed tank, pump 1 (Lowara-4HMS3/A) was used to provide pressure and top up flow to the membrane system. Following the pump was a filter cartridge housing (Pentek-3G housing) containing a string wound polypropylene 1.0 micron pre filter (Prosep) used to protect the NF or RO membrane from particulate debris. A second pump (Lowara-4HMS3/A) was used to provide circulation flow through the NF or RO membrane. After the second pump, a pressure gauge (Wika 232.50 and L990.22 Sanitary Seal) indicated the pressure of the membrane feed. A paddle flowmeter (Burkert S030 DN25 and Burkert 8035) allowed the flow rate through the membrane to be recorded, a parameter needed for calculating the cross-flow velocity needed for understanding the membrane performance throughout the trial. Both membranes were contained within a custom stainless steel housing (Axium Process Ltd., Hendy, UK). A second pressure gauge after the membrane indicated the pressure of the membrane retentate, in conjunction with the first pressure gauge the transmembrane pressure can be calculated to allow the calculation of the membrane flux per bar across the membrane. A flowmeter (Omega Engineering) was placed on the membrane retentate return to the OWS. This flowmeter was switched out between a FL7205 and FL2098 depending on operation of the rig, with the majority of retentate being recycled to the Swansea feed tank or back to the OWS, respectively. In addition, a flowmeter (Omega Engineering-FL7201) on the membrane permeate outlet was used to monitor the permeate flow rate, a parameter used to monitor the membrane performance by calculating the membrane flux using the known surface area of the membrane. A schematic of the pilot plant is included in Figure 3 and photographed in Figure 4.

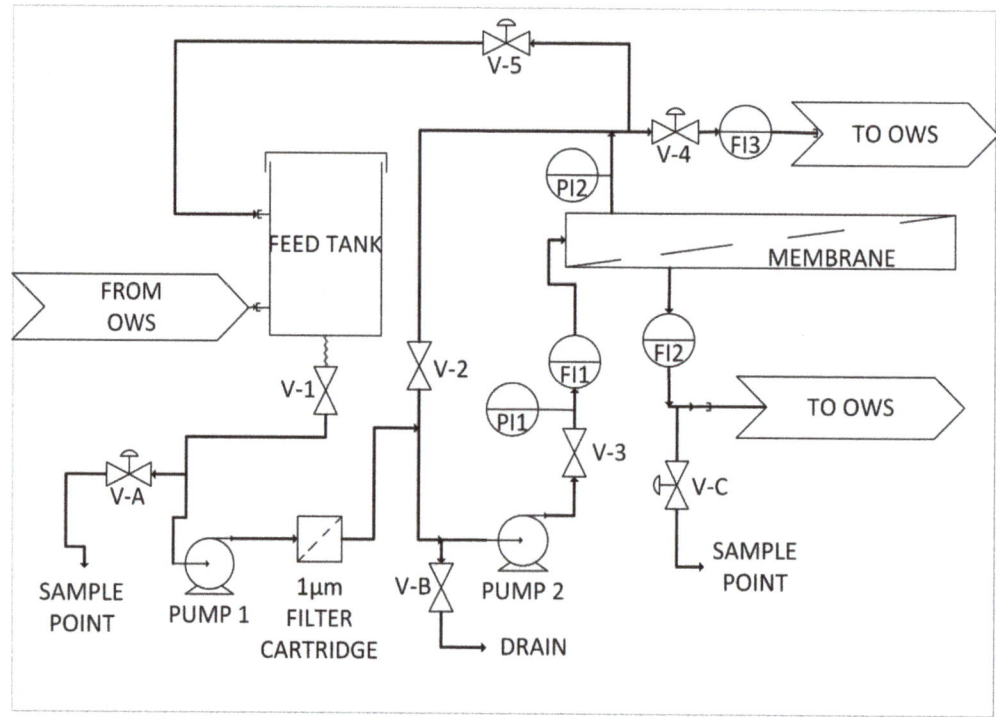

Figure 3. A schematic drawing of the pilot membrane rig used for the Bedfordshire site trials.

Figure 4. Pilot membrane rig used for the Bedfordshire site trials. Left: front view of the pilot. Right: connection to the existing Bedfordshire OWS unit.

All materials processed during the site trial were removed from the existing OWS within the groundwater facility and returned to the OWS following processing, i.e., there was no risk of non-treatment leading to non-compliance of the site during the trial period.

Throughout the trial, either total organic carbon (TOC) analysis conducted by Swansea University (Shimadzu, TOC-LPH, Milton Keynes, UK) or gas chromatography mass spectrometry (GC–MS) conducted by Natural Resources Wales (NRW, Llanelli, UK) using a method equivalent to that of the EPA 8260b for volatile organic compounds (VOCs) was used to analyze the separation performance.

The NF membrane used was a GE Osmonics DK series (model: DK2540F1073). The RO membrane used was a GE Osmonics AK series (model: AK2540TM).

2.3. Laboratory Analysis

2.3.1. General Laboratory Filtration Trials

Salt rejection was measured at the previously stated pressures using sodium chloride (NaCl) (Fisher Scientific, Loughborough, UK) at a concentration of 2000 ppm, and the conductivity of feed and permeate samples was measured using a conductivity probe (Jenway Model 3450). Toluene rejection was measured at the previously stated pressures using toluene (Reagent grade, Fisher Scientific, Loughborough, UK) at a concentration of 100 ppm, and the concentrations of feed and permeate samples were measured using a total organic carbon analyzer (Shimadzu TOC-LCPH, Shimadzu Corporation, Milton Keynes, UK). TCE rejection was measured on its own at the previously stated pressures using TCE (Puriss $\geq$ 99.5% (GC), Sigma, Gillingham, UK) at a concentration of 100 ppm, and the concentrations of feed and permeate samples were measured using a total organic carbon analyzer (Shimadzu TOC-LCPH, Shimadzu Corporation, Milton Keynes, UK). A feed VOC mixture of TCE, PCE, DCE and DCP (Sigma, Gillingham, UK) was created with a concentration of 100 ppm for all solvents. For all feed batches, the VOCs were initially dissolved in 50 mL of methanol (Fisher Scientific, Loughborough, UK) and added to a 5 L flask. A further 50 mL of methanol was added to the measuring cylinder and transferred to the 5 L flask to avoid any residual VOCs clinging to the glassware. VOC rejection was measured at the previously stated pressures. Feed and permeate samples were measured using a headspace GC–MS system at Swansea University (Agilent GC 6850, MS 5977A, HS 7697A 12-vial) not previously available during the preliminary and pilot-scale trial. Two calibrations were conducted—a ppb calibration and ppm calibration—with the following standards: blank, 10, 50, 100, 250, 500, and 1000 ppb and 1, 10, 50, 100, and 150 ppm, respectively. Data acquisition and analyses were performed using the MassHunter Workstation software with quantification using the selected ion monitoring (SIM) method. See Table 2 for GC–MS HS method.

Table 2. GC–MS HS method settings.

Headspace Parameters	Agilent 7697A HS—12 vial
Temperature settings	
Oven temperature	70 °C
Loop temperature	85 °C
Transfer line temperature	120 °C
Timing settings	
Vial equilibration	10 min
Injection duration	0.3 min
GC cycle time	22 min
Vial settings	
Vial size	20 mL
Vial pressurization	15 psi
Loop size	1.0 mL
Extraction time	0.3 min
Mode	Single extraction
Transfer line flow	20 mL/min
Transfer line	Agilent p/n 160-2535-5
Line type	Fused silica, deactivated
Line diameter	0.53 mm
Vial and cap	20 mL, PTFE/silicone septa

Table 2. *Cont.*

Headspace Parameters	Agilent 7697A HS—12 vial		
GC Parameters	Agilent 6850 series II GC		
Inlet settings			
Heater	On—150 °C		
Pressure	On—6.4 psi		
Total flow	On—42.6 mL/min		
Run time	6.8 min		
Gas saver	Off		
Split ratio	40:1		
Split flow	40 mL/min		
Oven settings			
Oven ramp	°C/min	Next °C	Hold
Initial		30	0.3 min
Ramp 1	5	55	0 min
Ramp 2	10	70	0 min
Total run time	15.8 min		
Equilibration time	0.5 min		
Oven Max temperature	260 °C		
Column	Agilent J & W HP-5		
Length	30 m		
Diameter	0.25 mm		
Film thickness	0.25 µm		
Mode	Constant flow		
Pressure	6.4 psi		
Nominal initial flow	1 mL/min		
Inlet	Front		
Outlet	MSD		
Outlet pressure	Vacuum		
MSD Parameters	Agilent 5977A MSD		
MSD settings			
Solvent delay	0.0 min		
Sim ions	4		
Quantitation ions	61, 77, 130, 166 M/Z		
Sim dwell	50 msec/ion		
Quad temperature	150		
Source temperature	230		
Transfer line temperature	250		
Gain factor	5		

All rejection measurements analyzed by TOC were based on 20 mL of permeate once the initial 5 mL of permeate was discarded, with 25 mL removed in total. All GC–MS rejection measurements were based on 10 mL of permeate once the initial 5 mL of permeate was discarded, with 15 mL removed in total. After each rejection experiment, the membrane was rinsed with DI water to remove any residual solvent. Following each experiment set—NaCl, toluene, TCE and VOC mixture, respectively—a clean water flux was conducted at 20 bar to ensure the membrane was not deteriorating or fouling.

2.3.2. Contact Angle Measurements

Contact angles of the dry membranes were measured using a VCA Optima contact angle analyzer (AST Products Inc., Billerica, USA) using the static sessile drop method. A droplet of ultrapure water obtained from the Millipore Elix 5 was delivered onto the dry membrane surface and a static image of the droplet was taken after contact with the surface. Contact angle measurements performed using the VCA Optima software for each membrane at 5 different locations were recorded and the average taken. Contact angles and standard deviations are included in Table 3a,b.

Table 3. Characteristics and supporting information for the membranes used in this study.

a. Properties of NF membranes used in this study.

Membrane	DK	DL	NF90	NF270
Manufacturer	GE Osmonics	GE Osmonics	Dow Filmtec	Dow Filmtec
Support material	Polysulfone	Polysulfone	Polysulfone	Polysulfone
Surface material	TFC PA	TFC PA	PA	PA
Maximum operating temperature	50 °C	50 °C	45 °C	45 °C
Maximum operating pressure	41 bar	41 bar	41 bar	41 bar
pH range	3–9	3–9	2–11	2–11
Flux (GFD)/psi	22/100	28/220	46–60/130	72–98/130
MWCO	~150–300	~150–300	~200–400	~200–400
Contact angle	26.36 ± 0.48°	27.60 ± 0.38°	27.40 ± 0.89°	21.22 ± 0.88°

b. Properties of RO membranes used in this study.

Membrane	AK	AG	BW30	BW30LE	BW30XFR
Manufacturer	GE Osmonics	GE Osmonics	Dow Filmtec	Dow Filmtec	Dow Filmtec
Support material	Polysulfone	Polysulfone	Polysulfone	Polysulfone	Polysulfone
Surface material	TFC PA	TFC PA	PA	PA	PA
Maximum operating temperature	50 °C	50 °C	45 °C	45 °C	45 °C
Maximum operating pressure	27 bar	41 bar	41 bar	41 bar	41 bar
pH range	4–11	4–11	2–11	2–11	2–11
Flux (GFD)/psi	26/115	26/225	26/225	37–46/225	28–33/225
MWCO	~0	~0	~100	~100	~100
Contact angle	70.21 ± 1.09°	72.21 ± 2.52°	55.86 ± 0.67°	67.58 ± 0.30°	56.61 ± 0.66°

2.3.3. Membranes Used

The membranes used in this study were obtained in either flat sheet format for laboratory trials or spiral wound format for pilot and deployment trials. Four NF membranes were used in total; namely, the DK and DL membranes sourced from GE Osmonics and the NF90 and NF270 sourced from Dow Filmtec. Five RO membranes were used and these were the AK and AG (GE Osmonics) and the BW30, BW30LE and BW30XFR (Dow Filmtec). Full details of the NF and RO membranes are illustrated in Table 3a,b, respectively.

2.4. Rejection Theory

The experimental rejection characteristics of a membrane are usually defined by the observed rejection:

$$R_{obs} = 1 - \frac{C_P}{C_F} \quad (1)$$

where C_F and C_P are the concentrations of the feed and permeate, respectively. However, due to concentration polarization, the concentration at the membrane surface, C_W is higher than that of the bulk feed concentration, C_F. Therefore, real rejection of the solute, R_{real}, which is always equal to or greater than R_{obs} is defined as:

$$R_{real} = 1 - \frac{C_P}{C_W} \quad (2)$$

The concentration at the wall, C_W, can be calculated indirectly using a suitable model for concentration polarization [33–35]. The approach to concentration polarization taken in this study is that of the infinite rejection method first reported by Nakao and Kimura [36] and given as:

$$\exp\left(\frac{J_v}{k}\right) = \frac{C_W - C_P}{C_F - C_P} \quad (3)$$

where k is the mass transfer, defined as:

$$k = \frac{D_{eff,\infty}}{\delta} \quad (4)$$

and $D_{eff,\infty}$ is the diffusion coefficient at infinite dilution and δ is the thickness of the concentration polarization layer.

The mass transfer coefficient may be determined experimentally by the substitution of Equations (1) and (2) into Equation (3), yielding:

$$\ln\left(\frac{1 - R_{obs}}{R_{obs}}\right) = \frac{J_v}{k} + \ln\left(\frac{1 - R_{real}}{R_{real}}\right) \quad (5)$$

In this case, the mass transfer coefficient may be represented as

$$k = a\omega^n \quad (6)$$

where a and n are predetermined constants and ω is the stirrer speed. For the Membranology cell, these constants are 2.993×10^{-6} and 0.415, respectively [34].

3. Results and Discussion

3.1. Preliminary Experiments

As previously alluded to, our understanding of the modified ceramic separation capabilities changed during the preliminary investigations into their use for this study. Previous and ongoing work using frack-produced water has demonstrated that the membranes are capable of cleaning water at the molecular level [32]. These waters are heavily contaminated far beyond solubility limits, resulting in biphasic mixtures and oil–water emulsions. Under these conditions, the modified ceramics are capable of separating small hydrocarbons such as toluene. However, upon attempting to use these membranes on a saturated toluene solution in just DI water, 1 g L^{-1} at 20 °C, no separation occurred. Therefore, the modified ceramics would not be suitable for the Bedfordshire trial, or at least not the only type of filter, hence exploratory NF laboratory studies were conducted to assess potential. The same toluene–water concentration solution that was showing 0% rejection with the ceramic membranes was rejecting between 70% and 93% of toluene using a NF membrane with a transmembrane pressure (TMP) of 2.5–20 bar, respectively. See Table 4.

Table 4. TOC data for dead-end filtration of the Bedfordshire site water.

	1 g/L Toluene–Water Dead-End Filtration	
Sample Name	TOC (mg/L)	Rejection
Desal DK feed	508.6	
Desal DK 2.5 bar permeate	34.15	93.29%
Desal DK 5 bar permeate	108.5	78.67%
Desal DK 10 bar permeate	127.1	75.01%
Desal DK 20 bar permeate	154.2	69.68%

With a successful separation, a further experiment using a 350 mL sample of the Bedfordshire site water was run. The sample was prefiltered by a 0.2 micron ceramic membrane prior to feed analysis and NF filtration occurring. It was first discovered during this experiment that the total organic carbon content of the feed water was considerably lower than expected. The results of the test are shown below in Table 5.

Table 5. TOC data for dead-end filtration of the Bedfordshire site water.

Company a Water Dead-End Filtration		
Sample Name	TOC (µg/L)	Rejection
Desal DK feed	3913	
Desal DK 2.5 bar permeate	316.4	91.91%
Desal DK 5 bar permeate	37.5	99.04%
Desal DK 10 bar permeate	0	100.00%
Desal DK 20 bar permeate	0.11	~100.00%

The NF DK series showed excellent overall organic carbon removal even at relatively low pressures. Unfortunately, the GC–MS equipment was not available at this stage to test specific volatile organic removal capabilities of the membrane.

3.2. Pilot Scale

3.2.1. Nanofiltration

The initial setup employed a GE DK series NF membrane, set to operate with most the retentate (concentrated dirty water) from the membrane directed back to the Swansea feed tank, whilst a small bleed of retentate and clean water permeate was returned to the OWS. In this configuration, if the membrane performed well and separated the volatile organic carbons, then the concentration of these organics in the Swansea feed tank would steadily increase. Further, the configuration removes only a small quantity of liquids from the OWS to replenish the lost permeate from the pilot rig to avoid increased mixing within the OWS. After 4 weeks of operation, the configuration was changed to return both the retentate and the permeate back to the OWS. In this mode, the membrane process is no longer a recycle of retentate or concentration process, but a once-through continuous process more suitable for larger-scale operations. During the once-through trial period, an observation was made that the prefilter experienced considerable fouling compared to the recycle mode of operation. See Figure 5. This was most likely the result of the increased flow of liquid from the OWS causing settled sediment within the OWS to be dispersed and transferred to the pilot system. There is also the possibility that during this period of operation, the well pumps were sending more silt to the OWS. Prior to deployment, it was recommended that a prefilter be installed in the system as a precaution despite the OWS water containing 'minimal solids.' Throughout the trial, the solids concentration of the OWS water was above what would be considered minimal for a NF membrane system under normal operation—a permanent system would have to cope with this level of solids more appropriately than weekly filter changes such as the incorporation of the functionalized ceramic microfiltration membrane. The decision was made to return to the original configuration to prevent starving the second pump or deadheading the first pump should the prefilter block completely.

The feed to the pilot from the Bedfordshire OWS contained five major compounds in varying quantities, totaling between 7358 and 1360 ppb across the trial period. The GC–MS data of the five main compounds are shown graphically in Figure 6a–e for VC, DCE, DCP, TCE and PCE, respectively.

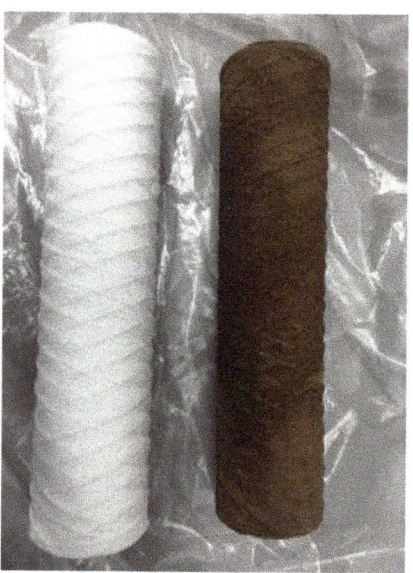

Figure 5. An example of a virgin prefilter (**left**) and a fouled prefilter (**right**).

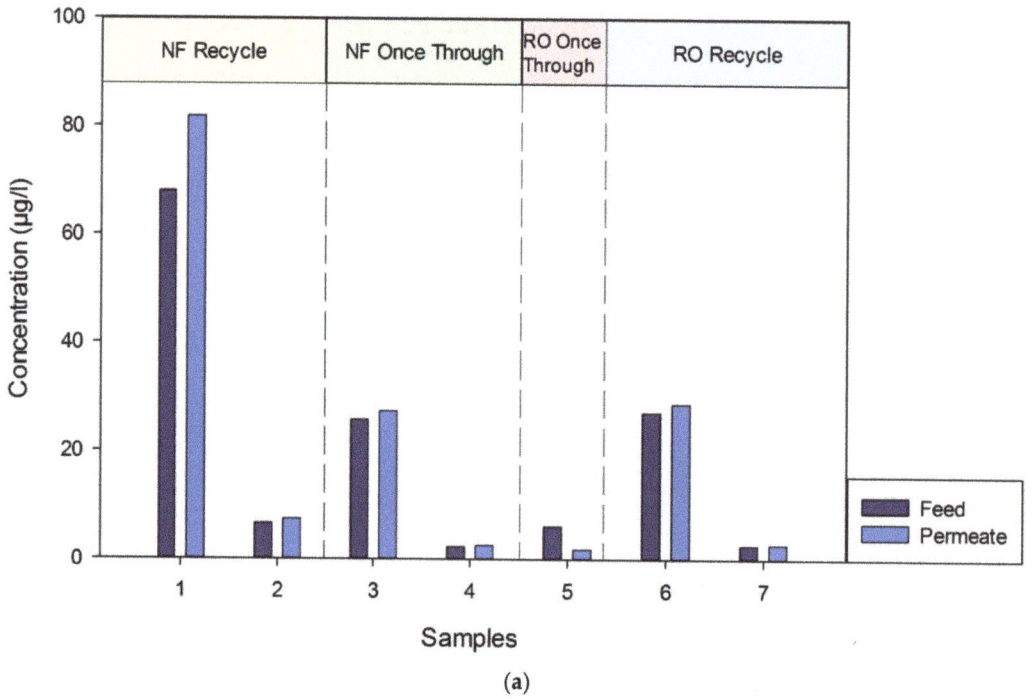

(**a**)

Figure 6. *Cont.*

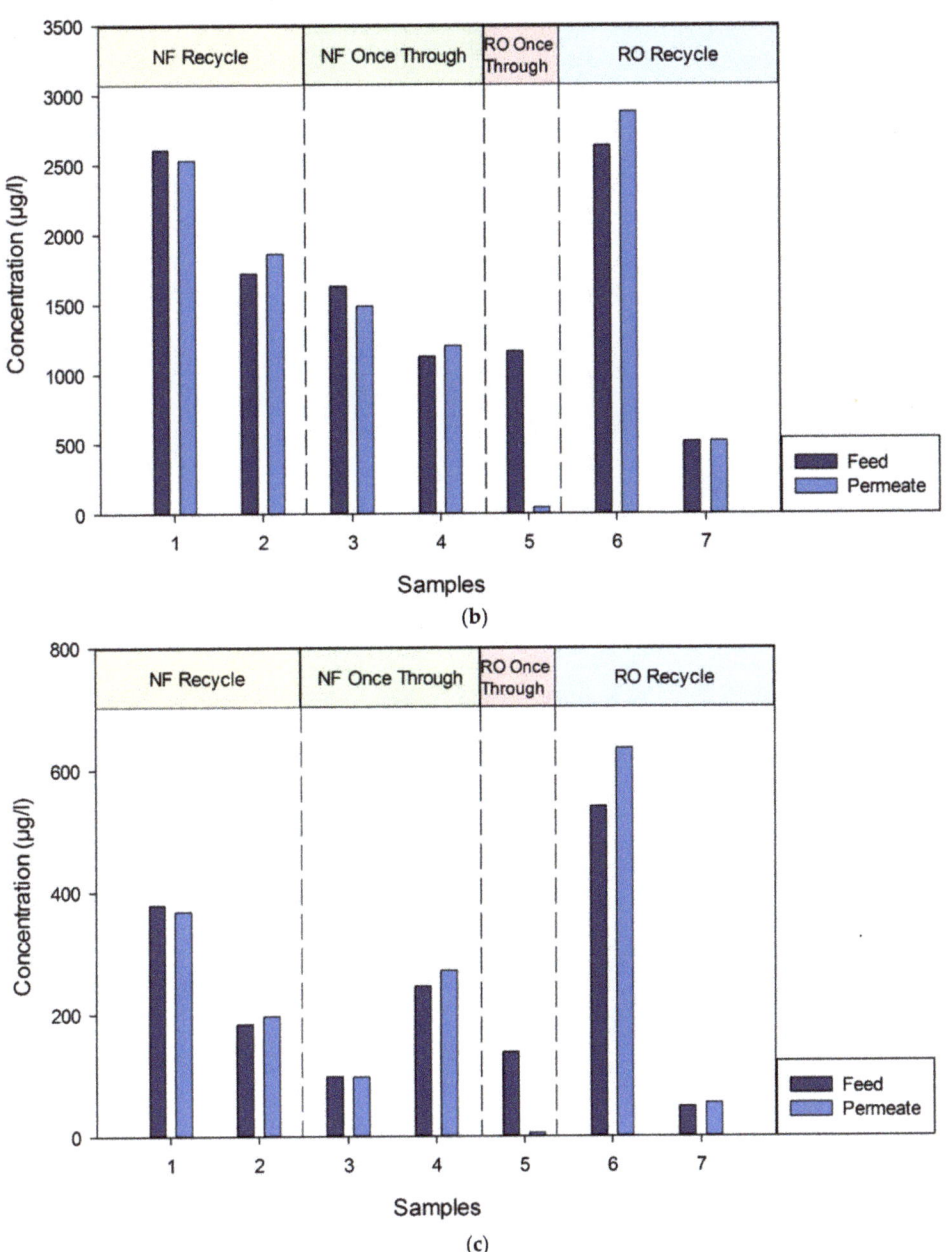

Figure 6. Cont.

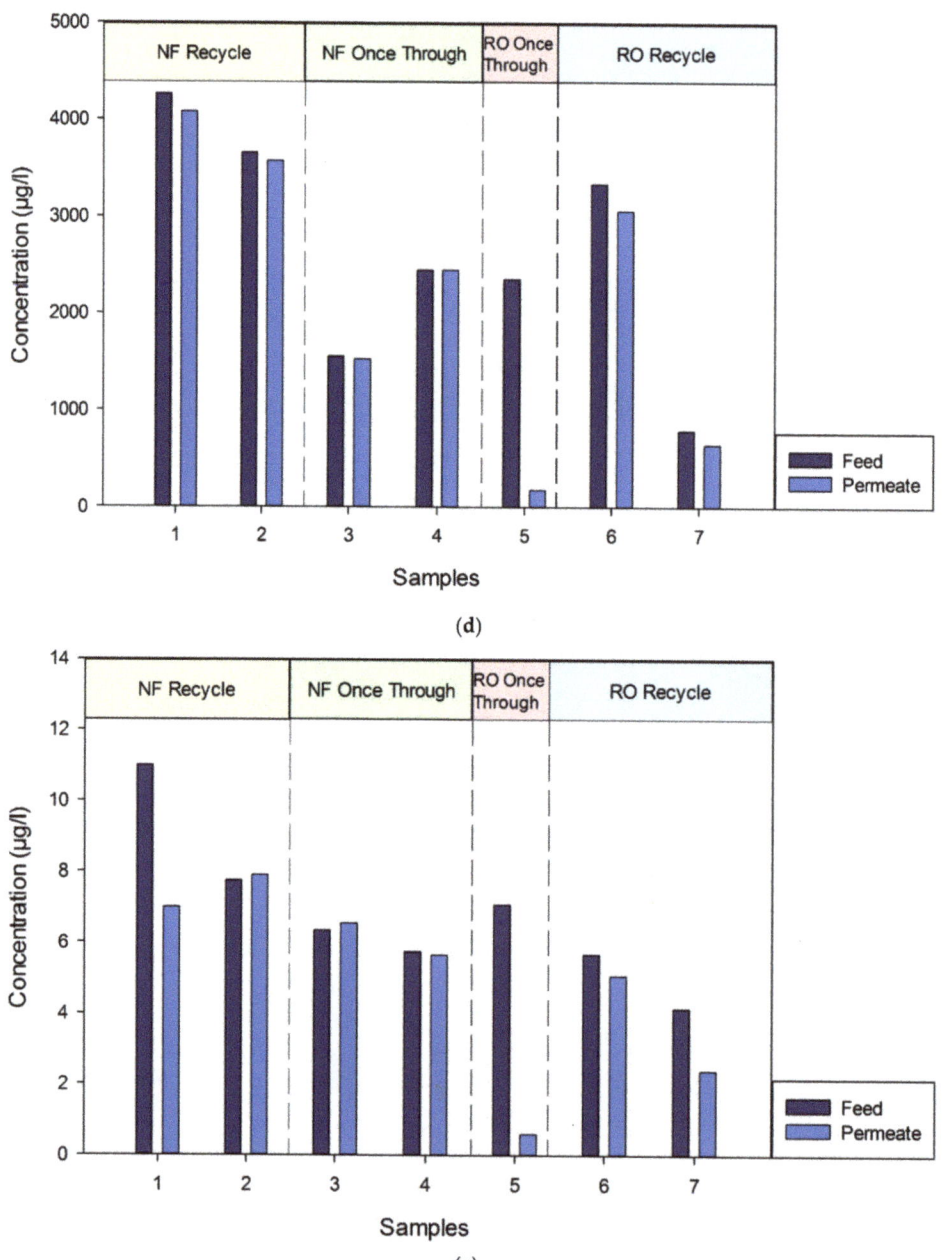

Figure 6. (**a**) Graphical representation of GC–MS data for vinyl chloride throughout the pilot trial. (**b**) Graphical representation of GC–MS data for cis-1,2-dichlorothylene throughout the pilot trial. (**c**) Graphical representation of GC–MS data for 2,2-dichloropropane throughout the pilot trial. (**d**) Graphical representation of GC–MS data for trichloroethylene throughout the pilot trial. (**e**) Graphical representation of GC–MS data for tetrachloroethylene throughout the pilot trial.

For both operating configurations, the GC–MS method for VOCs detected very little or no separation occurring during the NF trial, with the feed and permeate samples being comparable. As an example, sample data from 27th May of the five main compounds, shown in Table 6, clearly show no separation of the VOCs occurring.

Table 6. GC–MS data of five main components detected for the 27th May samples.

Compound	Feed 1 (µg/L)	Feed 2 (µg/L)	Permeate 1 (µg/L)	Permeate 2 (µg/L)
Total	5476	5684	5662	5639
VC	6.6	6.5	7.5	7.3
DCE	1687	1758	1974	1749
DCP	177.4	188.6	212	180.1
TCE	3594	3720	3457	3691
PCE	7.9	7.6	8.1	7.7

The results for the NF were disappointing but not completely unexpected if the membranes were to follow normal separation principles. It is well known from literature [37] that rejection of a solute is dependent on the applied pressure for steric rejection—the higher the applied pressure, the higher the rejection. At low pressures, or Bedfordshire trial operating pressures, the rejections are considerably lower than the rejection expected at typical operating pressures of NF and RO membranes. During the NF stage of the trial, TOC analysis was conducted in conjunction with the GC–MS analysis. As shown in Table 7, on average, 75% of the total organic carbon was removed by the NF membrane.

Table 7. Swansea University TOC data for NF trial.

Date	Sample	Total Carbon (µg/L)	Inorganic Carbon (µg/L)	Total Organic Carbon (µg/L)	Average (µg/L)	Rejection
13/05/2015	Feed	89,796	82,635	7161	7598	57.19%
	Feed	90,388	82,353	8035		
	Permeate	64,208	60,934	3274	3253	
	Permeate	63,759	60,527	3232		
27/05/2015	Feed	84,059	76,947	7112	7314	70.30%
	Feed	83,069	75,553	7516		
	Permeate	65,731	63,566	2165	2172	
	Permeate	65,714	63,535	2179		
09/06/2015	Feed	80,894	74,534	6360	6360	78.30%
	Permeate	62,659	61,279	1380	1380	
	Feed	79,405	73,265	6140	6140	79.10%
	Permeate	59,334	58,051	1283	1283	
23/06/2015	Feed	75,762	70,445	5317	5317	71.96%
	Permeate	60,341	58,850	1491	1491	

A rejection of 75% organic carbon is a much more positive result than the equivalent VOC rejection—the 1 µm prefilter (microfilter) will absorb only a fraction, if any, dissolved organic carbon. This is clear evidence that the NF membrane is still separating some organics, if not quite as efficiently as the laboratory conditions—the use of an NF is not entirely redundant as the removal of organics will assist in preventing an RO membrane downstream from being fouled. When trying to understand why the membrane displayed reduced organic carbon, the harsher conditions found in real-world trial conditions and the damage potentially inflicted on the membrane module must be considered. If the membrane had been damaged or pore sizes enlarged by the water contaminants, then the permeate flux would have been much greater than a flux of 5 LMH/bar, as suggested by the manufacturer. Upon closer examination of the recorded permeate flow rates and operating pressures during the trial, the membrane flux performance changed very little throughout the trial, with the specific flux averaging 5.45 LMH/bar ± 0.5 during operation. See Table 8.

Table 8. Nanofiltration membrane performance data.

Date	Inlet (bar)	Outlet (bar)	TMP (bar)	Recycle (l/s)	Permeate Flow (usgph)	Permeate Flow (l/h)	Flux (LMH)	Specific Flux (LMH/bar)
13th May	2.85	1.65	2.25	1.04	4.50	17.03	10.65	4.73
19th May	2.60	1.40	2.00	1.05	3.40	12.87	8.04	5.75
19th May	3.00	1.80	2.40	1.05	4.20	15.90	9.94	5.52
27th May	2.60	1.40	2.00	1.04	3.20	12.11	7.57	5.41
27th May	3.00	1.80	2.40	1.05	4.60	17.41	10.88	6.05
1st June	2.95	2.40	2.68	1.04	4.00	15.14	9.46	3.94
9th June	2.80	1.60	2.20	1.02	4.00	15.14	9.46	5.91
9th June	3.00	1.80	2.40	1.03	4.20	15.90	9.94	5.52
23rd June	2.50	1.30	1.90	1.02	3.00	11.36	7.10	5.46
23rd June	3.00	1.80	2.40	1.03	4.40	16.66	10.41	5.78
8th July	2.00	0.80	1.40	1.03	2.00	7.57	4.73	5.91

Therefore, the membrane was not inadvertently damaged by dissolved solids or other materials. Despite the nanofiltration membrane not removing the VOCs, it effectively removed a large percentage of other organic carbons in the water source. As previously noted, the operating pressure of 3 bar is very low for a NF membrane of this type. A minimum operating pressure would usually be 8 bar(g), with a preferred operating pressure upwards of 20–30 bar(g). Increasing the operating pressure could potentially improve the rejection characteristics of the membrane. However, the lack of rejection of the VOCs even at 3 bar(g) suggests that VOC removal may not be viable with an NF membrane and only a trial using more appropriate equipment could confirm this fact. However, the initial proposal of using a ceramic microfilter required a reduced pressure system. To confirm this conclusion further in-house laboratory testing was necessary to gather rejection data at higher pressures (5–30 bar).

Upon analyzing the used pilot membrane, the membrane surface was considerably fouled see Figure 7. This fouling layer does not appear to have affected the permeate flow rate due to the pore size of the membrane being considerably smaller than the expected pore size of the foulant. However, this filter cake could have significantly affected the separation properties of the membrane, negating the surface charge effects [38,39].

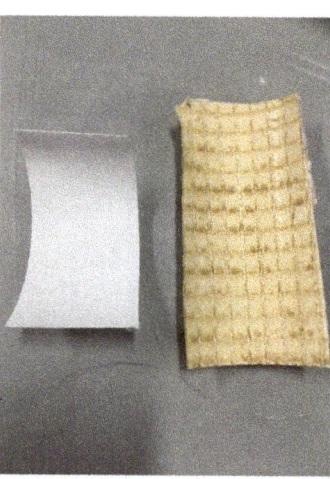

Figure 7. New DK membrane (**left**) vs. post trial DK membrane (**right**).

3.2.2. Reverse Osmosis

Prior to the Bedfordshire deployment, there were concerns that the nanofiltration membrane would not be fully effective at removing the VOCs to below the levels desired due to the pore size and reduced pressures being used. As a result, a low-energy RO membrane was purchased ready to switch out the nanofiltration membrane during the trial if necessary. A low-energy RO is designed to operate at a much lower pressure, 7 to 10 bar(g), than normal RO pressures (>60 bar) while separating 98% of NaCl versus 99.5% NaCl, respectively. Despite the Bedfordshire pilot rig pressure being limited to 3 bar, the low-energy RO membrane was anticipated to yield better separation than the NF membrane and considerably better separation than a standard RO membrane.

Due to circumstances outside of our control, TOC analysis was unavailable during the RO trial period. The GC–MS results for VOCs showed 92.8% removal on the first sample taken after an hour of operation. See Table 9.

Table 9. GC–MS data of five main components detected for the 8th July samples.

Compound	Feed (µg/L)	Permeate (µg/L)	Rejection (%)
Total	3670.0	264.0	92.8
VC	6.1	1.8	70.3
DCE	1165.0	43.5	96.3
DCP	137.2	5.1	96.3
TCE	2351.7	175.8	92.5
PCE	7.1	0.6	91.7

This was a very promising result considering the membrane was being operated at a pressure considerably less than optimum, suggesting that a further increase in pressure would increase the separation capabilities beyond 93% rejection of the VOCs. However, the following sample (two weeks later) and those beyond showed no VOC separation. See Table 10.

Table 10. GC–MS data for five main components detected on 21th July.

Compound	Feed (µg/L)	Permeate (µg/L)
Total	6552	6613
VC	27.1	28.7
DCE	2636.9	2881.8
DCP	539.8	635.2
TCE	3340.8	3058.7
PCE	5.7	5.1

Once again, if the membrane had been damaged, then the permeate flux would have been much greater than the initial flux when the membrane was first installed. The results indicate that the specific flux did not change significantly during the sample period. See Table 11.

Table 11. RO membrane pilot-scale performance data.

Time	Inlet (bar)	Outlet (bar)	TMP (bar)	Recycle Flow (L/s)	Permeate Flow (L/h)	Flux (LMH)	Specific Flux (LMH/bar)
8th July	2.00	1.40	1.70	0.67	2398	2.12	1.51
21st July	3.00	1.60	2.30	1.03	3708	2.42	1.51
04th Aug	3.00	1.40	2.20	0.68	2441	1.97	1.41
18th Aug	3.00	1.40	2.20	*	*	1.51	1.08
1st Sept	2.60	0.70	1.65	*	*	0.30	0.43
1st Sept	3.00	1.50	2.25	*	*	1.67	1.11

* Unable to read totalizer.

The membrane does appear to have experienced considerable fouling towards the end of the trial as the specific flux drops to approximately 29% of the original value. Membrane fouling can affect the quality of rejection, the surface charge of the membrane can be severely impacted by the build up of a fouling layer, and this may well explain the results seen [40,41]. Also of note during this period of operation, the prefilter experienced considerable fouling by silt. It is possible that the silt saturated the microfilter, and was forced through to the RO membrane and that this then had a detrimental effect on the membrane performance. This silt contamination was confirmed post trial. See Figure 8a,b. In addition, the dissolved organic material rejected by the NF membrane would also be rejected (and to a higher extent) by the RO membrane and could possibly cause fouling. However, without GC–MS data for the latter stages of the RO trial and no TOC data throughout, a categorical confirmation that fouling is the reason behind this reduced separation is not possible from the pilot-scale trials.

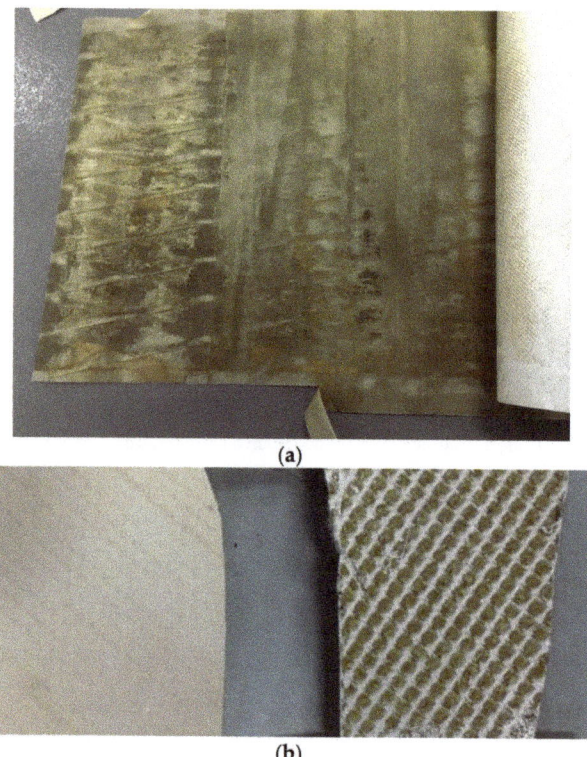

Figure 8. Illustration of the fouling that occurred for the RO membrane. (**a**) Fouled backing layer; (**b**) virgin membrane surface [left] vs. post trial membrane surface [right].

As with the NF trial, further testing on the RO membranes across an increased pressure range, including TOC and GC–MS analysis at the Bedfordshire site conditions and the operating conditions suggested by the manufacturer, was required. At the same time, an investigation into the fouling effects on the membrane under more stringent monitoring conditions should be conducted. This will determine whether fouling caused the rejection of the VOCs to drop so drastically. As a final note on the pilot-scale trial, during decommissioning of the system, it was discovered that one of the side port ferrules had been severely rusted on the stainless steel feed tank. Further, multiple pin holes were found on the side of the tank. See Figure 9. The tank was returned to the manufacturer for diagnostic analysis. The manufacturer's report indicated that the tank was made of an inferior quality stainless steel, 304, while all other components of the system were 316 L. The ferrule issue was also a result of the stainless steel used and the thickness of the tank. The ferrule was 316 L grade, and as a result, 316 L weld filler was used, which requires a higher temperature than 304 weld filler. When heated to the higher welding temperatures, the chromium combines with the carbon, leaving the steel short of chromium. Therefore, any further systems should be fully constructed from 316 L as a minimum. However, the use of a suitable plastic such as PVC or HDPE would be recommended to considerably reduce the cost of any subsequent system.

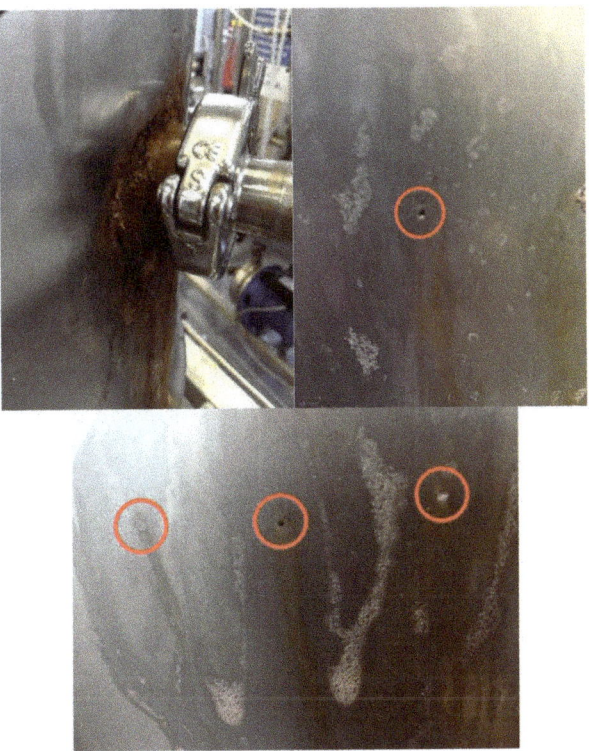

Figure 9. Swansea feed tank corrosion. Top left, rusting around the tri-clamp connection. Top right and bottom, pitting on the vessel surface.

3.2.3. Contact Angle

The contact angle was measured for the fouled DK and AK pilot-scale trial membranes, and the DK NF membrane hydrophobicity increased significantly to 72.88 ± 0.78°. The AK RO membrane, on the other hand, behaved in the opposite manner—the hydrophilicity increased to 48.76 ± 2.52°. Similar phenomena have been reported previously [39,42].

3.3. Laboratory Scale

3.3.1. TOC Analysis

Additional testing under laboratory conditions was necessary to understand the results of the pilot-scale trial. As well as the GE DK membrane used in the pilot scale, another three NF membranes were tested, as listed in Table 3a. As for RO, together with the pilot-scale RO membrane, GE Osmonics AK, an additional four RO membranes were assessed, as shown in Table 3b. The first experiment was investigating salt rejection, and the reason for this test was twofold. Firstly, membrane manufacturers supply NaCl rejection data for RO at a specific operating condition, and therefore this testing allows comparison with the manufacturer's published results as a secondary check that the membrane is operating as expected. Secondly, the treated groundwater will eventually be released back into the local water supply, and since a change in groundwater salinity may have an adverse effect on the local environment, this must be monitored and balanced. The second set of experiments using toluene, an immiscible aromatic hydrocarbon, were used to replicate a very common groundwater contamination of a water–oil biphasic mixture with extremely low solubility, 0.52 g/L at 20 °C. The final set of experiments were on a TCE–water mixture, the predominant contaminate found in the groundwater at the Bedfordshire site. TOC analysis was used to analyze the rejection as the GC–MS was initially unavailable. Results from all three experimental sets conducted for nine membranes can be found in Table 12a–i.

Table 12. Laboratory separation assessment trials using nanofiltration and reverse osmosis membranes.

a. Dow NF90 membrane flux (LMH) and observed and real rejection data for NaCl, toluene and trichloroethylene from conductivity and TOC measurements.

Membrane	NaCl			Toluene			TCE		
	Flux	Obsd	Real	Flux	Obsd	Real	Flux	Obsd	Real
NF90 5 bar	35.5	50.5%	69.2%	40.2	90.4%	95.8%	45.9	72.7%	88.1%
NF90 10 bar	80.4	58.6%	89.4%	85.8	70.5%	94.1%	86.1	42.6%	83.4%
NF90 15 bar	123	60.6%	95.9%	131.1	58.6%	96.3%	126.3	32.3%	88.7%
NF90 20 bar	172.2	52.4%	98.1%	179.5	49.7%	98.1%	163.9	32.6%	94.8%
NF90 25 bar	213.1	54.8%	99.3%	225.6	49.5%	99.3%	203.8	39.3%	98.4%
NF90 30 bar	236.5	32.1%	98.9%	275.6	51.7%	99.8%	241.1	32.7%	99.0%

b. Dow NF270 membrane flux (LMH) and observed and real rejection data for NaCl, toluene and trichloroethylene from conductivity and TOC measurements.

Membrane	NaCl			Toluene			TCE		
	Flux	Obsd	Real	Flux	Obsd	Real	Flux	Obsd	Real
NF270 5 bar	45.9	48.6%	72.3%	54.5	83.5%	94.4%	47.8	53.9%	77.2%
NF270 10 bar	94.7	49.2%	88.8%	108.4	57.1%	93.6%	98.8	36.4%	83.7%
NF270 15 bar	137.8	50.4%	95.6%	157.7	45.6%	96.5%	155	29.2%	92.8%
NF270 20 bar	186	48.4%	98.3%	215.3	36.4%	98.6%	199.3	23.2%	96.2%
NF270 25 bar	230.4	38.1%	99.0%	259.4	36.0%	99.4%	258.4	21.5%	98.8%
NF270 30 bar	282.9	41.8%	99.7%	303.2	39.9%	99.8%	296.3	28.5%	99.7%

c. GE DK membrane flux (LMH) and observed and real rejection data for NaCl, toluene and trichloroethylene from conductivity and TOC measurements.

Membrane	NaCl			Toluene			TCE		
	Flux	Obsd	Real	Flux	Obsd	Real	Flux	Obsd	Real
DK 5 bar	27.8	38.3%	53.5%	28.5	87.8%	93.1%	26.6	72.8%	82.9%
DK 10 bar	49.8	42.0%	68.6%	55.2	66.5%	87.1%	49.7	58.8%	81.2%
DK 15 bar	72.6	48.2%	82.4%	80.4	59.0%	89.6%	75.7	47.7%	83.1%
DK 20 bar	103.3	50.4%	91.0%	109.6	53.3%	92.8%	96.2	47.4%	88.4%
DK 25 bar	120.6	50.8%	93.8%	129.2	53.0%	95.2%	115.6	52.2%	93.4%
DK 30 bar	146.2	51.4%	96.5%	155	51.8%	97.1%	137.8	48.2%	95.2%

d. GE DL membrane flux (LMH) and observed and real rejection data for NaCl, toluene and trichloroethylene from conductivity and TOC measurements.

Membrane	NaCl			Toluene			TCE		
	Flux	Obsd	Real	Flux	Obsd	Real	Flux	Obsd	Real
DL 5 bar	31.6	33.7%	50.6%	32.7	77.5%	87.7%	29.4	77.8%	87.0%
DL 10 bar	64.6	34.7%	69.0%	60.8	58.4%	84.4%	59.2	50.9%	79.4%
DL 15 bar	91.7	39.1%	83.1%	90	47.7%	87.0%	84.9	38.9%	80.7%
DL 20 bar	117.6	40.7%	90.3%	114.4	50.2%	92.7%	111.6	35.5%	86.7%
DL 25 bar	140.3	41.2%	94.0%	136.4	50.5%	95.5%	137.8	35.6%	92.2%
DL 30 bar	174.5	40.9%	97.1%	160.7	53.1%	97.6%	155	34.5%	94.3%

Table 12. *Cont.*

e. GE AK membrane flux (LMH) and observed and real rejection data for NaCl, toluene and trichloroethylene from conductivity and TOC measurements.

Membrane	NaCl			Toluene			TCE		
	Flux	Obsd	Real	Flux	Obsd	Real	Flux	Obsd	Real
AK 5 bar	6.9	93.2%	94.1%	9.2	98.3%	98.6%	8.5	79.1%	82.1%
AK 10 bar	15.3	95.3%	96.6%	16.7	94.9%	96.4%	17.2	73.4%	80.2%
AK 15 bar	21.8	95.8%	97.3%	24.1	96.5%	97.9%	26.3	70.4%	81.0%
AK 20 bar	30.7	96.6%	98.3%	32.1	95.4%	97.7%	34.4	73.1%	85.4%
AK 25 bar	38.2	97.0%	98.7%	40.6	94.0%	97.5%	41.2	72.2%	86.6%
AK 30 bar	47.1	97.0%	98.9%	48.1	93.8%	97.8%	51.7	69.7%	87.9%

f. GE AG membrane flux (LMH) and observed and real rejection data for NaCl, toluene and trichloroethylene from conductivity and TOC measurements.

Membrane	NaCl			Toluene			TCE		
	Flux	Obsd	Real	Flux	Obsd	Real	Flux	Obsd	Real
AG 5 bar	10.9	96.8%	97.5%	15.4	98.9%	99.3%	14.4	78.2%	83.2%
AG 10 bar	27.8	98.9%	99.4%	28.8	95.0%	97.3%	28.5	63.1%	76.3%
AG 15 bar	46.5	99.2%	99.7%	41.7	95.9%	98.3%	42.3	68.9%	85.0%
AG 20 bar	60.3	99.3%	99.8%	56.5	94.1%	98.2%	57.4	69.9%	89.3%
AG 25 bar	76.8	99.4%	99.9%	70.2	92.9%	98.4%	70.1	69.8%	91.6%
AG 30 bar	95.5	99.3%	99.9%	85.1	92.3%	98.8%	84	68.0%	93.2%

g. Dow BW30 membrane flux (LMH) and observed and real rejection data for NaCl, toluene and trichloroethylene from conductivity and TOC measurements.

Membrane	NaCl			Toluene			TCE		
	Flux	Obsd	Real	Flux	Obsd	Real	Flux	Obsd	Real
BW30 5 bar	11.1	93.4%	94.7%	14	95.4%	96.6%	14.6	84.2%	88.1%
BW30 10 bar	24.4	94.9%	96.9%	27.7	96.7%	98.2%	27.3	79.0%	87.3%
BW30 15 bar	37.3	96.1%	98.2%	40.2	93.0%	97.0%	40.3	63.6%	81.1%
BW30 20 bar	52.6	96.6%	98.9%	53.5	92.0%	97.4%	54.5	52.2%	78.6%
BW30 25 bar	64.9	96.7%	99.2%	66	91.3%	97.9%	66	66.7%	89.7%
BW30 30 bar	77.5	96.9%	99.4%	78.9	90.7%	98.2%	79.6	65.1%	91.6%

h. Dow BW30LE membrane flux (LMH) and observed and real rejection data for NaCl, toluene and trichloroethylene from conductivity and TOC measurements.

Membrane	NaCl			Toluene			TCE		
	Flux	Obsd	Real	Flux	Obsd	Real	Flux	Obsd	Real
BW30LE 5 bar	15.2	91.5%	93.8%	20.7	96.0%	97.5%	22.6	64.9%	75.3%
BW30LE 10 bar	35.9	94.3%	97.4%	41.1	91.2%	96.3%	43.2	47.6%	70.3%
BW30LE 15 bar	53.9	96.1%	98.8%	60.5	84.3%	95.4%	63.2	48.8%	79.5%
BW30LE 20 bar	77.5	96.1%	99.3%	79.8	83.1%	96.7%	87.2	48.5%	86.7%
BW30LE 25 bar	94.7	96.4%	99.5%	103.3	82.0%	97.8%	105.3	49.3%	91.0%
BW30LE 30 bar	112	96.6%	99.7%	112	77.0%	97.6%	127.4	65.2%	96.9%

i. Dow BW30XFR membrane flux (LMH) and observed and real rejection data for NaCl, toluene and trichloroethylene from conductivity and TOC measurements.

Membrane	NaCl			Toluene			TCE		
	Flux	Obsd	Real	Flux	Obsd	Real	Flux	Obsd	Real
BW30XFR 5 bar	10	94.8%	95.8%	13.6	95.4%	96.5%	13	68.9%	74.7%
BW30XFR 10 bar	23	96.0%	97.6%	26.9	98.1%	98.9%	26.6	83.6%	90.2%
BW30XFR 15 bar	35.7	96.8%	98.5%	40.5	97.0%	98.8%	40.2	76.0%	88.5%
BW30XFR 20 bar	48.4	96.9%	98.9%	53.9	95.7%	98.7%	51.7	74.3%	90.1%
BW30XFR 25 bar	60.3	97.0%	99.2%	66	95.0%	98.8%	63.2	76.5%	93.0%
BW30XFR 30 bar	68.9	97.1%	99.4%	80.4	94.7%	99.1%	75.2	76.8%	94.6%

The rejection of NaCl for NF membranes is not widely published by the manufacturer as NF is primarily used as a water softener, typically removing divalent salts not monovalent. However, literature values suggest that a NaCl rejection of 5–95% can be expected for NF depending on feed salinity and operating conditions [43]. Clean water fluxes obtained from the new membranes confirmed that the membranes were operating effectively. Toluene is a relatively inexpensive solvent, making it ideal as an initial solvent to test the performance of the NF membranes. As previously discussed, toluene has a low solubility with water at room temperature. When the concentration exceeds the solubility limit, the solution forms a biphasic mixture with a top layer of toluene. Due to this nature of biphasic mixtures, representative sampling can be particularly difficult to achieve. To avoid these difficulties, the feed concentration selected, 0.1 g L^{-1}, removes the potential for inaccurate feed concentration samples. For all four NF membranes, the observed rejection decreases as the applied pressure increases. For example, for Dow NF90, the observed rejection is 90.4% to 51.7%, between 5 and 30 bar, respectively—this is counter to the real rejection, which takes concentration polarization at the membrane surface into account. The real rejection ranges between 88.1% and 99%, respectively, and the greatest difference between the observed and real rejection, $\geq$60%, is at the highest pressures. This occurrence is sensible, since as the pressure increases, the flux rate increases, and the resulting mass transfer is governed by convective transport of solute to the membrane surface. This convective flux to the membrane surface is significantly higher than the mixing rate removing solute from the membrane surface and thus concentration polarization is inevitable. The rejection data obtained for TCE demonstrate a similar pattern. However, the observed and real rejections are both lower when compared to the equivalent toluene results. If nanofiltration separation was based solely on steric (size) exclusion, then we would expect to see toluene rejecting less than TCE due to its lower MW. However, previous studies have proven that NF exhibits a combination of separation mechanisms from both UF and RO, steric and Donnan (charge) exclusion or ionic diffusion, respectively [44–46]. All four membranes exhibit a similar decline in observed rejection from 5 to 15 bar, to varying degrees. Between 15 and 30 bar, the rejection appears to reach a plateau, with only some slight fluctuations observed.

The rejection of NaCl for RO membranes is published by the manufacturer, allowing a cheap and relatively simple way of testing the membranes integrity. GE Osmonics market the AK as a low-energy brackish water reverse osmosis (BWRO) with a minimum NaCl rejection of 98% using a 500 ppm NaCl solution, with operating conditions of 8 bar pressure, 25 °C and pH 7.5. As shown in Table 12e, the observed rejection was less than the advertised minimum. However, this can be explained as the feed solution used was 4-fold above the manufacturer test solution, 2000 ppm. A higher feed NaCl concentration increases the salt gradient between the feed and permeate or osmotic difference. An increase in osmotic gradient requires a higher applied pressure to overcome the osmotic pressure to maintain the permeate flux. With a constant pressure, the water flux decreases, and therefore salt passage increases. GE Osmonics tests on the AG series use a 2000 ppm NaCl solution, with operating conditions set as 15.5 bar operating pressure, 25 °C and pH 7.5. The AG series is marketed as a standard BWRO with a minimum NaCl rejection of 99%. It is unclear why GE use two different feed solutions for comparable membranes. As shown in Table 12f, the observed rejection was 99.2% at 15 bar. Dow Filmtec publicize the BW30 as a low-energy BWRO with a minimum NaCl rejection of 99% using a 2000 ppm NaCl solution, with operating conditions of 15.5 bar pressure, 25 °C and pH 8. As shown in Table 12g, BW30 only obtained a salt rejection of 96.1%. BW30LE, a low-energy BWRO, has a minimum NaCl rejection of 98% using a 2000 ppm NaCl solution, with operating conditions of 10 bar pressure, 25 °C and pH 8. As shown in Table 12h, BW30LE only obtained an observed salt rejection of 94.3%. The final Dow Filmtec membrane, BW30XFR, is an optimized extra fouling-resistant BWRO membrane. Dow claim a minimum salt rejection of 99.4%, and operational conditions are the same as those used in BW30. When tested, the membrane attained a salt rejection of 96.8%, as noted in Table 12i. It is unclear whether the manufacturers' stated minimum rejections are observed or calculated real rejections. It is assumed that the published data are observed, but if the results are real, the results in this study are closer to the manufacturers' values. The RO membranes exhibit the same separation tendencies as the NF membranes for toluene and TCE, but with higher observed rejections for all the RO membranes. As reverse osmosis membranes have no physical pores, size exclusion is no longer a separation mechanism. Solute passage is determined purely by solution diffusion, thus, the increased rejection for the RO membranes is entirely logical. Comparing the observed rejection and calculated real rejection, the variances between them are less pronounced compared to the NF membranes. This fully agrees with the NF filtration pressure increase vs. observed/real rejection theory—increased convective flux results in increased concentration polarization. RO membranes have a greatly reduced specific flux when evaluated against NF membranes. A lower permeate flux produces less convection to the membrane surface, which in turn minimizes concentration polarization, hence, less discrepancy between observed and real rejection.

3.3.2. GC–MS Analysis

The TOC system at Swansea university was not equipped with the optional purgeable organic carbon (POC) analyzer. VOCs are easily purged from a sample via sparging, the TOC of the sample can be determined by the addition of the POC and non-purgeable organic carbon (NPOC). Despite the TOC method used not including a sparge of the sample, volatiles can be difficult to quantify through the subtraction method compared to the addition method, TOC = TC − IC and TOC = POC + NPOC, respectively. The analysis of volatile organic compounds (VOCs) in environmental water samples is usually performed by either headspace (HS) or purge and trap (P&T), with separation by gas chromatography (GC) and detection by mass spectrometry (MS). The P&T method uses a carrier gas remove volatiles from the solution and these are caught in an adsorbent trap. The trap is then heated which releases the volatiles into the GC-MS for subsequent analysis. This method provides excellent sensitivity as the total sample is extracted. However, in comparison to other methods, P&T methods are generally more complicated to operate and maintain. They can also suffer from a degree of water carry over, which may lead to a loss in sensitivity, a loss of peak shape and, in some cases, sample cross contamination. The obvious alternative is the HS method which uses a closed sample arrangement. In this case, the sample vial containing the total solution is heated (and agitated in some systems) in order to drive the volatiles out of the solution and into the headspace of the vial. When this is the case, an equilibrium forms between the volatile contained in the solution and the headspace. This equilibrium can be shifted by the addition of salt to the sample. After a specified time, a portion of the headspace is transferred onto the GC–MS via a valve with a sample loop. This technique is robust and experiences few carryover problems as less water is transferred to the GC–MS. The GC-MS HS system was installed at Swansea university after the pilot-scale trials, but during the laboratory trial. Therefore, a mixture of VOCs was produced. See Section 3.3.1. This VOC mixture underwent the same experimental pressures as the TOC analysis. The rejection results are shown in Figure 10a–i.

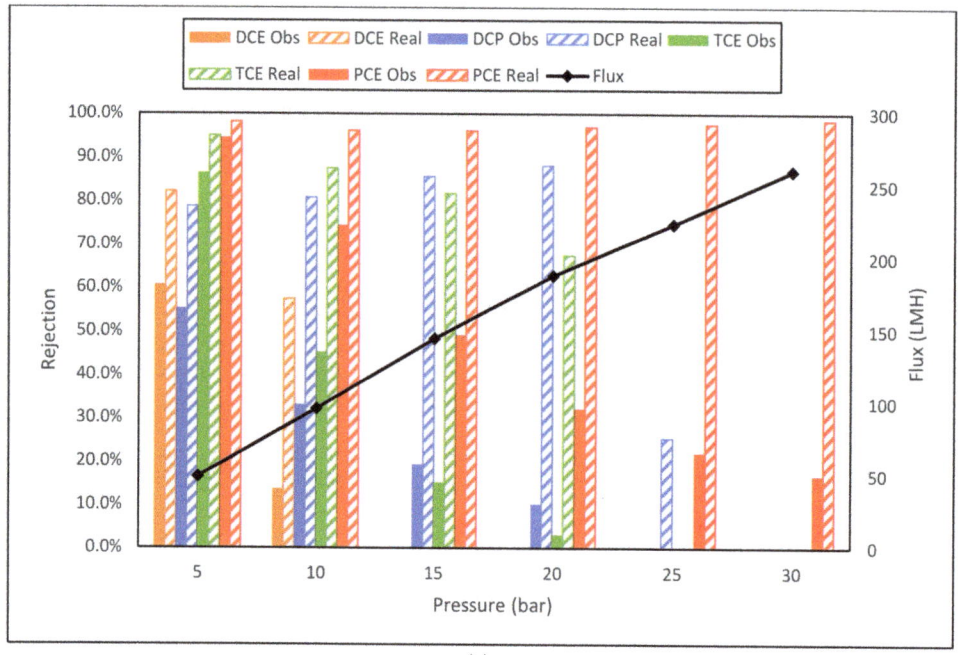

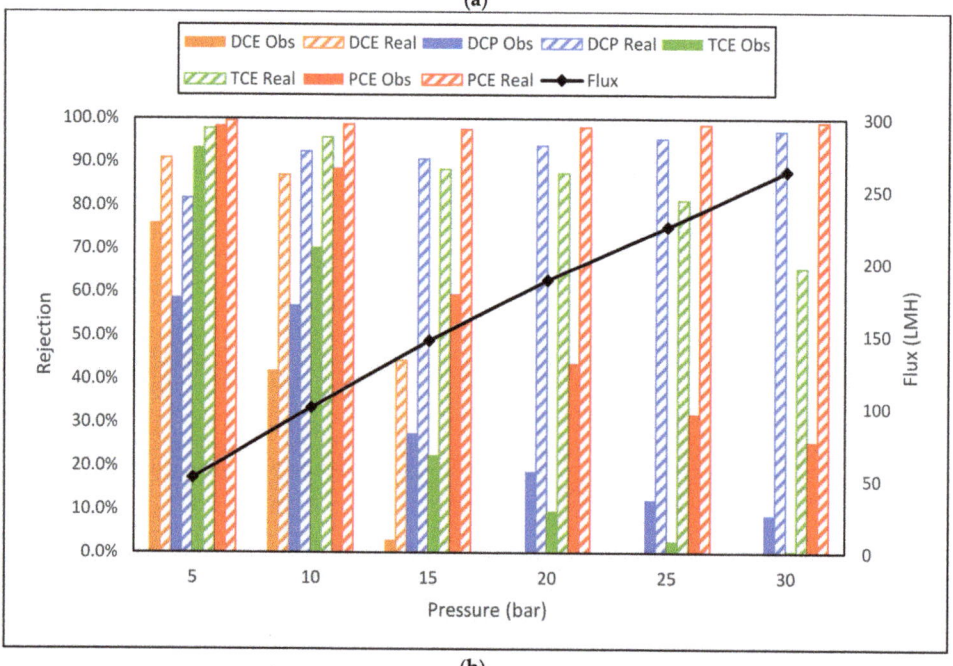

Figure 10. *Cont.*

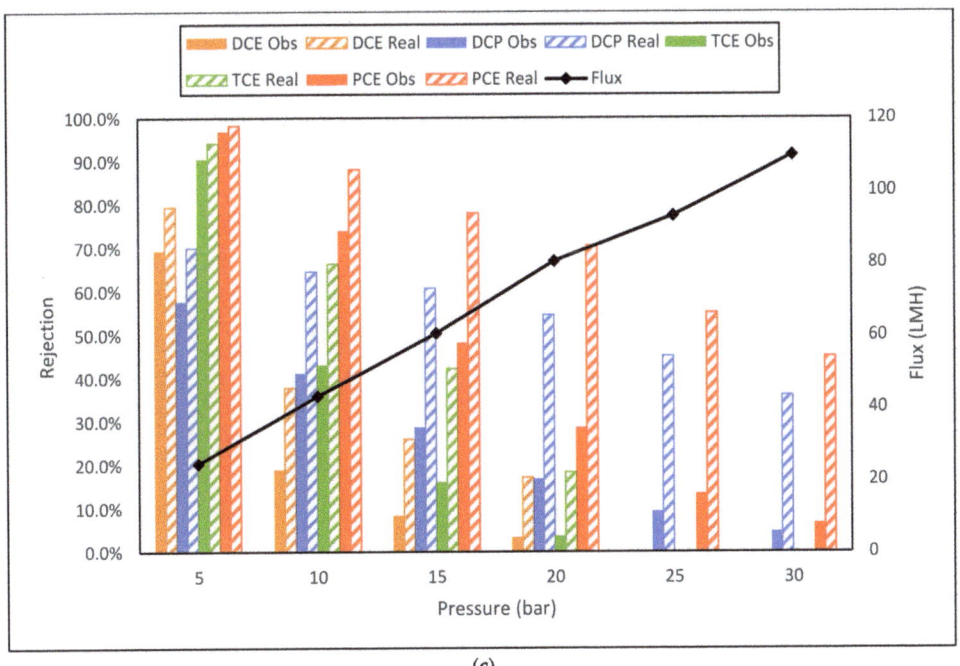

(c)

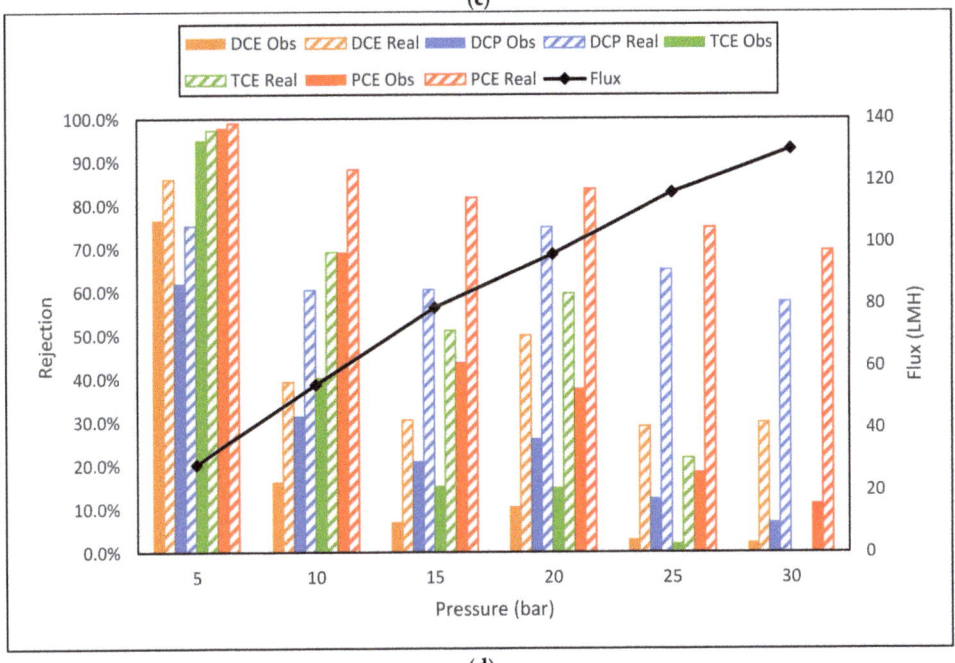

(d)

Figure 10. *Cont.*

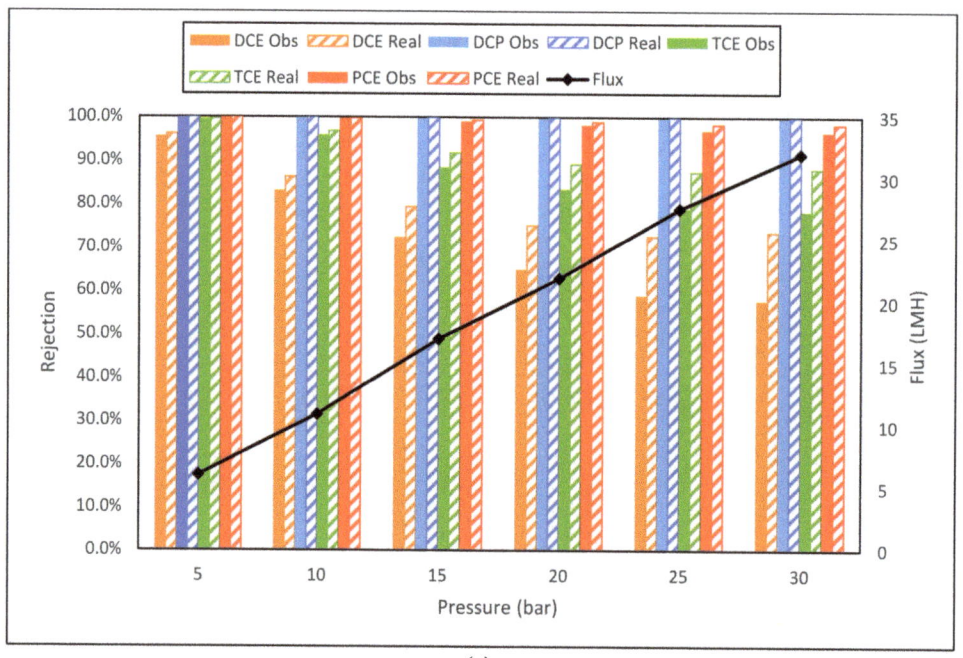

(e)

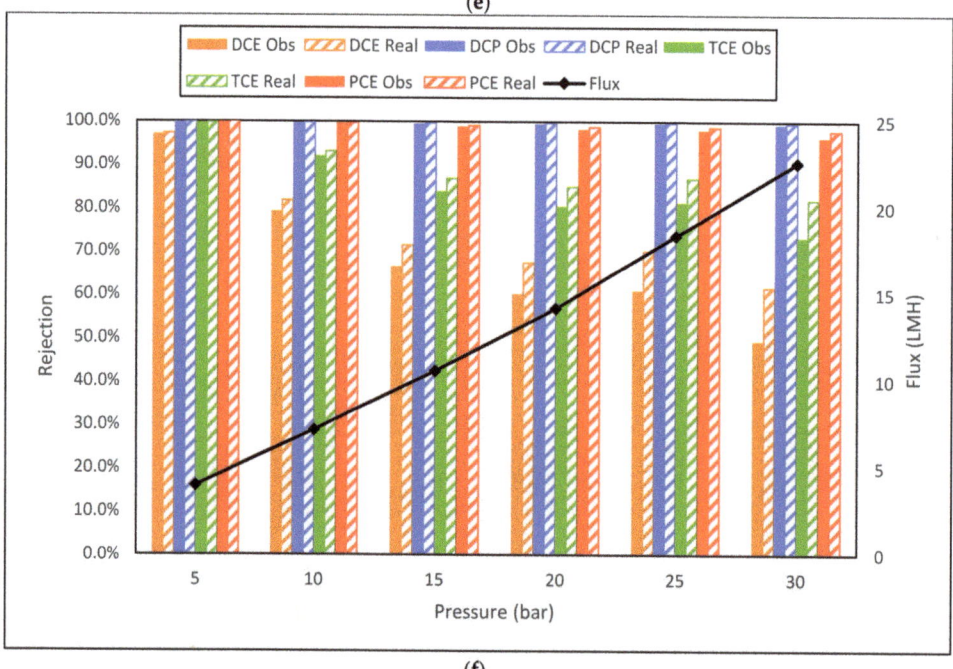

(f)

Figure 10. *Cont.*

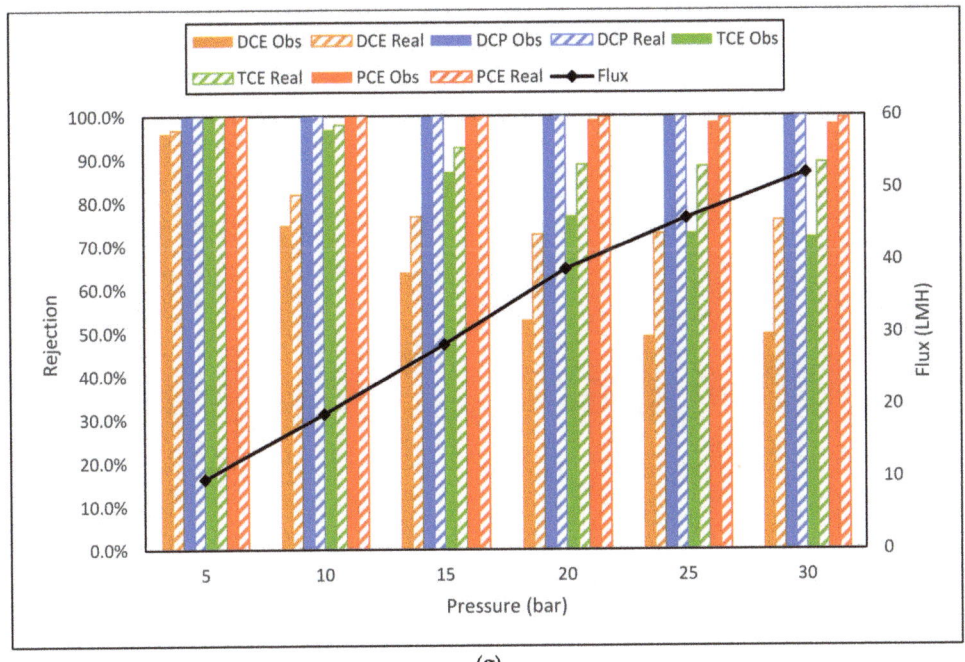

(g)

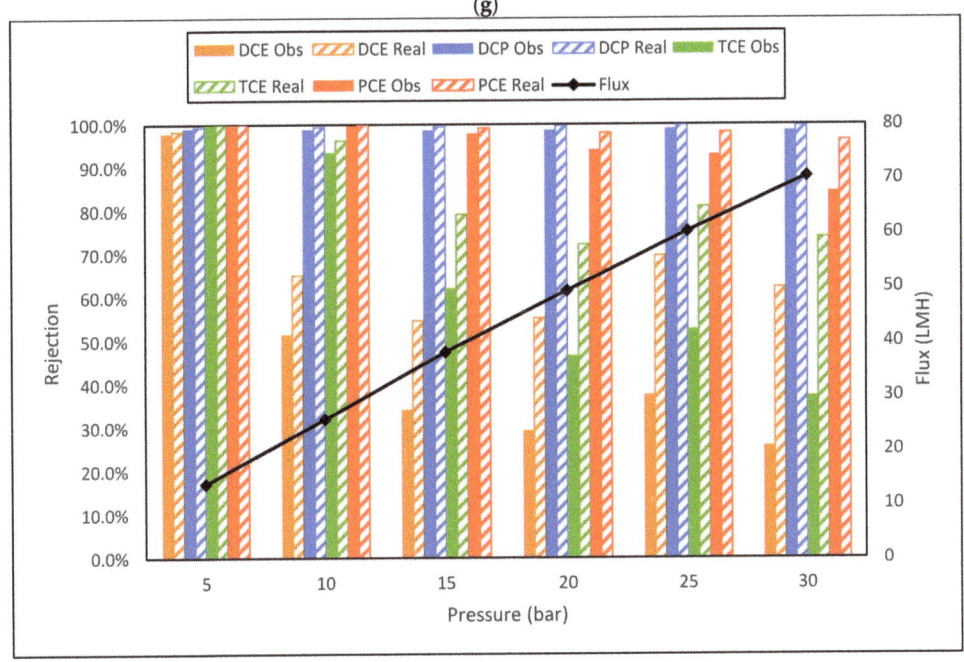

(h)

Figure 10. *Cont.*

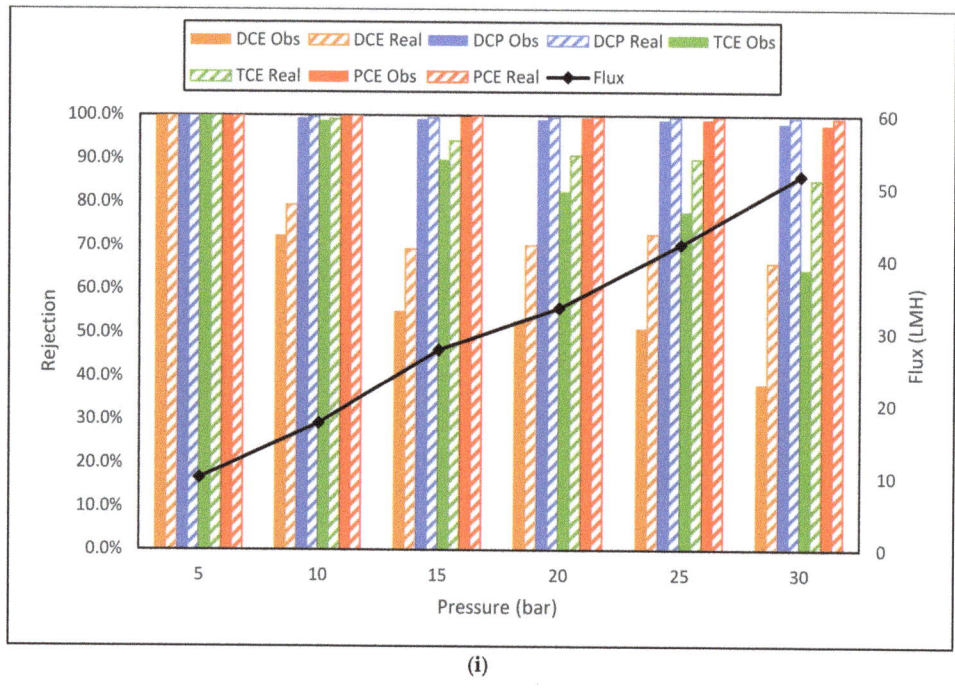

(i)

Figure 10. (**a**) Dow NF90 membrane flux (LMH) and observed and real rejection data for VOC solution using GC–MS HS analysis. (**b**) Dow N270 membrane flux (LMH) and observed and real rejection data for VOC solution using GC–MS HS analysis. (**c**) GE DK membrane flux (LMH) and observed and real rejection data for VOC solution using GC–MS HS analysis. (**d**) GE DL membrane flux (LMH) and observed and real rejection data for VOC solution using GC–MS HS analysis. (**e**) GE AK membrane flux (LMH) and observed and real rejection data for VOC solution using GC–MS HS analysis. (**f**) GE AG membrane flux (LMH) and observed and real rejection data for VOC solution using GC–MS HS analysis. (**g**) Dow BW30 membrane flux (LMH) and observed and real rejection data for VOC solution using GC–MS HS analysis. (**h**) Dow BW30LE membrane flux (LMH) and observed and real rejection data for VOC solution using GC–MS HS analysis. (**i**) Dow BW30XFR membrane flux (LMH) and observed and real rejection data for VOC solution using GC–MS HS analysis.

The membrane fluxes recorded with the VOC mixture feed show an interesting difference within the NF membranes tested and the RO membranes compared to the fluxes documented during the TCE TOC analysis tests. The GE NF90 behaved differently compared to all the other membranes—the flux on average, increased 11% when the mixture of VOC was filtered. All the other investigated NF membranes exhibited a decrease in flux—on average, 4%, 16% and 10% for NF270, DK and DL, respectively. The RO membranes suffered a far greater deterioration in flux—BW30 and BW30LE lost 31% of their permeate flux. Dow BW30XFR flux was down by 41% and 34% for the GE Osmonics AK membrane when compared to the TOC trials. The greatest drop in permeate flow rate was suffered by the AG membrane—a 74% loss. Upon examining the rejections from the GC–MS quantitation data, the GC–MS results generally agree with the TOC analysis—an increase in pressure decreases the solute retention. The reduction in rejection for the nanofiltration membranes was more severe than the equivalent TOC results for TCE—the lowest TCE rejection for TOC was between 21% and 47% for the four NF membranes. The equivalent GC–MS rejections observed were 0% for all four membranes at the highest operating pressure. This large discrepancy demonstrates the inaccuracies of using TOC as a method of substantiating the levels of VOCs in an aqueous sample. Concentration polarization is evident once again for the NF membranes. The largest difference of observed to real rejection was seen with DCP using NF270 at 30 bar, 8.8% to 97.1%, respectively—an

88.3% change. The NF membranes show good separation of TCE and PCE at 5 bar. DCE and DCP separation, however, would be considered poor across the pressure range. All the RO membranes demonstrate very good separation of PCE across the pressure range. TCE separation is good for observed rejection, and once concentration polarization is considered, the rejection increases to very good. Observed and real rejection both decrease as the pressure increases. DCE rejection remains very poor above 5 bar applied pressure for all RO membranes, particularly the low-energy BWRO Dow Filmtec BW30LE. In fact, the low-energy BW30LE performs worse compared to the standard RO membrane BW30 for all tests. Low-energy RO membranes are designed to operate at a significantly reduced operating pressure, whilst maintaining the separation capabilities and permeate flow rates of a standard RO operating at high pressure, this allows vastly reduced operating costs, making RO desalination much more appealing. This study has proven that this is not strictly true—there is a clear loss in rejection proficiency with an increase in permeate flow rate when comparing a standard RO to a low-energy RO at the same operating pressure. Previous and more recently published research [47,48] have suggested that solvent passage through an NF or RO membrane is due to the convection of the solvent to the membrane surface—adsorption of the VOCs through the membrane matrix followed by desorption from the membrane into the low concentration permeate. From the results, this study would confirm this theory. It has also been suggested that pressure has little to no effect on the rejection with this theory [47]. However, the experimental data presented in this study show this not to be the case. An increase in pressure increases the permeate flow rate, as a result the convection of solute towards the membrane surface increases, increasing concentration polarization. This increase in concentration polarization and permeate flow rate increases the rate of adsorption and desorption through the membrane matrix, respectively. Despite the continuous stirring within the cell, concentration polarization is made more likely due to the use of a frontal or dead-end filtration cell. A cross-flow filtration cell, where the feed is passed across the membrane surface rather than directly downwards onto surface, would reduce the possibility of concentration polarization occurring. The variations of adsorption kinetics have been attributed to the hydrophilicity or hydrophilicity of the membrane surface and the solvent molecule [48,49]. Solvents considered hydrophilic have a higher logarithm octanol–water partitioning coefficient, XLogP3 > 2.5, hydrophobic VOCs, XLogP3 < 2.5. The six highlighted VOCs throughout this paper are listed in Table 1, from most hydrophobic to hydrophilic. If the theory of hydrophilicity rejection is correct, then the rejection of the GC–MS results should be as follows (lowest to highest rejection) DCE < DCP < TCE < PCE. However, the rejection data from this study contradict this theory—DCP rejections are higher than TCE and PCE across the pressure range for all five RO membranes. DCP was not a VOC investigated by the previous researchers. The referenced study [48] also researched the rejection characteristics over a prolonged period of time, and it was determined that the concentration of VOCs in the permeate steadily increases until no separation occurs after 24 h. This would agree with the adsorption theory and the decrease in rejection at higher pressures observed in this study. An increase in pressure is simply speeding up the rate of adsorption and desorption that is experienced over a longer timeframe at a low pressure. All the membranes tested in this study were hydrophilic, having a contact angle <90 °C, the use of a membrane with a hydrophobic surface may present better selectivity of VOCs. Conversely, a hydrophobic membrane will experience a significant reduction in membrane flux, permeate flow rate, compared to the analogous hydrophilic membrane. Hence, a compromise between selectivity and permeate production rate must be balanced—a common situation to contemplate when selecting a suitable membrane for an application.

4. Conclusions

The results reported in this study demonstrate that an RO membrane has the potential to separate some VOCs that can be found in a contaminated groundwater source. The use of an extra fouling-resistant membrane, BWRO, from Dow Filmtec BW30XFR, produced promising results for the removal of all tested VOCs at 5 bar, with 100% observed rejection. The use of RO membranes for VOC removal is, however, an issue at high pressures and over a prolonged period of operation. Rejection proficiencies decline as the permeate flux increases, which in turn increases the convection of the VOCs to the membrane surface, causing concentration polarization at the membrane wall. An inevitable increase in concentration polarization results in a higher probability of adsorption of VOCs onto the membrane. The same principal explains the decline in VOC retention over time, due to adsorption of VOCs through the membrane matrix. The use a frontal filtration in the study compounds this problem. DCE was a compound that was inefficiently separated for all membranes tested at pressures above 5 bar. Hydrophobic VOCs generally appear to be more susceptible to adsorption than their counterparts, the hydrophilic VOCs. However, DCP would either discredit this correlation or is simply an anomaly to the rule. Further, in-depth study of VOCs is required to determine the validity of the suggested hypothesis.

Author Contributions: Conceptualization and methodology, A.R.B. and D.L.O.-R.; formal analysis, T.J.A.; investigation, T.J.A. and D.L.O.-R.; resources, A.R.B.; data curation, T.J.A. and D.L.O.-R.; writing—original draft preparation, T.J.A.; writing—review and editing, A.R.B. and D.L.O.-R.; visualization, T.J.A.; supervision, A.R.B. and D.L.O.-R.; project administration, A.R.B.; funding acquisition, A.R.B. All authors have read and agreed to the published version of the manuscript.

Funding: Financial support was provided by Lockheed Martin Corporation, the Robert A. Welch Foundation, the BioInnovation Wales projects which are part funded by the European Social Fund (ESF) and the Welsh government Sêr Cymru program.

Data Availability Statement: The data presented in this study are available in article.

Conflicts of Interest: The authors declare no conflict of interest.

References

1. Houlihan, M.F.; Lucia, P.C. Remediation of Contaminated Groundwater. In *The Handbook of Groundwater Engineering*, 1st ed.; Delleur, J., Ed.; CRC Press: Boca Raton, FL, USA, 1999; pp. 29–39.
2. Henze, M.; Harremoes, P.; la Cour Jansen, J.; Arvin, E. *Wastewater Treatment: Biological and Chemical Processes*, 3rd ed.; Springer: Berlin/Heidelberg, Germany, 2002.
3. Rahman, K.S.M.; Thahira-Rahman, J.; Lakshmanaperumalsamy, P.; Banat, I.M. Towards efficient crude oil degradation by a mixed bacterial consortium. *Bioresour. Technol.* **2002**, *85*, 257–261. [CrossRef]
4. Peña, A.A.; Hirasaki, A.G.J.; Miller, C.A. Chemically Induced Destabilization of Water-in-Crude Oil Emulsions. *Ind. Eng. Chem. Res.* **2005**, *44*, 1139–1149. [CrossRef]
5. Li, X.; Zhang, L.; Yang, Z.; Wang, P.; Yan, Y.; Ran, J. Adsorption materials for volatile organic compounds (VOCs) and the key factors for VOCs adsorption process: A review. *Sep. Purif. Technol.* **2020**, *235*, 116213. [CrossRef]
6. KP Veerapandian, S.; De Geyter, N.; Giraudon, J.M.; Lamonier, J.F.; Morent, R. The use of zeolites for VOCs abatement by combining non-thermal plasma, adsorption, and/or catalysis: A review. *Catalysts* **2019**, *9*, 98. [CrossRef]
7. Fritzmann, C.; Löwenberg, J.; Wintgens, T.; Melin, T. State-of-the-art of reverse osmosis desalination. *Desalination* **2007**, *216*, 1–76. [CrossRef]
8. Greenlee, L.F.; Lawler, D.F.; Freeman, B.D.; Marrot, B.; Moulin, P. Reverse osmosis desalination: Water sources, technology, and today's challenges. *Water Res.* **2009**, *43*, 2317–2348. [CrossRef]
9. Mohammad, A.W.; Teow, Y.H.; Ang, W.L.; Chung, Y.T.; Oatleyradcliffe, D.L.; Hilal, N. Nanofiltration membranes review: Recent advances and future prospects. *Desalination* **2015**, *356*, 226–254. [CrossRef]
10. Costa, A.R.; De Pinho, M.N. Performance and cost estimation of nanofiltration for surface water treatment in drinking water production. *Desalination* **2006**, *196*, 55–65. [CrossRef]
11. Schaep, J.; Van Der Bruggen, B.; Uytterhoeven, S.; Croux, R.; Vandecasteele, C.; Wilms, D.; Van Houtte, E.; Vanlerberghe, F. Removal of hardness from groundwater by nanofiltration. *Desalination* **1998**, *119*, 295–301. [CrossRef]
12. Van der Bruggen, B.; Vandecasteele, C. Removal of pollutants from surface water and groundwater by nanofiltration: Overview of possible applications in the drinking water industry. *Environ. Pollut.* **2003**, *122*, 435–445. [CrossRef]
13. Van Der Bruggen, B.; Everaert, K.; Wilms, D.; Vandecasteele, C. Application of nanofiltration for removal of pesticides, nitrate and hardness from ground water: Rejection properties and economic evaluation. *J. Membr. Sci.* **2001**, *193*, 239–248. [CrossRef]
14. Hilal, N.; Al-Zoubi, H.; Darwish, N.A.; Mohamma, A.W.; Abu Arabi, M. A comprehensive review of nanofiltration membranes:Treatment, pretreatment, modelling, and atomic force microscopy. *Desalination* **2004**, *170*, 281–308. [CrossRef]
15. Hong, S.; Elimelech, M. Chemical and physical aspects of natural organic matter (NOM) fouling of nanofiltration membranes. *J. Membr. Sci.* **1997**, *132*, 159–181. [CrossRef]

16. Mondal, S.; Wickramasinghe, S.R. Produced water treatment by nanofiltration and reverse osmosis membranes. *J. Membr. Sci.* **2008**, *322*, 162–170. [CrossRef]
17. Dias, J.M.; Alvim-Ferraz, M.C.; Almeida, M.F.; Rivera-Utrilla, J.; Sánchez-Polo, M. Waste materials for activated carbon preparation and its use in aqueous-phase treatment: A review. *J. Environ. Manag.* **2007**, *85*, 833–846. [CrossRef]
18. Lee, L.J.-H.; Chan, C.-C.; Chung, C.-W.; Ma, Y.-C.; Wang, G.-S.; Wang, J. Health risk assessment on residents exposed to chlorinated hydrocarbons contaminated in groundwater of a hazardous waste site. *J. Toxicol. Environ. Health Part A* **2002**, *65*, 219–235. [CrossRef]
19. Fan, C.; Wang, G.-S.; Chen, Y.-C.; Ko, C.-H. Risk assessment of exposure to volatile organic compounds in groundwater in Taiwan. *Sci. Total. Environ.* **2009**, *407*, 2165–2174. [CrossRef]
20. Huang, B.; Lei, C.; Wei, C.; Zeng, G. Chlorinated volatile organic compounds (Cl-VOCs) in environment—Sources, potential human health impacts, and current remediation technologies. *Environ. Int.* **2014**, *71*, 118–138. [CrossRef]
21. Cooney, D.O.; Nagerl, A.; Hines, A.L. Solvent regeneration of activated carbon. *Water Res.* **1983**, *17*, 403–410. [CrossRef]
22. Tamon, H.; Saito, T.; Kishimura, M.; Okazaki, M.; Toei, R. Solvent regeneration of spent activated carbon in wastewater treatment. *J. Chem. Eng. Jpn.* **1990**, *23*, 426–432. [CrossRef]
23. Kim, J.-H.; Ryu, Y.-K.; Haam, S.; Lee, C.-H.; Kim, W.-S. Adsorption and steam regeneration of n-hexane, mek, and toluene on activated carbon fiber. *Sep. Sci. Technol.* **2001**, *36*, 263–281. [CrossRef]
24. San Miguel, G.; Lambert, S.D.; Graham, N. The regeneration of field-spent granular-activated carbons. *Water Res.* **2001**, *35*, 2740–2748. [CrossRef]
25. Suzuki, M.; Misic, D.M.; Koyama, O.; Kawazoe, K. Study of thermal regeneration of spent activated carbons: Thermogravimetric measurement of various single component organics loaded on activated carbons. *Chem. Eng. Sci.* **1978**, *33*, 271–279. [CrossRef]
26. Moreno-Castilla, C.; Rivera-Utrilla, J.; Joly, J.P.; López-Ramón, M.V.; Ferro-García, M.A.; Carrasco-Marín, F. Thermal regeneration of an activated carbon exhausted with different substituted phenols. *Carbon* **1995**, *33*, 1417–1423. [CrossRef]
27. Sabio, E.; González, E.; González, J.F.; González-García, C.M.; Ramiro, A.; Ganan, J. Thermal regeneration of activated carbon saturated with p-nitrophenol. *Carbon* **2004**, *42*, 2285–2293. [CrossRef]
28. United States Environmental Protection Agency (USEPA). *Wastewater Technology Fact Sheet: Granular Activated Carbon Adsorption and Regeneration*; EPA 832-F-00-017; UESPA: Washington, DC, USA, 2000.
29. Álvarez, P.M.; Beltrán, F.J.; Gómez-Serrano, V.; Jaramillo, J.; Rodríguez, E.M. Comparison between thermal and ozone regenerations of spent activated carbon exhausted with phenol. *Water Res.* **2004**, *38*, 2155–2165. [CrossRef]
30. Ledesma, B.; Román, S.; Álvarez-Murillo, A.; Sabio, E.; González, J. Cyclic adsorption/thermal regeneration of activated carbons. *J. Anal. Appl. Pyrolysis* **2014**, *106*, 112–117. [CrossRef]
31. Maguire-Boyle, S.J.; Barron, A.R. A new functionalization strategy for oil/water separation membranes. *J. Membr. Sci.* **2011**, *382*, 107–115. [CrossRef]
32. Maguire-Boyle, S.J.; Huszman, J.E.; Ainscough, T.J.; Oatley-Radcliffe, D.; Alabdulkarem, A.A.; Al-Mojil, S.F.; Barron, A.R. Superhydrophilic functionalization of micro-filtration ceramic membranes enables separation of hydrocarbons from frac and produced waters without fouling. *Sci. Rep.* **2017**, *7*, 12267. [CrossRef]
33. Oatley, D.L.; Williams, S.R.; Barrow, M.S.; Williams, P.M. Critical appraisal of current nanofiltration modelling strategies for seawater desalination and further insights on dielectric exclusion. *Desalination* **2014**, *343*, 154–161. [CrossRef]
34. Oatley-Radcliffe, D.L.; Williams, S.R.; Ainscough, T.J.; Lee, C.; Johnson, D.J.; Williams, P.M. Experimental determination of the hydrodynamic forces within nanofiltration membranes and evaluation of the current theoretical descriptions. *Sep. Purif. Technol.* **2015**, *149*, 339–348. [CrossRef]
35. Nicolas, S.; Balannec, B.; Béline, F.; Bariou, B. Ultrafiltration and reverse osmosis of small non-charged molecules: A comparison study of rejection in a stirred and an unstirred batch cell. *J. Membr. Sci.* **2000**, *164*, 141–155. [CrossRef]
36. Nakao, S.-I.; Kimura, S. Analysis of solutes rejection in ultrafiltration. *J. Chem. Eng. Jpn.* **1981**, *14*, 32–37. [CrossRef]
37. Spiegler, K.S.; Kedem, O. Thermodynamics of hyperfiltration (reverse osmosis): Criteria for efficient membranes. *Desalination* **1966**, *1*, 311–326. [CrossRef]
38. Verliefde, A.R.; Cornelissen, E.R.; Heijman, S.G.J.; Petrinić, I.; Luxbacher, T.; Amy, G.L.; Van Der Bruggen, B.; Van Dijk, J.C. Influence of membrane fouling by (pretreated) surface water on rejection of pharmaceutically active compounds (PhACs) by nanofiltration membranes. *J. Membr. Sci.* **2009**, *330*, 90–103. [CrossRef]
39. Bellona, C.; Marts, M.; Drewes, J.E. The effect of organic membrane fouling on the properties and rejection characteristics of nanofiltration membranes. *Sep. Purif. Technol.* **2010**, *74*, 44–54. [CrossRef]
40. Xu, P.; Drewes, J.E.; Kim, T.-U.; Bellona, C.; Amy, G. Effect of membrane fouling on transport of organic contaminants in NF/RO membrane applications. *J. Membr. Sci.* **2006**, *279*, 165–175. [CrossRef]
41. Agenson, K.O.; Urase, T. Change in membrane performance due to organic fouling in nanofiltration (NF)/reverse osmosis (RO) applications. *Sep. Purif. Technol.* **2007**, *55*, 147–156. [CrossRef]
42. Chang, E.E.; Chang, Y.C.; Liang, C.H.; Huang, C.P.; Chiang, P.C. Identifying the rejection mechanism for nanofiltration membranes fouled by humic acid and calcium ions exemplified by acetaminophen, sulfamethoxazole, and triclosan. *J. Hazard. Mater.* **2012**, *221*, 19–27. [CrossRef]
43. Hilal, N.; Al-Zoubi, H.; Mohammad, A.W.; Darwish, N.A. Nanofiltration of highly concentrated salt solutions up to seawater salinity. *Desalination* **2005**, *184*, 315–326. [CrossRef]

44. Bowen, W.R.; Mohammad, A.W.; Hilal, N. Characterisation of nanofiltration membranes for predictive purposes—Use of salts, uncharged solutes and atomic force microscopy. *J. Membr. Sci.* **1997**, *126*, 91–105. [CrossRef]
45. Oatley, D.L.; Llenas, L.; Pérez, R.; Williams, P.M.; Martínez-Lladó, X.; Rovira, M. Review of the dielectric properties of nanofiltration membranes and verification of the single oriented layer approximation. *Adv. Colloid Interface Sci.* **2012**, *173*, 1–11. [CrossRef] [PubMed]
46. Oatley, D.L.; Llenas, L.; Aljohani, N.H.; Williams, P.M.; Martínez-Lladó, X.; Rovira, M.; De Pablo, J. Investigation of the dielectric properties of nanofiltration membranes. *Desalination* **2013**, *315*, 100–106. [CrossRef]
47. Ducom, G.; Cabassud, C. Interests and limitations of nanofiltration for the removal of volatile organic compounds in drinking water production. *Desalination* **1999**, *124*, 115–123. [CrossRef]
48. Altalyan, H.N.; Jones, B.; Bradd, J.; Nghiem, L.D.; Alyazichi, Y.M. Removal of volatile organic compounds (VOCs) from groundwater by reverse osmosis and nanofiltration. *J. Water Process. Eng.* **2016**, *9*, 9–21. [CrossRef]
49. Kiso, Y.; Nishimura, Y.; Kitao, T.; Nishimura, K. Rejection properties of non-phenylic pesticides with nanofiltration membranes. *J. Membr. Sci.* **2000**, *171*, 229–237. [CrossRef]

Article

The Removal of Selected Inorganics from Municipal Membrane Bioreactor Wastewater Using UF/NF/RO Membranes for Water Reuse Application: A Pilot-Scale Study

Mujahid Aziz *[ ] and Godwill Kasongo

Environmental Engineering Research Group (*EnvERG*), Department of Chemical Engineering, Faculty of Engineering and the Built Environment, Cape Peninsula University of Technology, Bellville, P. O. Box 1906, Cape Town 7535, South Africa; godwillkas26@gmail.com
* Correspondence: azizm@cput.ac.za; Tel.: +27-(0)21-460-4292

Abstract: Membrane technology has advanced substantially as a preferred choice for the exclusion of widespread pollutants for reclaiming water from various treatment effluent. Currently, little information is available about Ultrafiltration (UF)/Nanofiltration (NF)/Reverse Osmosis (RO) performance at a pilot scale as a practical engineering application. In this study, the effluent from a full-scale membrane bioreactor (MBR) municipal wastewater treatment works (MWWTWs) was treated with an RO pilot plant. The aim was to evaluate the effect of operating conditions in the removal of selected inorganics as a potential indirect water reuse application. The influent pH, flux, and membrane recovery were the operating conditions varied to measure its influence on the rejection rate. MBR/RO exhibited excellent removal rates (>90%) for all selected inorganics and met the standard requirements for reuse in cooling and irrigation system applications. The UF and NF reduction of inorganics was shown to be limited to meet water standards for some of the reuse applications due to the high Electron Conductivity (EC > 250 μS·cm^{-1}) levels. The MBR/NF was irrigation and cooling system compliant, while for the MBR/UF, only the cooling system was compliant.

Keywords: membrane bioreactor (MBR); secondary effluent; ultrafiltration (UF); inorganics; Nanofiltration (NF); reverse Osmosis (RO); chemical oxygen demand (COD); municipal wastewater treatment works (MWWTWs); flux; reuse

Citation: Aziz, M.; Kasongo, G. The Removal of Selected Inorganics from Municipal Membrane Bioreactor Wastewater Using UF/NF/RO Membranes for Water Reuse Application: A Pilot-Scale Study. *Membranes* **2021**, *11*, 117. https://doi.org/10.3390/membranes11020117

Received: 15 November 2020
Accepted: 10 December 2020
Published: 6 February 2021

Publisher's Note: MDPI stays neutral with regard to jurisdictional claims in published maps and institutional affiliations.

Copyright: © 2021 by the authors. Licensee MDPI, Basel, Switzerland. This article is an open access article distributed under the terms and conditions of the Creative Commons Attribution (CC BY) license (https://creativecommons.org/licenses/by/4.0/).

1. Introduction

Water reclamation is substantial to contribute to the increasing demand for water due to climate change, population growth, and over-consumption [1]. Municipal wastewater treatment constitutes a more reliable and significant source for reclaimed water [2,3]. Membrane bioreactor (MBR) technology has drawn much attention for the treatment of municipal wastewater due to its advantages, which include a better effluent quality compared to parallel processes, absolute control of solids, and hydraulic retention times, as well as a smaller footprint [4]. However, in many cases, high-quality effluent provided for discharge by MBR systems is still not able to be used directly as irrigation and process water, because it does not meet the recommended final pollutant concentrations for reuse [5].

Membrane technology has been accepted as the most effective technique for the removal of inorganic and organic pollutants due to its outstanding performance [3]. Reverse Osmosis (RO) has been mostly used for desalination, the purification of brackish [6], seawater [7], and wastewater [8,9], due to its ability to achieve high particulate rejection levels [5]. Ultrafiltration (UF), Nanofiltration (NF) and RO are tertiary pressure-driven membrane processes that can potentially eradicate dissolved species not removed by the MBR effluent [10]. Acero et al. (2010) [2] reported that treated municipal wastewater effluent is considered a source to produce reclaimed water [11,12] and can help inhibit the harmful effects of algal blooms and eutrophication in urban water systems [13]. Numerous authors concur that the MBR process, combined with tertiary treatment, is found to be suitable for the purification of municipal wastewater to produce high-quality water for reuse [10–12]. Some studies concluded that a combined MBR–NF/RO system could be considered as a possible alternative for treated wastewater recycling for irrigation purposes [6,14].

Membranes were operating at pH ranges between 6 and 8, which are perfect separation conditions for conventional emerging contaminants (CEC) such as pharmaceuticals, pesticides, industrial chemicals, surfactants, and personal care products. Findings by Xu et al. (2005) [15] specified that NF and extra-low energy (XLE) RO membranes, with molecular weight cut-off (MWCO) of 200 Da and less, perform similarly to conventional RO membranes when removing CEC. The membrane surface charge in high-pressure (HP) membranes are more important for the rejection than the MWCO [15]. De Souza et al. (2020) [13] concur with Xu et al. (2005) [15] regarding the MWCO, but they reiterate that the separation mechanisms applicable on the membrane surfaces are adsorption, steric hindrance, and electrostatic effects. According to Ezugbi and Rathilal (2020) [16], membrane technology has the potential of connecting the reliability and economic gap due to its accessibility and environmental sustainability.

In this study, the performances of three different membranes, namely Ultrafiltration (UF), Nanofiltration (NF), and Reverse Osmosis (RO), were evaluated in the removal of inorganic compounds and chemical oxygen demand (COD) from secondary municipal sewage wastewater treatment plant (MSWWTP) MBR effluent. The objective of this study was to assess the effects of operating condition parameters, such as pH, permeate flux, and system recovery, for reuse application. The reuse of a secondary MSWWTP MBR effluent for cooling system application and agricultural irrigation could increase agricultural production as well as water availability. The reduction of over-abstraction of surface and groundwater due to integrated usage of water resources will decrease water scarcity.

2. Materials and Methods

2.1. Full-Scale MBR and RO Pilot-Plant Hybrid System

The wastewater treatment works (WWTWs) is equipped to treat the wastewater from the largest informal settlement and its surroundings in the province. It is situated in the City of Cape Town in the Western Cape, South Africa. The MBR system incorporates ZeeWeed® 500 ultrafiltration membranes (GE Zenon, trading as Suez Technologies and Solution, Trevose, PA, USA), producing 18 megaliters of effluent per day. The pilot plant consisted of three different thin film composite (TFC) polyamide (PA) membrane modules, in parallel, which was subjected to various experimental running conditions (Table 1). The secondary MBR effluent (Table 2) was used to feed into the UF/NF/RO pilot plant (Figure 1). Batches, 8 h, once through experimental mode runs, were conducted with individual membranes at any given time.

A frequency converter controlled the influent flow rate and the operational pressure through the inlet and high-pressure pumps. Two bag filters with pore sizes of 5.0 and 1.0 µm were installed between these two pumps to prevent potential damage to the membranes by large particles. The pressures, flow rates, and temperature of influent, effluent, and brine were all monitored by online pressure gauges, rotameters, and thermometers. Online monitoring instruments measured the pH and conductivity of influent, effluent, and brine. Two online automatic dosing systems for pH and antiscalant are included. Phosphonic acid ($H_2O_3P^+$) Vitec 3000, which is a broad spectrum antiscalant and dispersant liquid obtained from Avista Technologies, Inc., South Africa, was used to minimize fouling.

The feedwater pH fluctuation (6.5 and 7) affected the membrane surface charge and the state of the solute, rendering the rejection of pollutants complex. The experimental tests were conducted with both constant pH and uncontrolled pH to describe the effects on salt rejection and the removal of inorganics and COD. Sulfuric acid (H_2SO_4) was used to keep the pH constant at 6.5.

The RO system was also equipped with an online membrane cleaning system by flushing an industrial biocide; Hydrex 7000, obtained from Veolia Water Technologies (Pty Ltd.), Paarl, South Africa; daily after an operation.

Table 1. Pilot plant operating conditions [8].

Parameters	Operating Conditions		
Membrane module	XLE	NF270	UA60
Recovery (%)	50; 75	75	75
Flux (L·m^{-2}h^{-1})	25; 30	30	30
pH	uncontrolled; 6.5	uncontrolled	uncontrolled

Table 2. The physicochemical characteristics of the membrane bioreactor (MBR) effluent.

Parameter	Units	Average MBR Effluent	Limit [1]
Electron conductivity (EC)	mS/m	56	75 [1]
pH		6.5	5.5–9.5 [1]
Total Dissolved Solids (TDS)	mg/L	360	450 [1]
Chemical oxygen demand (COD)	mg/L	<20	75 [1]
Ammonium (NH_4^{2-})	mg/L	<0.4	3.0 [1]
Phosphate (PO_4)	mg/L	2.6	10 [1]
Nitrate (NO_3)	mg/L	13	15 [1]
Turbidity	NTU	0.25	-
Temperature	°C	25	35 [1]

[1] Department of Water and Forestry (DWAF) 2010 guideline [9].

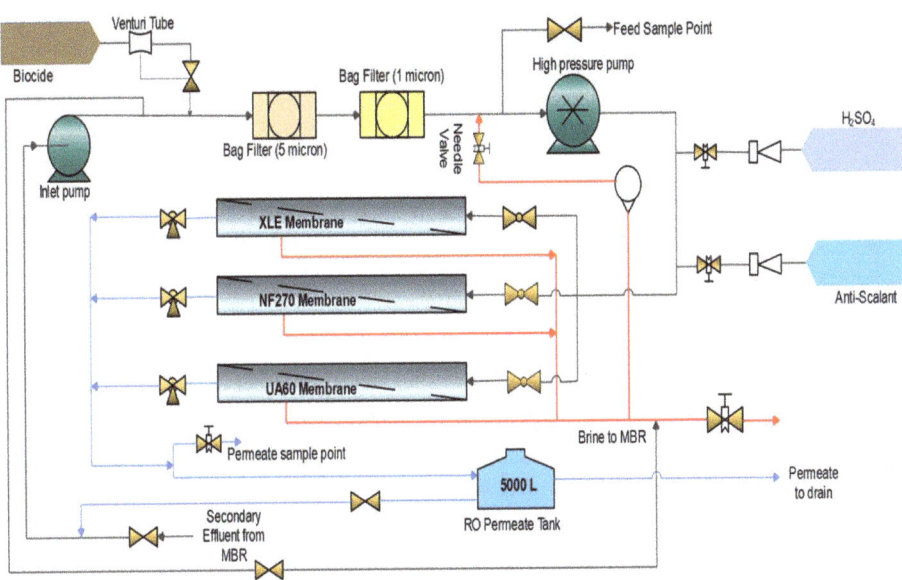

Figure 1. Process flow diagram of the Ultrafiltration (UF)/Nanofiltration (NF)/Reverse Osmosis (RO) pilot plant.

2.2. UF/NF/RO Membranes

Pilot plant experimental runs were carried out with three commercial spiral wound membranes: (1) XLE, a polyamide extra low energy RO membrane from DOW-Filmtec (Midland, MI, USA), with an MWCO of approximately 200 Da [17]; (2) NF270, a polyamide loose NF membrane from DOW-Filmtec (Midland, MI, USA), with an MWCO of approximately 400 Da [17]; and (3) UA60, a piperazine loose UF membrane from TriSep (Goleta, CA, USA) with an MWCO of approximately 1000 Da [17]. These membranes were chosen because they have MWCOs covering the MW range (200–1000 Da) of most inorganics reported in the literature [16]. The characteristics of these membranes are presented in Table 3.

Table 3. Properties of three membrane modules.

Membrane Component [1]	Texture [1]	Type [1]	Rejection [1] %	Effective Area [1] (m²)	MWCO [1] (Da)	Maximum Pressure [1] (bar)	Maximum Temperature [1] (°C)	Maximum Permeate Flowrate [1] (m³/h)
RO	TFC Polyamide	Filmtec XLE−4040	99% NaCl	8.1	<200	6.9	45	9.8
NF	TFC Polyamide	Filmtec NF270-4040	>97% MgSO4	7.6	400	4.8	45	9.5
UF	TFC Piperazine	TriSep 4040-UA60-TSA	80% MgSO4	8.2	1000	7.6	45	11.4

[1] Obtained from the literature [18].

2.3. Membrane Energy Consumptions

The energy usage of membranes systems contributes to nearly 35–45% of the total permeate production cost [19,20]. Therefore, it is imperative to evaluate the energy consumption of membranes and study the effects of operating conditions such as percentage recovery and flux on the consumption of energy. The sources of energy consumption of membranes systems include the feedwater intake and pre-treatment; high-pressure pumps; membrane type; and post-treatment [19]. The principal source of energy consumption is the high-pressure pump, which is essential to drive water flux across the membrane [21]. The pump energy usage can be expressed as specific energy consumption (SEC), which can be obtained using the following equation [20]:

$$SEC = \frac{\Delta P}{\gamma} = \frac{\Delta P \times Q_F}{Q_P} \qquad (1)$$

where ΔP is the transmembrane pressure (Pa), v is the percentage recovery, Q_F is the feed flowrate, and Q_P the permeate flow rate.

2.4. Analytical Methods and Water Analysis

The influent, effluent, and brine were all collected to investigate the operation performance of the RO system. The physicochemical parameters assessed were Electrical Conductivity (EC), Total Dissolved Solids (TDS), pH, Temperature (T), Turbidity, Ammonium (NH_4^{2+}), Phosphate (PO_4), Nitrate (NO_3), and Chemical Oxygen Demand (COD). Sampling was carried out from four different sampling points: (1) MBR effluent; (2) permeate of UF; (3) permeate of NF; (4) permeate of RO element. To avoid frequent fluctuations in concentrations, each sample taken from the pilot plant was an 8-h composite sample taken for the duration of each experimental run. All water samples were collected in amber glass bottles (2.5 L) prewashed with nitric acid and rinsed thoroughly with distilled water. Samples were filtered through 1.0 µm pore size glass fiber filter paper (Whatman GF/B); then, the filtrates were stored at 4 °C and analyzed within 24 h of collection. All equipment and meters for the on-site measurements were calibrated and checked according to the manufacturer's instruction. EC, T, and TDS were measured using a Crison CM 35+ hand-held meter (Merck Pty Ltd., Bellville, Cape Town, South Africa). The pH measured with Jenway 3510 Bench pH/mV Meter and Turbidity with an HF Scientific Micro TPI Infrared Turbidity Meter. COD samples were digested in a Thermo reactor Model HI839800-02 (Hanna Pty Ltd., Bellville, Cape Town, South Africa) and measured using a COD Meter and Multiparameter photometer Model HI83214-02 (Hanna Pty Ltd., Bellville, Cape Town, South Africa). The concentration levels of NH_4^{2+}, PO_4, and NO_3 (Hanna Pty Ltd., Bellville, Cape Town, South Africa) were analyzed using the Multiparameter photometer Model HI83214-02 according to the Standard Methods.

2.5. Statistical Analysis

The data presented were analyzed with statistical calculations to approve the significance of the data obtained. Analyses of variance (ANOVA) and T-test with a significance level of 0.05 were applied to evaluate correlations between membrane type (UF, NF, and RO) and pH (controlled and uncontrolled), respectively.

3. Results and Discussion

3.1. Salt Rejection and Total Dissolved Solids (TDS)

The performance of membranes was assessed by measuring the physicochemical parameters, salt rejection and total dissolved solids (TDS), with the pilot plant operation condition of flux, recovery at 30 $L \cdot m^{-2} h^{-1}$, 75% respectively, as well as control and uncontrolled pH. Figure 2 shows the permeate salt rejection (Figure 2A) and TDS (Figure 2B) as a function of time, obtained for all three membranes (UF/NF/RO) during experimental runs on the pilot plant. The RO (XLE) membrane rejection was the highest between 94.4–96.6% at controlled (6.5) and 89.2–91.4%, uncontrolled pH. The UF (UA60) and NF (NF270) membranes performed as expected and had better rejection at a controlled pH. Although there was MBR effluent (real feed) with concentration variation into the UF/NF/RO pilot plant, as can be seen with the TDS, the results still indicate that the performance of the membranes was stable throughout the experimental study. The RO salt rejection usually is high due to its membrane design characteristics, where the skin layer is much denser than the other two, UF and NF. The mechanisms that can be attributed to the rejection of ionic species in the water are size exclusion, charge, and ionic electrostatic interactions of the ions with the surface of the membrane. It is reported that monovalent ions in the feed water can generally pass through the membrane more easily than divalent ions due to size exclusion [18]. The UF and NF membranes have similar separation characteristics; however, the membrane parameters are quite different.

Garcia-Aleman et al. (2004) [22] state that the transport and selectivity of NF membranes are mainly due to steric/hydrodynamic effects and charge repulsion. The relative size of the ions causes the steric effect to the membrane pores, and the repulsion effect is caused by the charged nature of the membrane and electrolytes. The NF270 is a loose NF membrane, but it is tighter than the UF, with relatively high permeability and charge density. The UA60 membrane has a larger pore radius, is less permeable, and has a higher surface charge density; thus, steric effects are not as applicable. During salt separation, the UA60 membrane depends exclusively on Donnan exclusion [22]. According to Üstün et al. (2011) [23], the UF and NF membrane surfaces are negatively charged at pH values higher than 4. The pH of the MBR-UF/NF/RO influent in this study was between 6 and 7, thus presenting a negative charge density on the membrane surface. The primary mechanism of ion rejection by these (UF and NF) membranes is the sieving mechanism [24]. A solution–diffusion model describes the XLE membrane transport mechanism because of the nominal pore size, where diffusion dominates over convection [25].

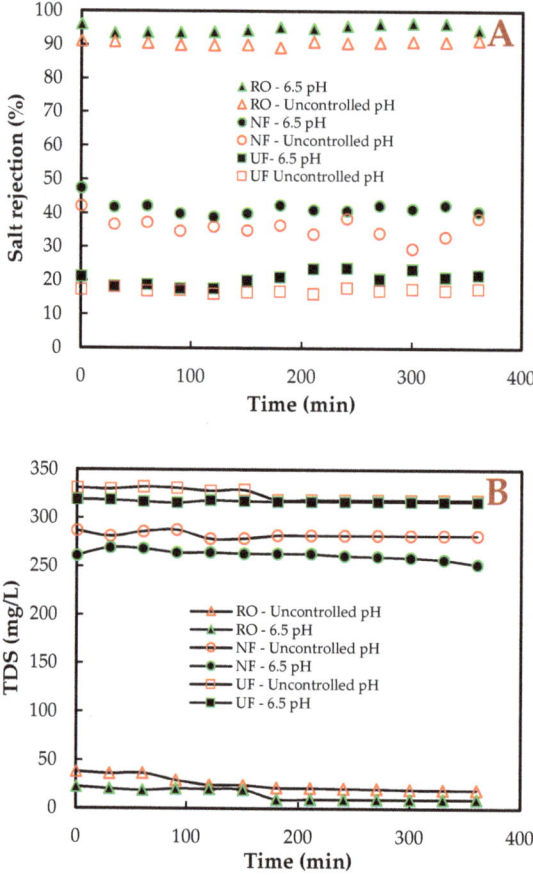

Figure 2. The measured effluents of the UF, NF, and RO membrane treatment at uncontrolled and controlled 6.5 pH pilot plant condition. Salt rejection (**A**) and TDS (**B**), all at a flux of 30 L·m^{-2}h^{-1}, and 75% recovery.

3.2. Chemical Oxygen Demand (COD)

The pH level of water defines its application for different purposes. Low or high pH has a poisonous effect on marine life and alters the solubility of other chemical pollutants and elements in the water. The South African limit for pH in the water for reuse is 6 to 9 [17]. Chemical oxygen demand is described as the amount of strong oxidant required to break down both organic and inorganic substances in water. The removal of COD with all membranes is presented in Figure 3, where the RO percentage removal of 92% and 99% for uncontrolled and controlled pH, respectively, were significantly higher than the UF and NF membranes ($p = 0.018$ for UF, $p = 0.013$ for NF and $p = 0.009$ for RO; $\alpha = 0.05$). This is consistent with a similar study of MBR/NF and MBR/RO membrane effluent rejection [26]. MBR is considered a relatively improved treatment process for the exclusions of COD compared to conventional activated sludge processes as a pre-treatment for NF and RO reuse [10].

The effect of pH on the COD removal with the NF and UF membranes appeared to have the opposite effect as compared to the RO membrane, where the higher percentage was achieved. The pH range with uncontrolled pH experimental runs was between 6.7 and 7.1, while experimental runs with controlled pH maintained the latter at 6.5. The COD removal has been reported to increase with increasing the pH, which was in part attributed to the rise in the hydroxide ions concentration, increasing the production of hydroxyl free radicals [21]. Therefore, this may suggest that the increase in pH was a predominant factor in the removal of COD when using the UF and NF membranes. Other researchers have reported changes in properties such as pH to affect contaminant removal, which was found to be substantially lower when operated without pH control [22]. The COD percentage removals for UF were 80 and 72; NF were 85 and 82. This is in the range of a study of Xu et al. (2020) [27] where the NF membrane showed a COD percentage removal of 90. The best removal of COD achieved with a controlled pH when using the RO membrane may be explained by the fact that a controlled pH results in a higher and more sustained osmotic flow, which caused a more significant COD removal [28], as well as the surface of the membrane, which became less negative with the decrease in pH as compared to experimental with no adjustment of pH [29].

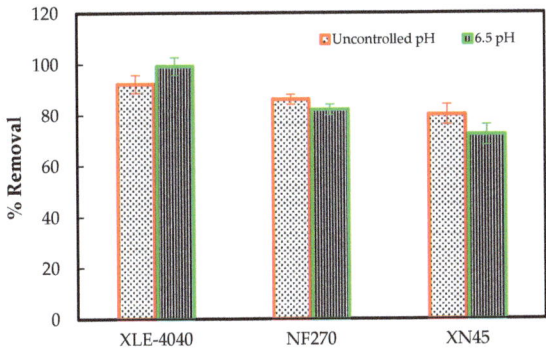

Figure 3. The measured effluents of the UF, NF, and RO membrane treatment at uncontrolled and controlled 6.5 pH pilot plant condition of chemical oxygen demand (COD) removal.

3.3. Inorganics Removal

The permeate quality of the MBR/UA60, MBR/NF270 and MBR/XLE units of the selected inorganics are summarized in Figure 4. Shad et al. (2019) [30] confirmed that inorganics are found in municipal wastewater originating from domestic and industrial products such as pesticides, preservatives, surfactants, perfluorochemicals, pharmaceutical residues, and steroidal hormones, which are all found in excreted human waste. These salts were selected to evaluate the correlation of anionic, neutral, and cationic solutes with the membrane-type and pH. It can be seen in Figure 4A and B the percentage removal of phosphate: 40, 89 and 94% and phosphorous: 58, 90.5, and 96% with the UA60, NF270, and XLE, respectively.

There was a significant difference in the removal of selected inorganics observed with the three membranes ($p = 0.001$ for uncontrolled pH, $p = 0.043$ for 6.5 pH at $\alpha = 0.05$). The phosphorus removal is visibly higher than the phosphate, which is due to the size exclusion and chemical charge. Phosphorous is a neutral molecule that differs from phosphate, which is a multivalent anion that may increase electrostatic repulsion with the surface of the membrane [31]. The reduction of both phosphate and phosphorous with pH change indicates that pH adjustment affects the removal of these physicochemical properties slightly ($p > 0.05$ for both phosphorous and phosphate). Contrary, the adjustment of pH had a significant effect on ammonia percentage removal with the NF270 and XLE membranes ($p = 0.018$ at $\alpha = 0.05$). This ammonia percentage removal is shown in Figure 4C, where it increased from 62 to 99% and 52 to 87%, respectively, when changing from 6.5 pH to uncontrolled. This could be explained by the fact that the pH adjustment (pH 6.5) shifted the equilibrium of ammonia, resulting in higher removal and permeance of cations than the anions due to the deprotonated carboxylic groups of the polyamide membrane [32]. According to Chu et al. (2017) and Pagès et al. (2017) [31,32], ionizable functional groups can affect water and solute permeation due to the production of pH-dependent charges on the active membrane layer. Sert et al. (2017) [8] reiterated that the higher rejection of these monovalent ions by the XLE membrane is due to its dense surface layer without pores. The UA60 and NF270 with higher MWCO are classified as loose membranes that reject monovalent ions with lower percentages by electrostatic interaction mechanism.

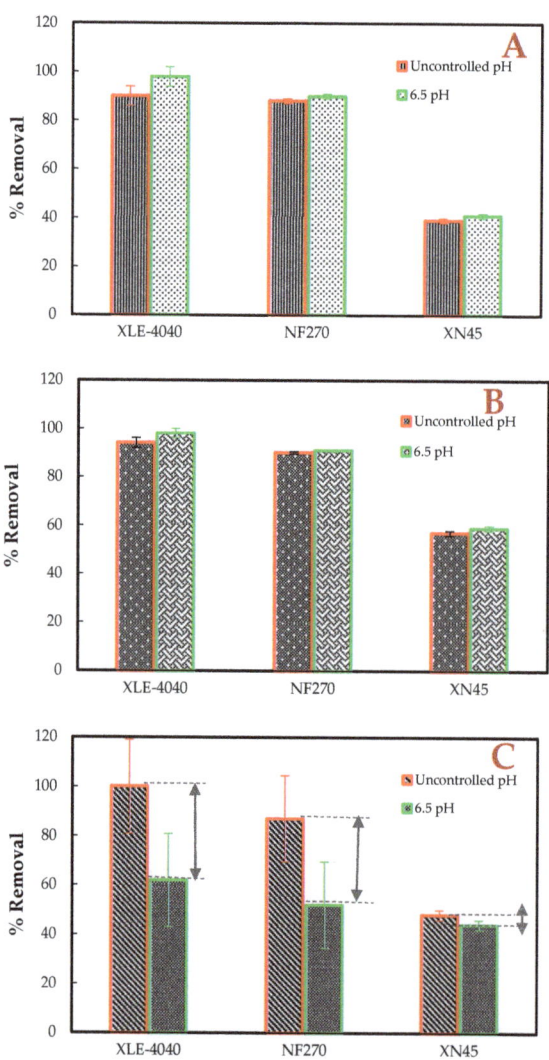

Figure 4. The measured effluents of the UF, NF, and RO membrane treatment at uncontrolled and controlled 6.5 pH pilot plant condition. Phosphate (**A**), Phosphorous (**B**), and Ammonia removal (**C**).

3.4. Membranes Energy Consumption Comparison and Effect of RO Operating Conditions

Energy usage of the membrane system was calculated using the specific energy consumption of pumps (SEC), as 50 to 75% of the energy consumed by RO systems emanates from the high-pressure pumps of the system [19]. Table 4 shows the SEC of the

RO membranes at different conditions and compared to the UF and NF SEC calculated. The RO membrane used in this study indicated a low consumption of energy compared to other studies conducted, where a minimum SEC of 1.37 to 2 kWh·m^{-3} could be obtained at 50% recovery [20,33].

Table 4. Energy consumption comparison of RO membrane operating conditions with NF and UF membranes.

	RO				NF	UF
	30 L·m^{-2} h^{-1} Flux		25 L·m^{-2} h^{-1} Flux		30 L·m^{-2} h^{-1} Flux	
	% Recovery (Y)				75% Recovery (Y)	
Parameter	75	50	75	50		
ΔP (kPa)	384.2	423.7	385.7	389.2	298.7	269.7
Q_B (m^3·h^{-1})	0.081	0.243	0.068	0.204	0.063	0.205
Q_P (m^3·h^{-1})	0.243	0.243	0.2024	0.204	0.19	0.205
Q_F (m^3·h^{-1})	0.324	0.486	0.270	0.405	0.253	0.41
SEC (kWh·m^{-3})	0.142	0.235	0.143	0.216	0.111	0.100

Furthermore, it shows that a change in percentage recovery has more effect on the SEC than water flux change when comparing the different experimental conditions of the RO membrane. At a lower recovery of water (50%), the SEC required was higher than the SEC at 75% recovery. This can be explained by the decline in differential pressure drop across the membrane as a function increased velocity in the concentrate stream [20]. When comparing SEC of membranes, results show that the RO membrane consumed more energy than NF membranes, and the latter consumed more than the UF membrane. This is again due to the membranes' design characteristics difference; hence, the RO required more pressure to drive the solvent across the dense surface of the membrane.

The SEC difference between the NF and the UF is not significant, as the two membranes have similar separation characteristics. However, the energy consumption only increased by a factor of 1.42 when shifting from UF or NF to RO. No energy recovery device (ERD) was used when using the RO membrane. Therefore, the use of an ERD would help sensibly reduce the energy consumption of the RO, as suggested by several researchers [19–21]. It is indicated that ERD can help reduce the SEC up to 16% [19]. Figure 5 shows the specific energy consumed versus permeate total dissolved solids (TDS). In order to achieve a lower, permeate TDS concentration, more energy was required from the pump. The low TDS concentration is directly proportional to the characteristic of the membranes, which in this case required more energy usage when moving from UF to NF and RO, respectively.

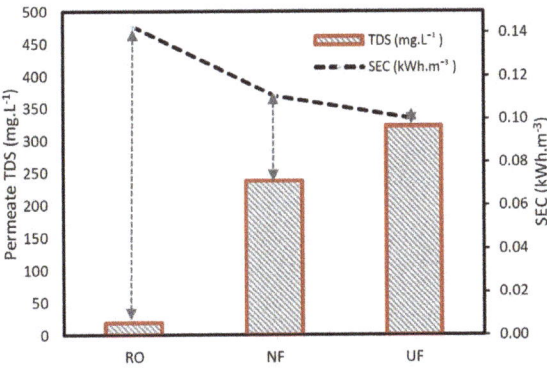

Figure 5. Comparison of membranes systems specific energy consumption concerning the permeate total dissolved solids (TDS) at a fixed permeate flux of 30 L·m^{-2}h^{-1} and recovery of 75%.

3.5. Reuse Application for Wastewater Effluent

The effluent samples of the MBR/UF, MBR/NF, and MBR/RO were compared to the water quality requirements for reuse in cooling and irrigation applications, as summarized in Table 5. This table shows essential changes in the average inorganic and COD concentration. The MBR/RO effluent satisfied all reuse conditions required for cooling and irrigation application, and this is consistent with similar studies [23,24,33,34]. The physiochemical properties of the MBR/NF effluent are suitable for reuse in industrial cooling applications. However, they may be restricted to specific irrigation applications, because some of the parameters such as the EC (355 $\mu S \cdot cm^{-1}$) are outside the required range (<250 $\mu S \cdot cm^{-1}$) for unrestricted irrigation water quality. As expected, the MBR/UF effluent with an EC of 471 $\mu S \cdot cm^{-1}$ is only suitable for cooling system reuse application. Falizi et al. (2018) [26] cautions that water salinity (measured by EC) is the primary factor threatening crop productivity and quality with the usage of irrigation water at a pH between 6 and 9. The earth may appear wet, but if the EC is high, then the available water to the vegetations will be less. The acceptable EC limits for effluent discharge and domestic water supply usage, according to South African guidelines, are 250 $\mu S \cdot cm^{-1}$ and 70 $\mu S \cdot cm^{-1}$, respectively [26].

Table 5. Characteristics of UF/NF/RO effluent average water quality with reuse criteria for wastewater in different applications.

Parameter	Irrigation [23,24]	Cooling System [34,35]	UF	NF	RO
COD (mg·L^{-1})	<50	<30	16	10	2
NH$_3$ (mg·L^{-1})	<6.08	<1	0.62	0.28	0.17
P (mg·L^{-1})	<1.5	-	1.8	0.79	0.21
PO$_4$ (mg·L^{-1})	<2	<7	2.07	0.91	0.45
TDS (mg·L^{-1})	<200	-	300	255	19
pH	6.5–8.4	6.8–7.2		6.5–7.05	
EC ($\mu S \cdot cm^{-1}$)	<250	<1445	471	355	37
Turbidity (NTU)	<2	<36	-	-	0.08

The results in Table 5 show that the removal of phosphate, phosphorous, and ammonia with MBR/NF270 and MBR/XLE membranes is within the specification guidelines for cooling and irrigation applications. The rejection of the phosphates and ammonia with the XLE membrane is due to size exclusion but with the UA60 and NF270 (both negatively charged membranes) charge effects. Van Voorthuizen et al. (2005) [25] explained that the difference in rejection between ammonia and phosphate for the negatively charged membranes (UA60 and NF270) could be explained by the hydrogen–phosphate ion (HPO$_4^{2+}$), which is bigger than the bicarbonate ion (HCO^{3+}). The hydrogen–phosphate ion has a larger negative charge and will repel much more assertively with the UA60 and NF270 membrane. NH$_4^+$ cation enters and is retained by membrane pores when the hydration energy of 407 kJ·mol^{-1} and the ionic radius of 0.095, due to surface forces [36].

3.6. Effect of Operation Conditions on RO Membrane Rejection

3.6.1. Selected Inorganic Rejection

The properties of operating conditions such as pH, permeate flux and system percentage recovery were evaluated using the RO (XLE) membrane. The parameters in the effluent are lower than those obtained in the influent, as shown in Table 6. For the experimental runs conducted at a constant pH of 6.5, the best results were attained at a permeate flux of 25 L/m^2·hr and a system percentage recovery of 75. The highest average percentage reductions obtained for ammonia, nitrate, nitrite, phosphate and phosphorous were 98%, 100%, 83%, 97%, and 98%, respectively. Although the findings did not match expectations suggesting a slight decrease in permeate inorganics concentration when increasing flux, on the assumption that ions leakage across the membrane remains reasonably constant [30], the phenomenon may be explained by the increase in percentage recovery (75%), which allows for the mass of ions at the surface of the membrane to be blended with more permeate, resulting in a lower concentration of inorganics in the permeate.

Table 6. Selected inorganics percentage removal using RO membrane.

Operating Conditions		30 L/m²·h Flux		25 L/m²·h Flux	
		% Recovery			
Water pH	Inorganic	75	50	75	50
6.5	NH_3	92	97	98	97
	NO_3	100	87	80	100
	NO_2	63	60	83	82
	PO_4	90	97	90	97
	P	92	97	98	97
Uncontrolled	NH_3	80	87	94	92
	NO_3	68	76	63	63
	NO_2	61	55	86	71
	PO_4	98	86	98	88
	P	80	87	94	92

The percentage reduction of the different inorganics tested, increase from nitrite, nitrate, phosphate, ammonia to phosphorous removal. Higher rejection of multivalent ions can be explained by the size of multivalent ions, which is larger than monovalent ones. Therefore, an increase in ion charge causes an increase in electrostatic interactions with membranes, which determine the contaminant removal mechanism [33]. The lower rejection of nitrate compared with the ammonia rejection measured with the same membranes at the same operating conditions is notable. NH_4^+ is regarded as a single-charged trace cation that was better rejected than the anions (NO_3^-) [32]. Nitrate rejection for the RO membrane follows the behavior described by Donnan exclusion theory where the co-ions (NO_3^-) are rejected due to electrostatic repulsion. In some studies, the nitrates in the effluent (permeate) increases. This may be due to the oxidation of ammonia and nitrite to nitrate by nitrite-oxidizing bacteria [37].

3.6.2. Chemical Oxygen Demand (COD) rejection

The effects of several operating conditions such as pH, permeate flux and system percentage recovery was evaluated using the RO (XLE) membrane with the removal of COD. Figure 6 describes the results where the best COD percentage removal of 99 was obtained when the system operated at 75% recovery and 25 $L·m^{-2}·h^{-1}$ of flux, with a controlled pH of 6.5. The change in operating conditions of flux and system recovery does not have a significant effect on COD rejection, except for pH. The average COD rejection increases by 5% for a pH variable with a minimum of 96 and a maximum of 99%, respectively. This suggests that the accumulation of organic matter in the treatment of the effluent with an RO membrane can be effectively reduced with a controlled pH and by adjusting the flux and recovery. According to Paugum et al. (2004) [36], the degree of membrane ionization is a function of the effluent pH where the isoelectric point corresponds to the pH value for which the electric charge of the fixed cations neutralizes that of the anions.

3.6.3. Water Turbidity Using the RO System

The effects of several operating conditions such as pH, permeate flux and system percentage recovery were evaluated using the RO (XLE) membrane with the measuring of turbidity. Figure 7 describes the results for the lowest turbidity obtained at a flux of 25 $L·m^{-2}·h^{-1}$ of flux, with a controlled pH of 6.5 and system recovery of 75% (0.08 NTU) and 50% (0.09 NTU), respectively.

The highest turbidity measurement in the permeate (0.57 NTU) was obtained at 50% recovery and 25 $L·m^{-2}·h^{-1}$. The high turbidity observed in the samples indicates the presence of finely divided organic and inorganic matter, soluble colored organic compounds and microscopic organisms. Studies have shown that too much turbidity in water can lead to interference with water treatment techniques and increase the cost.

When turbid water is chlorinated, then, a possible rise in trihalomethane (THM) precursor formation is possible [37]. All the turbidity influent or effluent results are below 1 NTU except for one influent (25 L·m^{-2} h^{-1} permeate flux, 75% recovery) at 1.05 NTU. The slightly high permeate turbidity of 0.57 NTU obtained at these operating conditions is explained by the high turbidity (1.05) of the RO influent, which influenced the turbidity of the effluent. There is no South African guideline for turbidity in effluent discharge, although the South African Target Water Quality Range for turbidity in water for domestic water supply is 0–1 NTU, while the World Health Organization (WHO) standard is 5 NTU [38].

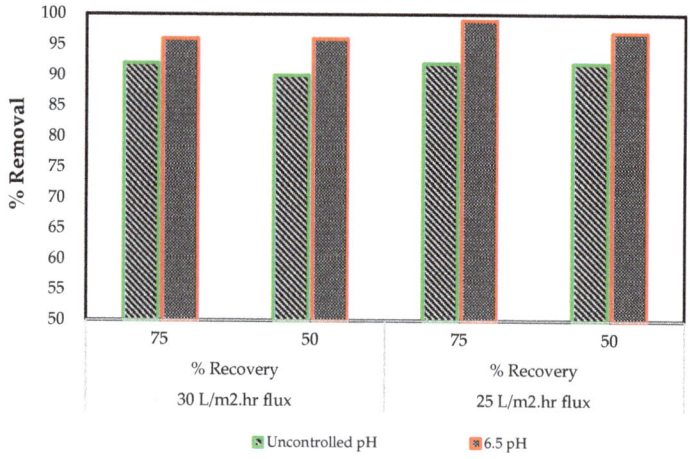

Figure 6. The measured effluents of the RO membrane treatment at uncontrolled and controlled 6.5 pH pilot plant conditions of COD removal.

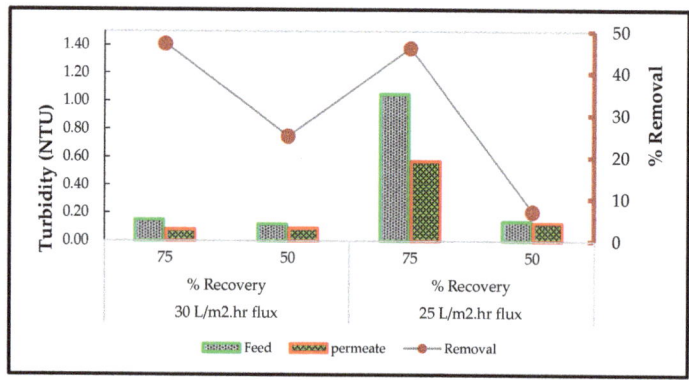

Figure 7. Turbidity removal using RO membrane with controlled pH at 6.5.

4. Conclusions

In this study, the performance of UF/NF/RO technology for treating municipal MBR wastewater on a pilot scale was investigated. RO (XLE) and NF (NF270) membranes exhibited exceptional removal rates of 90%, (UF > 40%) for COD, NH_4^{2+}, PO_4, and NO_3. The influence of pH, permeate flux, and percentage recovery had a visible effect on the rejection of these selected inorganics. The XLE membrane showed a 99% COD rejection at operational conditions where the pH, flux, and recovery were 6.5, 25 L·m^{-2}h^{-1} and 75%, respectively. Although the results show that the removal performance of inorganics and COD are significantly better with the XLE membrane, the energy consumption, however, increased by a factor of 1.42 than with UA60 or NF270 membranes. The MBR/RO comply with the standard requirements for potable and non-potable reuse applications.

Author Contributions: Conceptualisation, M.A.; data curation, M.A. and G.K.; formal analysis, M.A. and G.K.; funding acquisition, M.A.; investigation, M.A. and G.K.; methodology, M.A. and G.K.; project administration, M.A.; resources, M.A.; software, M.A.; validation, M.A.; visualization, M.A. and G.K.; writing—original draft, M.A. All authors have read and agreed to the published version of the manuscript.

Funding: This research was funded by the National Research Foundation (NRF-RSA).

Acknowledgments: The authors are thankful to the Environmental Engineering Research Group (*EnvERG*) in the Chemical Engineering Department at Cape Peninsula University of Technology (CPUT) and the National Research Foundation (NRF-RSA).

Conflicts of Interest: The authors declare no conflict of interest.

References

1. Kim, S.; Chu, K.H.; Al-Hamadani, Y.A.J.; Park, C.M.; Jang, M.; Kim, D.H.; Yu, M.; Heo, J.; Yoon, Y. Removal of contaminants of emerging concern by membranes in water and wastewater: A review. *Chem. Eng. J.* **2018**, *335*, 896–914. [CrossRef]
2. Acero, J.L.; Benitez, F.J.; Teva, F.; Leal, A.I. Retention of emerging micropollutants from UP water and a municipal secondary effluent by ultrafiltration and nanofiltration. *Chem. Eng. J.* **2010**, *163*, 264–272. [CrossRef]
3. Bunani, S.; Yörükoğlu, E.; Yüksel, Ü.; Kabay, N.; Yüksel, M.; Sert, G. Application of reverse osmosis for reuse of secondary treated urban wastewater in agricultural irrigation. *Desalination* **2015**, *364*, 68–74. [CrossRef]
4. Dialynas, E.; Diamadopoulos, E. Integration of a membrane bioreactor coupled with reverse osmosis for advanced treatment of municipal wastewater. *Desalination* **2009**, *238*, 302–311. [CrossRef]
5. Sahar, E.; David, I.; Gelman, Y.; Chikurel, H.; Aharoni, A.; Messalem, R.; Brenner, A. The use of RO to remove emerging micropollutants following CAS/UF or MBR treatment of municipal wastewater. *DES* **2011**, *273*, 142–147. [CrossRef]
6. Aziz, M.; Kasongo, G. Scaling prevention of thin film composite polyamide Reverse Osmosis membranes by Zn ions. *Desalination* **2019**, *464*, 76–83. [CrossRef]
7. Goh, P.S.; Lau, W.J.; Othman, M.H.D.; Ismail, A.F. Membrane fouling in desalination and its mitigation strategies. *Desalination* **2018**, *425*, 130–155. [CrossRef]
8. Rana, H.H.; Saha, N.K.; Jewrajka, S.K.; Reddy, A.V.R. Low fouling and improved chlorine resistant thin film composite reverse osmosis membranes by cerium(IV)/polyvinyl alcohol mediated surface modification. *Desalination* **2015**, *357*, 93–103. [CrossRef]
9. Kasongo, G.; Steenberg, C.; Morris, B.; Kapenda, G.; Jacobs, N.; Aziz, M. Surface grafting of polyvinyl alcohol (Pva) cross-linked with glutaraldehyde (ga) to improve resistance to fouling of aromatic polyamide thin film composite reverse osmosis membranes using municipal membrane bioreactor effluent. *Water Pract. Technol.* **2019**, *14*, 614–624. [CrossRef]
10. Sert, G.; Bunani, S.; Yörüko, E.; Kabay, N.; Egemen, Ö.; Yüksel, M.; Falizi, N.J.; Hacıfazlıoğlu, M.C.; Parlar, İ.; Kabay, N.; et al. Challenges of municipal wastewater reclamation for irrigation by MBR and NF/RO: Physico-chemical and microbiological parameters, and emerging contaminants. *Chem. Eng. J.* **2018**, *22*, 137959.
11. Malamis, S.; Katsou, E.; Takopoulos, K.; Demetriou, P.; Loizidou, M. Assessment of metal removal, biomass activity and RO concentrate treatment in an MBR-RO system. *J. Hazard. Mater.* **2012**, *209–210*, 1–8. [CrossRef] [PubMed]
12. Dolar, D.; Gros, M.; Rodriguez-Mozaz, S.; Moreno, J.; Comas, J.; Rodriguez-Roda, I.; Barceló, D. Removal of emerging contaminants from municipal wastewater with an integrated membrane system, MBR-RO. *J. Hazard. Mater.* **2012**, *239–240*, 64–69. [CrossRef] [PubMed]
13. de Souza, D.I.; Giacobbo, A.; Fernandes, E.d.S.; Rodrigues, M.A.S.; de Pinho, M.N.; Bernardes, A.M. Experimental design as a tool for optimizing and predicting the nanofiltration performance by treating antibiotic-containing wastewater. *Membranes* **2020**, *10*, 156. [CrossRef] [PubMed]
14. Bunani, S.; Yörükoğlu, E.; Sert, G.; Kabay, N.; Yüksel, Ü.; Yüksel, M.; Egemen, Ö.; Pek, T.Ö. Utilization of reverse osmosis (RO) for reuse of MBR-treated wastewater in irrigation-preliminary tests and quality analysis of product water. *Environ. Sci. Pollut. Res.* **2018**, *25*, 3030–3037. [CrossRef] [PubMed]
15. Xu, P.; Drewes, J.E.; Kim, T.U.; Bellona, C.; Amy, G. Effect of membrane fouling on transport of organic contaminants in NF/RO membrane applications. *J. Membr. Sci.* **2006**, *279*, 165–175. [CrossRef]
16. Ezugbe, E.O.; Rathilal, S. Membrane technologies in wastewater treatment: A review. *Membranes* **2020**, *10*, 89. [CrossRef]
17. Agoro, M.A.; Okoh, O.O.; Adefisoye, M.A.; Okoh, A.I. Physicochemical Properties of Wastewater in Three Typical South African Sewage Works. *Pol. J. Environ. Stud.* **2018**, *27*, 491–499. [CrossRef]

18. Aziz, M.; Ojumu, T. Exclusion of estrogenic and androgenic steroid hormones from municipal membrane bioreactor wastewater using UF/NF/RO membranes for water reuse application. *Membranes* **2020**, *10*, 37. [CrossRef]
19. Gude, V.G. Energy consumption and recovery in reverse osmosis. *Desalin. Water Treat.* **2011**, *36*, 239–260. [CrossRef]
20. Mazlan, N.M.; Peshev, D.; Livingston, A.G. Energy consumption for desalination—A comparison of forward osmosis with reverse osmosis, and the potential for perfect membranes. *Desalination* **2016**, *377*, 138–151. [CrossRef]
21. Qin, M.; Deshmukh, A.; Epsztein, R.; Patel, S.K.; Owoseni, O.M.; Walker, W.S.; Elimelech, M. Comparison of energy consumption in desalination by capacitive deionization and reverse osmosis. *Desalination* **2019**, *455*, 100–114. [CrossRef]
22. Garcia-Aleman, J.; Dickson, J.M. Mathematical modeling of nanofiltration membranes with mixed electrolyte solutions. *J. Membr. Sci.* **2004**, *235*, 1–13. [CrossRef]
23. Üstün, G.E.; Solmaz, S.K.A.; Çiner, F.; Bažkaya, H.S. Tertiary treatment of a secondary effluent by the coupling of coagulation-flocculation-disinfection for irrigation reuse. *Desalination* **2011**, *277*, 207–212. [CrossRef]
24. Emongor, V.E.; Khonga, E.B.; Ramolemana, G.M.; Marumo, K.; Machacha, S.; Motsamai, T. Suitability of treated secondary sewage effluent for irrigation of horticultural crops in Botswana. *J. Appl. Sci.* **2005**, *5*, 451–454. [CrossRef]
25. Van Voorthuizen, E.M.; Zwijnenburg, A.; Wessling, M. Nutrient removal by NF and RO membranes in a decentralized sanitation system. *Water Res.* **2005**, *39*, 3657–3667. [CrossRef]
26. Falizi, N.J.; Hacıfazlıoğlu, M.C.; Parlar, İ.; Kabay, N.; Pek, T.; Yüksel, M. Evaluation of MBR treated industrial wastewater quality before and after desalination by NF and RO processes for agricultural reuse. *J. Water Process Eng.* **2018**, *22*, 103–108. [CrossRef]
27. Xu, R.; Qin, W.; Zhang, B.; Wang, X.; Li, T.; Zhang, Y.; Wen, X. Nanofiltration in pilot scale for wastewater reclamation: Long-term performance and membrane biofouling characteristics. *Chem. Eng. J.* **2020**, *395*, 125087. [CrossRef]
28. Saichek, R.E.; Reddy, K.R. Effect of pH control at the anode for the electrokinetic removal of phenanthrene from kaolin soil. *Chemosphere* **2003**, *51*, 273–287. [CrossRef]
29. Chan, G.Y.S.; Chang, J.; Kurniawan, T.A.; Fu, C.X.; Jiang, H.; Je, Y. Removal of non-biodegradable compounds from stabilized leachate using VSEPRO membrane filtration. *Desalination* **2007**, *202*, 310–317. [CrossRef]
30. Farrokh Shad, M.; Juby, G.J.G.; Delagah, S.; Sharbatmaleki, M. Evaluating occurrence of contaminants of emerging concerns in MF/RO treatment of primary effluent for water reuse—Pilot study. *J. Water Reuse Desalin.* **2019**, *9*, 350–371. [CrossRef]
31. Chu, K.H.; Fathizadeh, M.; Yu, M.; Flora, J.R.V.; Jang, A.; Jang, M.; Park, C.M.; Yoo, S.S.; Her, N.; Yoon, Y. Evaluation of Removal Mechanisms in a Graphene Oxide-Coated Ceramic Ultrafiltration Membrane for Retention of Natural Organic Matter, Pharmaceuticals, and Inorganic Salts. *ACS Appl. Mater. Interfaces* **2017**, *9*, 40369–40377. [CrossRef] [PubMed]
32. Pagès, N.; Reig, M.; Gibert, O.; Cortina, J.L. Trace ions rejection tunning in NF by selecting solution composition: Ion permeances estimation. *Chem. Eng. J.* **2017**, *308*, 126–134. [CrossRef]
33. Elimelech, M.; Phillip, W.A. The future of seawater desalination: Energy, technology, and the environment. *Science* **2011**, *333*, 712–717. [CrossRef] [PubMed]
34. Hansen, E.; Rodrigues, M.A.S.; Aquim, P.M. de Wastewater reuse in a cascade based system of a petrochemical industry for the replacement of losses in cooling towers. *J. Environ. Manag.* **2016**, *181*, 157–162. [CrossRef]
35. Asano, T.; Mujeriego, R.; Parker, J.D. Evaluation of industrial cooling systems using reclaimed municipal wastewater. *Water Sci. Technol.* **1988**, *20*, 163–174. [CrossRef]
36. Paugam, L.; Taha, S.; Dorange, G.; Jaouen, P.; Quéméneur, F. Mechanism of nitrate ions transfer in nanofiltration depending on pressure, pH, concentration and medium composition. *J. Membr. Sci.* **2004**, *231*, 37–46. [CrossRef]
37. Egea-Corbacho Lopera, A.; Gutiérrez Ruiz, S.; Quiroga Alonso, J.M. Removal of emerging contaminants from wastewater using reverse osmosis for its subsequent reuse: Pilot plant. *J. Water Process Eng.* **2019**, *29*, 100800. [CrossRef]
38. Osode, A.N.; Okoh, A.I. Impact of discharged wastewater final effluent on the physicochemical qualities of a receiving watershed in a suburban community of the eastern Cape Province. *Clean Soil Air Water* **2009**, *37*, 938–944. [CrossRef]

Article

Valorization of Goat Cheese Whey through an Integrated Process of Ultrafiltration and Nanofiltration

Antónia Macedo [1,2,*], David Azedo [1], Elizabeth Duarte [2] and Carlos Pereira [3]

1. Polytechnic Institute of Beja, School of Agriculture, Rua Pedro Soares, Ap. 6158, 7801-908 Beja, Portugal; david_azedo@hotmail.com
2. LEAF—Linking Landscape, Environment, Agriculture and Food, Instituto Superior de Agronomia, University of Lisbon, Tapada da Ajuda, 1349-017 Lisboa, Portugal; eduarte@isa.utl.pt
3. Polytechnic Institute of Coimbra, School of Agriculture, 3045-601 Coimbra, Portugal; cpereira@esac.pt
* Correspondence: atmacedo@ipbeja.pt

Abstract: Goat cheese whey is a co-product that comes from goat cheese manufacture. Due to its high organic load, adequate treatment is necessary before its disposal. Additionally, the recent growing interest in caprine products, attributed to their specific nutritional and nutraceutical characteristics, such as the lower allergenicity of their proteins and higher content of oligosaccharides, compared with bovine products, made the recovery of goat cheese whey a challenge. In this study, an integrated process for the recovery of sweet goat whey components was carried out. It includes filtration, centrifugation and pasteurization, followed by sequential membrane processes, ultrafiltration/dilution, nanofiltration of ultrafiltration permeates in dilution mode and the concentration/dilution of nanofiltration retentates. Ultrafiltration was performed with membranes of 10 and 1 kDa. Membranes of 10 kDa have higher permeate fluxes and, in a single stage of dilution, allowed for better protein retention and higher lactose purity, with a separation factor of 14. The concentration of lactose by nanofiltration/dilution led to the retention of almost all the lactose in retentates and to a final permeate, whose application in cheese dairy plants will allow for the total recovery of whey. The application of this integrated process in small- or medium-sized goat cheese dairies can represent an important contribution to their sustainability.

Keywords: goat cheese whey; ultrafiltration; nanofiltration; dilution mode

1. Introduction

Goat cheese whey is a liquid co-product of goat cheese manufacture. It retains about 55% of the nutrients found in milk, including lactose, soluble proteins, bioactive peptides, lipids, minerals and vitamins [1]. In comparison with bovine and sheep cheese whey, goat cheese whey contains a higher concentration of oligosaccharides, namely sialic acid, that seems to promote the development of infants' brain [2]. Besides, it is rich in nonprotein nitrogen compounds, namely nucleotides and free amino acids, making it suitable for baby food or children with a cow's milk allergy [3,4]. Therefore, there is new and growing interest in producing caprine products due to the nutraceutical and hypoallergenic properties of caprine milk compared to cow's milk [5], which contributes to the increasing volumes of goat whey produced. Despite its nutritional and nutraceutical value, goat cheese whey is usually treated as waste, deposited in septic tanks or partially mixed with the wastewaters coming from cheese washing and the cleaning operations of equipment and from the cheese dairy; it is then delivered to wastewater plants. Its high values of chemical oxygen demand (COD) and biological oxygen demand (BOD$_5$), about 50–120 g L^{-1} and 27–60 g L^{-1}, respectively [6,7], can lead to the decline in treatment efficiency, transforming this co-product into one of the main environmental problems that the dairy industry has faced for decades. So, its reuse has the advantage of generating value-added products while mitigating its negative impacts on the environment.

Membrane technologies are the most common separation processes used for recovering valuable fractions from cheese whey. Ultrafiltration (UF) is well-established in the food industry to produce whey protein concentrates (WPC) from bovine cheese whey [8], using membranes with a cut-off equal to or higher than 10 kDa [9,10]. These membranes allow the retention of the predominant whey proteins (β-lactoglobulin (β-Lg), α-lactalbumin (α-La), immunoglobulin (Ig), serum albumin (SA)) and other minor proteins, such as lactoferrin (LF) and lactoperoxidase (LP), while lysozyme, glycomacropeptide (GMP) or bioactive peptides, with lower molecular weights or sizes [11], can mostly permeate through the membranes, together with lactose, other sugars and minerals. To increase the permeation of these smaller compounds, thus improving the separation of protein and lactose fractions, the dilution mode of UF retentates is also applied to obtain purified streams [12]. The recovery of permeates of ultrafiltration can be carried out by nanofiltration (NF), followed by the dilution mode of retentates [13]. NF, due to its specific characteristics, especially for the separation of the smaller solutes present in UF permeates, can be a valuable tool, retaining mostly lactose, the main component responsible for the environmental damage of whey [14]; bioactive peptides; oligosaccharides, free amino acids; and bivalent ions, originating a final permeate with a very low organic load [15,16]. The use of NF in dilution mode can enhance the permeation of salts, especially sodium and chloride, which can be high in UF permeates because of salt addition to milk during cheese manufacture. Therefore, the application of NF to UF permeates, in dilution mode, contributes to the reduction in osmotic pressure, a major drawback in these processes [17]. In addition, the purification of retentates may allow its application in food or pharmaceutical industries, due to their nutritional and nutraceutical characteristics, as cited above. This investigation aims to contribute to improvement in the sustainability of the artisanal production of goat cheese, carried out mainly in small- and medium-sized cheese dairy plants. This study has an innovative character in that it presents an integrated membrane process for the total recovery of a co-product with a very complex composition, such as goat cheese whey. Therefore, the main objectives are:

- to evaluate the performance of UF membranes of different cut-offs in the separation of the protein and lactose fractions of goat cheese whey;
- to investigate the influence of the dilution mode, in three stages, applied to UF retentates in separation efficiency;
- to assess the performance of the concentration process of nanofiltration, in dilution mode, of UF permeates;
- to study the influence of dilution, in three stages, applied to NF retentates for the removal of salts and to purify lactose;
- to produce a permeate with a low organic content.

2. Materials and Methods

2.1. Sampling and Pretreatment of Goat Cheese Whey

Six samples of goat cheese whey (GCW) were collected in the same cheese dairy, located close to Beja, Portugal. The GCW was produced during the cheese-making process, which involved the following steps: milk pasteurization; filtration of the milk; addition of salt to the milk cheese; rennet coagulation; syneresis, where whey was released; the new addition of salt to the curd; pressing and packing, before expedition. A volume of about 10 L of each sample was collected and carried out to our laboratory, keeping them refrigerated in ice during transportation. After arriving, samples were filtrated two times through cotton cloths, like those used in a traditional cheese dairy, to remove suspended solids and casein fines. After that, samples were skimmed in an Elecrem SAS, Fresnes, France, centrifuge, at a temperature of about 35 °C to remove most of the lipids and some minor residues of casein and bacteria. Since these samples have a high concentration of lipids, the reduction of their concentration is crucial to avoid membrane fouling. Finally, whey samples were subjected to a low pasteurization process, at 65 °C, for 30 min. When it was not possible to process the samples in the same day, they were immediately preserved at about 3 °C until the next day.

2.2. Permeation Experiments

All the permeation experiments were carried out in a plane-and-frame module, Lab Unit M20, from Alfa Laval, Navskov, Denmark. This filtration rig is a commercial installation with a membrane area ranging from a minimum of 0.036 m^2, which corresponds to two membrane sheets, to 0.72 m^2, the maximum area. The membranes were grouped in pairs, resting on the top and bottom of the same support plate. The plates were separated by spacers that acted as feed chambers 0.5 mm in height and that were divided into 30 channels to increase tangential velocity and thus, promote mass transfer in the adjacent layer near the membrane surface. In each support plate, there were individual collectors for the permeate. Support plates and spacers were made of polysulfone, and the module frame was made of stainless steel. The unit had a hydraulic system that allowed the flat plate module to be compressed, making it perfectly watertight.

These experiments included the UF of the pretreated goat cheese whey to remove residual protein and fat and to obtain a protein-containing retentate and a lactose-rich permeate. For a better recovery of lactose in permeates, dilution mode in ultrafiltration (DUF) of the retentates was also carried out. After that, all of the permeates resulting from UF and DUF were mixed and subjected to NF to recover the lactose fraction. For purifying this lactose-rich retentate, which may increase the possibilities for its use in various industries, dilution mode in NF was also applied to the retentates of the NF process (Figure 1).

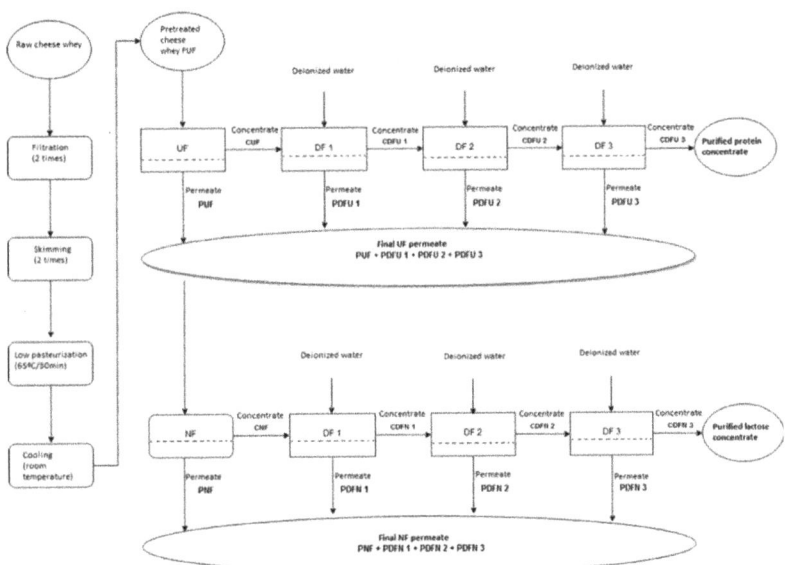

Figure 1. Experimental scheme.

Before each permeation test, the hydraulic permeability to pure water was determined by measuring the permeate fluxes at different transmembrane pressures at a feed circulation velocity of 0.94 ms^{-1} and using Equation (1). The hydraulic permeability of pure water is the slope of the linear regression obtained from the experimental water fluxes and corresponding transmembrane pressures.

$$J_w = \left(\frac{L_p}{\mu}\right)\Delta P \quad (1)$$

where J_w is the water permeate flux (ms^{-1}); (L_p/μ) is the hydraulic permeability to pure water (ms^{-1}Pa^{-1}); L_p is the intrinsic permeability of the membrane (m), related with its morphological characteristics; μ is the water viscosity (Pa·s), and ΔP is the applied transmembrane pressure (Pa).

After the tests, a cleaning and disinfection cycle was performed, according to the procedure shown in Table 1. To ensure that membrane's permeability characteristics were kept, the hydraulic permeability to pure water was again determined and, if it was at least 95% of the initial value, the same membranes were used in the following tests.

Table 1. Process of cleaning and disinfection of membranes [a].

Membrane parameters	Membranes			Time (min)
	RC70PP (UF)	ETNA01PP (UF)	NF	
pH	1–10	1–11	2–11	
Transmembrane pressure (MPa)	0.1–0.5	0.1–0.5	0.1–1	
Temperature (°C)	0–60	0–65	0–45	
Cleaning				
NaOH (% w/v)	0.05	0.05	0.05	15
Na-EDTA (% w/v)	0.20	0.20	0.20	15
HNO$_3$ (% w/v)	0.25	0.25	0.10	15
Citric acid (%w/v)	0.50	0.50	0.50	15
Disinfection				
H$_2$O$_2$ (mgL^{-1}) at 25 °C	1000	1000	1000	30

[a] Into tolerate pH limits of membranes; Na-EDTA, ethylenediaminetetra-acetic acid, sodium salt.

2.2.1. Ultrafiltration Experiments

Ultrafiltration experiments were carried out with two different kinds of membranes, one with an active layer made of regenerated cellulose acetate and a molecular weight cut-off of 10 kDa, designated as RC70PP, and another one with an active layer made of a composite fluoropolymer and a molecular weight cut-off of 1 kDa, named ETNA01PP. Both materials used to manufacture the membranes are hydrophilic in nature, which minimizes the effects of fouling by organic matter, particularly proteins, as described elsewhere [18]. The best operating conditions of transmembrane pressure (0.2 MPa) and feed circulation velocity (0.94 ms^{-1}) were selected, based on the results obtained in total recirculation UF experiments, carried out in the range of transmembrane pressure between 0.1–0.4 MPa, with the same membranes used in this study and established in previous works [18,19]. The highest permeate flux and relative flux (J_p/J_w) and the best separation between protein and lactose fractions were the criteria used for selection. The temperature varied from 16 °C to 22 °C and, to correct for the different viscosities of the permeates, all the permeate fluxes were converted to 25 °C [20]. The pressure drop along the module was about 0.1 MPa.

The first set of ultrafiltration experiments was done in three steps: preconcentration until the volume concentration factor, VFC = 2.0; dilution mode by adding deionized water; and postconcentration. This procedure allows the achievement of the concentration/purification of the protein fraction in the retentate and a better recovery of lactose in the permeate, thus contributing to improvements in the separation of these components.

Starting from an initial volume of 8.75 L of each sample, three UF experiments with GCW were performed in concentration mode until a volume concentration factor (VCF) of about two was reached with each of the membranes (10 kDa and 1 kDa), using a membrane area of 0.072 m^2. After concentration, the dilution (DF) of the final retentates was realized in three stages in a discontinuous mode. In each of them, a volume of deionized water, equal to the observed volume of the retentate in the tank, was added. After homogenization and stabilization at the same operating conditions of transmembrane pressure, feed circulation velocity and temperature, a new concentration process took place until the same volume of permeate was collected, thus maintaining the volume of the retentate. Samples of raw and pretreated GCW, retentates and permeates of UF and retentates and permeates of DF were taken for analyses.

The results obtained from this first set of experiments, with both membranes, were analyzed in terms of the following parameters: productivity, measured by volumetric permeate fluxes (J_p) and their evolution along the concentration processes, and the separation factor, α, for lactose and protein, which should be greater than 1; the higher it is, the better the separation between the two solutes.

The volumetric permeate fluxes were determined according to Equation (2):

$$J_p = \frac{\Delta V}{A_m \times \Delta t} \tag{2}$$

where J_p is the volumetric permeate flux; ΔV is the volume (m^3) of permeate collected during an interval of time Δt (s), and A_m is the total membrane area (m^2).

The separation factor, α, is defined as [21]:

$$\alpha = \frac{S_{microsolute}}{S_{macrosolute}} \tag{3}$$

where $S_{microsolute}$ is the sieving coefficient for the microsolute (lactose), and $S_{macrosolute}$ is the sieving coefficient for the macrosolute (protein). S_i, the sieving coefficient for a component i, is given by: $S_i = c_p/c_r$, in which c_r is the concentration of a solute in the bulk retentate, and c_p, the concentration of the solute in the bulk permeate.

Based on the results obtained in this first set of experiments, the best membrane in terms of productivity and the separation of protein and lactose fractions was selected to carry out NF experiments.

2.2.2. Nanofiltration Experiments

Nanofiltration experiments were realized with the mixture of permeates resulting from the UF (PUF+PDFU1+PDFU2+PDFU3) experiments (Figure 1). Permeates were mixed, homogenized and subjected to NF. NF experiments were performed with membrane NFT50 (NF), commercialized by Alfa Laval, Navskov, Denmark. These membranes have an active layer made of polyamide semi-aromatic (polipiperazine). The preconcentration of the feed was carried out until a VCF of about 2.0, at a transmembrane pressure of 2 MPa, a feed circulation velocity of 0.94 ms^{-1} and a membrane area of 0.072 m^2. The diafiltration of the final retentates of NF was performed in three stages by adding a volume of deionized water equal to that of the retentate in the tank, and, afterwards, the concentration process proceeded until an identical volume of permeate was collected. The experimental conditions used in this process were the same as those used for preconcentration by NF.

The performance of the process of concentration by NF, in dilution mode, followed by the diafiltration of the retentates obtained, was determined in terms of permeate fluxes (productivity), J_p; efficiency of the removal of salts, $\mu_{removal}$; and the evaluation of the quality of the final permeate for possible further application in cheese dairy plants.

The efficiency of removal of a certain solute is given by:

$$\mu_{removal} = \frac{c_{ri} - c_{rf}}{c_{ri}} \times 100 \tag{4}$$

where c_{ri} is the concentration of a solute in a retentate i, before a stage of diafiltration, and c_{rf} is its concentration in the retentate after that stage.

2.3. Cleaning and Disinfection Cycle

After the permeation experiments, samples were removed from the installation, and three flushes were carried out with water to ensure that no residues were present. The cleaning and disinfection cycle realized and the operating recommended conditions of the manufacturer for the membranes in this study are shown in Table 1. The cleaning procedure included four steps, in each one a different chemical was added, performed under recirculating conditions, which means that both permeate and retentate were recycled to the feed/retentate tank. A transmembrane pressure of 0.1 MPa, a feed circulation velocity of 0.92 ms^{-1} and a temperature of 25 °C were used during this operation. For NF membranes, a transmembrane pressure of 1 MPa was applied, maintaining the same values of feed circulation and temperature. Between each two cleaning solutions, water was permeated to remove the previous reagent, checking if the pH was already restored. After cleaning, a final disinfection step was carried out, as presented in Table 1, using the same transmembrane pressure, feed circulation velocity and temperature.

2.4. Physicochemical Characterization of the Samples

The samples (feed, retentates and permeates) were analyzed for: pH (by potentiometry); lactose, by determination of reducing sugars [22]; total solids, by gravimetry [23]; total nitrogen, by the Kjeldahl reference method; crude protein, obtained from total nitrogen multiplied by the factor 6.38 [24] and adapted for cheese whey; the fat content, determined by infrared spectroscopy using the equipment Milkoscan134B, previously calibrated for cheese whey with the standard method of Rose-Gottlied for milk and dairy products; sodium and potassium, by emission flame photometry, according to the procedure described in [25]; calcium and magnesium by atomic absorption spectrophotometry with air–acetylene flame [25]; chloride, by volumetric precipitation, according to the method of Charpentier-Volhard [26]; and phosphates, by the spectrophotometric method of ammonium molybdate [27].

3. Results and Discussion

3.1. Physicochemical Characterization of Raw and Pretreated Goat Cheese Whey

The average composition of raw and pretreated goat cheese whey is shown in Table 2.

Table 2. Average composition (mean ± standard deviation) of raw goat cheese whey and pretreated goat cheese whey [a].

Parameters	Raw Goat Cheese Whey	Pretreated Goat Cheese Whey (FUF)
pH (25 °C)	5.68 ± 0.72	5.90 ± 0.79
Total solids (% w/w)	9.24 ± 0.50	8.68 ± 0.47
Lactose (% w/w)	5.11 ± 0.33	5.12 ± 0.21
Lipids (% w/w)	1.06 ± 0.19	0.44 ± 0.09
Nkjeldahl (% w/w)	0.082 ± 0.005	0.078 ± 0.006
Crude protein (% w/w)	0.53 ± 0.03	0.50 ± 0.04
Ash (% w/w)	2.13 ± 0.03	1.81 ± 0.11
Cl (mg L^{-1})	9553 ± 217	8984 ± 336
P (mg L^{-1})	501 ± 98	392 ± 31
Ca (mg L^{-1})	171 ± 0.5	161 ± 0.3
Mg (mg L^{-1})	92.0 ± 0.3	74.4 ± 0.3
K (mg L^{-1})	1875 ± 3.4	1632 ± 7.6
Na (mg L^{-1})	6574 ± 235	5730 ± 186

[a] n (number of samples) = 6.

The goat cheese whey used in this study is classified as a sweet cheese whey because its pH is around 6.0 and is produced from milk coagulated by the enzymatic hydrolysis of casein through chymosin action, at a pH not lower than 5.6 [28]. Apart from water (around 90.8% w/w), the main components are lactose, followed by minerals, lipids and nitrogen compounds. Lactose, lipids, and nitrogen compounds are, in order of importance, primarily responsible for the high organic loading of these co-products, which are translated into high levels of COD and BOD, as stated in Section 1.

The pretreatment realized (filtration, centrifugation, pasteurization) allowed for a removal of about 58% of lipids, 5% of nitrogen compounds and 15% of ash, leading to a decrease of 6% of the total solids. These results suggest that a part of the organic matter present in the raw goat cheese whey, mainly related to its lipid content, was quickly removed during the pretreatment. This co-product, rich in lipids, after pasteurization, may eventually be reused in cheese dairies, added to milk cream to increase the yield of the manufacture process of goat butter and/or other types of spreads, and will be the subject of further study. However, lactose, most of the nitrogen fraction and around 42% of fat is still present in the pretreated goat cheese whey, thus contributing to its high content of organic matter.

Regarding the mineral composition, the most salient aspect is the very high concentrations of chloride and sodium, which are in contrast with goat milk, where the dominant minerals are potassium, chloride, calcium and phosphate [4]. This resulted from the addition of sodium chloride to the cheese milk, during the manufacture of goat cheese.

3.2. Permeation Experiments

3.2.1. Characteristics of Membranes

Before permeation experiments, the hydraulic permeability of membranes to pure water was determined (Table 3), according to the procedure described in Section 2.2. In Table 3 is displayed the hydraulic permeability of membranes, the intrinsic permeability and the MWCO of membranes, furnished by the supplier and determined, for NF membranes, according to the procedure described elsewhere [16].

Table 3. Hydraulic permeability of membranes (±95% confidence interval), intrinsic permeability and MWCO of membranes.

Membrane	Hydraulic Permeability to Pure Water (L_p/μ) (ms^{-1}Pa^{-1})	Intrinsic Permeability L_p(m) [1]	MWCO kDa
RC70PP (UF)	$1.73 \times 10^{-10} \pm 2.83 \times 10^{-11}$	1.54×10^{-13}	10
ETNA01PP (UF)	$4.85 \times 10^{-11} \pm 3.84 \times 10^{-13}$	4.32×10^{-14}	1
NFT50 (NF)	$1.48 \times 10^{-11} \pm 4.56 \times 10^{-13}$	1.32×10^{-14}	0.13 [2]

[1] The dynamic viscosity of pure water, at a temperature of 25 °C, used to calculate the intrinsic permeability of membranes, was 8.91×10^{-4} Pa·s; [2] In accord with [16], for the same set of membranes.

3.2.2. Performance of Ultrafiltration Experiments

- Permeate fluxes

The evolution of permeate fluxes along the process of concentration by UF, followed by dilution in UF mode with three stages for both types of membranes (RC70PP and ETNA01PP), is displayed in Figures 2 and 3, respectively. The horizontal line, in both figures, represents the water fluxes at the transmembrane pressure of 0.2 MPa, at which permeation experiments were carried out. As can be observed, until a VCF of about 2.5, the permeate fluxes obtained with samples and with water are close, which indicates that, in the experimental conditions used, fouling is negligible for both membranes. More experiments will be realized in the future on the highest VCFs to study flux behavior when protein concentration is increased. The effect of MWCO, and the corresponding mean pore radius, is evident, because permeate fluxes obtained with membranes of 1 kDa were around 50% of those produced with membranes of 10 kDa. Therefore, the use of membranes with higher MWCO allowed for higher permeate fluxes, as expected, probably because both membranes are made from hydrophilic materials, which are less susceptible to fouling by proteins, the component most responsible for this phenomena in the UF of cheese whey.

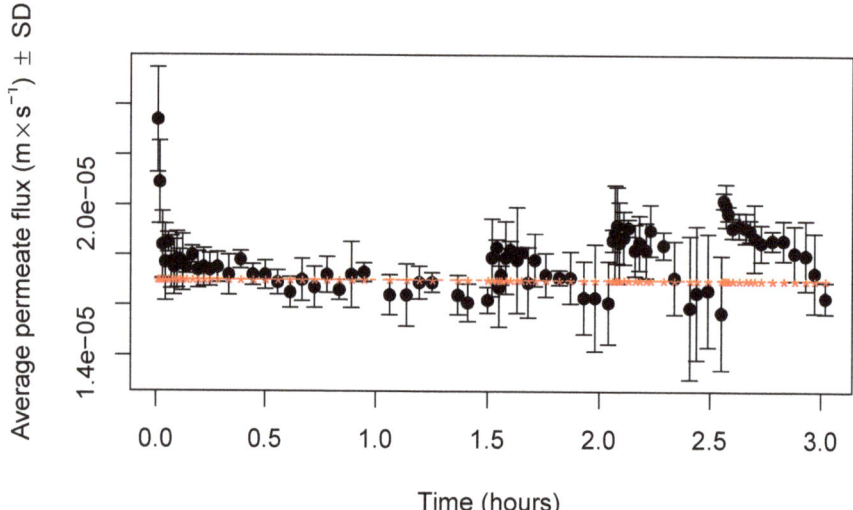

Figure 2. Variation of average permeate fluxes (±standard deviation) with time, during the process of UF/DF, obtained with membranes RC70PP (n = 3 experiments), at a transmembrane pressure of 0.2 MPa.

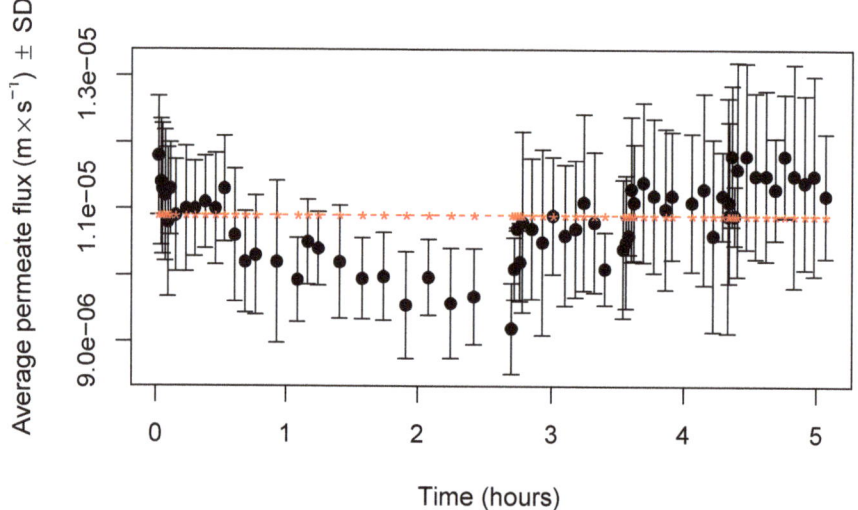

Figure 3. Variation of permeate fluxes (±standard deviation) with time, during the process of UF/DF, obtained with membranes ETNA01PP (n = 3 experiments), at a transmembrane pressure of 0.2 MPa.

In the case of UF experiments with the membrane RC70PP, at the very beginning, a decline in permeate fluxes was observed, probably due to polarization-concentration phenomena, which is more important when permeate fluxes are higher, due to the rapid accumulation of retained compounds near the membrane surface. However, after that, an average constant flux of about 1.73×10^{-5} ms^{-1} was reached. With membranes ETNA01PP (Figure 3), the initial decline in permeate fluxes was much less pronounced, because the lower average permeate flux, around 1.05×10^{-5} ms^{-1}, minimized the effect of the intensity of concentration-polarization phenomena [29].

Relative to the DF process carried out in three stages, we can observe in Figure 2 that the intensity of average permeate fluxes along the dilution processes were slightly higher than those observed during preconcentration by UF, because are the lower species, such lactose and minerals that are preferentially permeating UF membranes. During each stage, permeate fluxes were kept almost constant. The range of permeate fluxes during the dilution process realized with membranes RC70PP was between 1.79×10^{-5} ms^{-1} for DF1 and 2.00×10^{-5} ms^{-1} at the third stage (DF3). The same was true for the dilution process performed with membranes ETNA01PP (Figure 3), the permeate fluxes ranging from 1.08×10^{-5} ms^{-1} (DF1) up to 1.17×10^{-5} ms^{-1} (DF3). These results show that permeate fluxes, during the dilution processes, were not affected by the permeation of the lower species across UF membranes.

- Analysis of separation factors

The concentrations of lactose and protein in retentates and permeates, as well as the corresponding separation factors between these two components, along UF/DF processes, are shown in Tables 4 and 5 for membranes RC70PP and ETNA01PP, respectively.

Table 4. Separation factors [1] (α) between lactose and protein, along with UF/DF processes for membrane RC70PP.

Processes	C_p (Lac) (% w/w)	C_r (Lac) (% w/w)	C_p (prot) (% w/w)	C_r (prot) (% w/w)	α
UF	5.27 ± 0.68	5.31 ± 0.21	0.14 ± 0.06	0.84 ± 0.04	6.0 ± 0.4
DF1	3.42 ± 0.25	3.53 ± 0.26	0.06 ± 0.04	0.89 ± 0.16	14.4 ± 0.6
DF2	2.08 ± 0.37	2.24 ± 0.42	0.06 ± 0.05	0.93 ± 0.04	14.4 ± 0.5
DF3	1.26 ± 0.33	1.52 ± 0.45	0.10 ± 0.02	1.26 ± 0.11	10.4 ± 0.4

[1] α was determined at a transmembrane pressure of 0.2 MPa, feed circulation velocity of 0.92 ms^{-1} and temperature of 25 °C.

Table 5. Separation factors [1] (α) between lactose and protein, along with UF/DF processes for membrane ETNA01PP.

Processes	C_p (Lac) (% w/w)	C_r (Lac) (% w/w)	C_p (prot) (% w/w)	C_r (prot) (% w/w)	α
UF	5.08 ± 0.13	5.91 ± 0.08	0.09 ± 0.03	1.05 ± 0.05	10.0 ± 0.1
DF1	3.20 ± 0.11	3.83 ± 0.01	0.11 ± 0.02	1.10 ± 0.04	8.4 ± 0.2
DF2	2.26 ± 0.01	2.77 ± 0.07	0.10 ± 0.02	1.08 ± 0.05	8.8 ± 0.3
DF3	1.52 ± 0.07	2.11 ± 0.07	0.09 ± 0.01	1.08 ± 0.05	8.6 ± 0.2

[1] α was determined at a transmembrane pressure of 0.2 MPa, feed circulation velocity of 0.92 ms^{-1} and temperature of 25 °C.

The observation of the data displayed in Table 4 allows the conclusion that the use of dilution in ultrafiltration mode for UF retentates led to a large increase in the separation factor between lactose and protein, from 6 to 14, right after the first stage of dilution. This is mainly due to the permeation of lactose into permeate streams, as can be confirmed by the decrease in its concentration in UF retentates. However, after the first stage (DF1), the separation factor between those components remained or even declined. Therefore, since dilution mode involves water consumption, which should be minimized for economic and environmental reasons, the use of a second and third stage in dilution mode will be dependent on the desired purification of the final protein fraction.

For preconcentration by UF, the membranes of lower MWCO (1 kDa) led to a better separation between lactose and protein, because α is around 10 and, for the other membranes, it is 6 (Table 5). This is likely due to the greater accumulation of the protein fraction in retentates and, consequently, the lower loss of protein into the permeates, probably the lower-molecular-weight whey proteins, such as glycomacropeptide (GMP), with a molecular weight of 6.80 kDa, and bioactive peptides, as described elsewhere [9]. However, unlike what was observed with membranes of 10 kDa, during the dilution process in three stages, the separation factor decreased to about 9 and was kept constant until the end of this process. This may be due to the higher accumulation of the protein fraction in retentates that may have hampered the removal of lactose into the permeate stream, leaving it retained within the protein fraction, which can be confirmed by its higher concentration in retentates. Therefore, relative to the separation factor, despite the fact that membranes of lower MWCO allowed, in a single UF operation, the obtention of a better separation between protein and lactose fractions, the final decision as to which of the membranes should be selected will depend on the intended application of protein retentates.

3.2.3. Performance of Nanofiltration Experiments

- Variation of permeate fluxes with VCF

Nanofiltration experiments were carried out with the permeates resulting from the UF/DF of membranes with the higher MWCO because their separation between the protein and lactose fractions was better. During the process of dilution in the nanofiltration mode of UF permeates, it was observed a sharp decline in the average permeate flux of about 45%, ranging from 1.59×10^{-5} ms^{-1} (57.24 Lh^{-1}m^{-2}) to 8.33×10^{-6} ms^{-1} (29.99 Lh^{-1}m^{-2}), and until the VCF of 2.34 was attained. A similar pattern was observed during the DF of the NF retentates performed in three stages. Permeate fluxes were only slightly higher than those measured during the preconcentration process by NF, ranging from 1.74×10^{-5} to 8.72×10^{-6} ms^{-1}, and the decline of permeate fluxes varied from 39 to 47%. This behavior is explained by the fact that UF permeates are mainly composed of the smaller solutes of cheese whey, like lactose and minerals, especially sodium, chloride and potassium, that mostly contribute to its higher osmotic pressure [11], leading to a decrease in effective membrane pressure and thus, to the decrease of permeate fluxes. One possibility to overcome this disadvantage will be the reduction of the amount of sodium chloride added to the milk and curd during the manufacturing process, which will also be beneficial for human health. Another factor that can also contribute to the decrease in NF productivity is the possible formation of insoluble salts of calcium or magnesium phosphates near the membrane surface due to their high retention by NF membranes.

- Physicochemical characterization of NF and DF/NF samples and removal efficiency

Table 6 shows the physicochemical characterization of the following samples (Figure 1): final UF permeate, which is the feed for nanofiltration; CNF, the concentrate of nanofiltration; PNF, the permeate of nanofiltration; CDNF3, the concentrates of 3rd stage of diafiltration; and PDNF3, the corresponding permeates.

Table 6. Physicochemical characterization of NF and DF/NF samples.

Parameters	Feed	CNF	PNF	CDNF3	PDNF3
pH (25 °C)	6.28 ± 0.02	6.27 ± 0.01	6.20 ± 0.30	6.33 ± 0.05	5.99 ± 0.09
Total solids (%w/w)	6.93 ± 0.39	12.49 ± 0.04	1.68 ± 0.79	12.74 ± 0.87	1.16 ± 0.04
Lactose (%w/w)	4.67 ± 0.10	9.49 ± 0.46	n.d. [1]	10.01 ± 0.14	n.d.
Lipids (%w/w)	0.043 ± 0.01	0.057 ± 0.02	0.046 ± 0.05	0.070 ± 0.01	0.061 ± 0.02
$N_{Kjeldahl}$ (%w/w)	0.023 ± 0.003	0.034 ± 0.004	0.021 ± 0.007	0.042 ± 0.002	0.010 ± 0.002
Crude protein (%w/w)	0.15 ± 0.02	0.22 ± 0.02	0.13 ± 0.03	0.27 ± 0.05	0.05 ± 0.02
Ash (%w/w)	1.29 ± 0.21	2.23 ± 0.25	1.59 ± 0.72	0.89 ± 0.04	0.25 ± 0.02
Cl (mg L^{-1})	9351.3 ± 961.1	12,338.1 ± 1760.0	12,835.8 ± 754.3	2702.3 ± 497.8	2015.0 ± 249.7
P (mg L^{-1})	203.23 ± 26.32	215.33 ± 16.82	n.d.	209.93 ± 24.25	n.d.
Ca (mg L^{-1})	112.80 ± 0.18	163.20 ± 0.51	n.d.	168.40 ± 0.35	n.d.
Mg (mg L^{-1})	87.20 ± 0.17	128.00 ± 0.38	n.d.	146.60 ± 0.44	n.d.
K (mg L^{-1})	163.20 ± 0.18	138.89 ± 0.25	126.74 ± 0.14	78.13 ± 0.12	53.82 ± 0.11
Na (mg L^{-1})	5447.89 ± 234.13	4240.64 ± 265.12	4200.40 ± 186.71	1946.88 ± 256.21	699.40 ± 38.24

[1] Not detectable.

In Table 6, it can be observed that the main components of the feed of the DF/NF process (mixture of all the permeates from ultrafiltration) are, apart from water, lactose and minerals, nitrogen compounds at the lowest concentration (0.023% w/w). During the previous ultrafiltration, most nitrogen compounds were retained by UF membranes, in accord with the sieving coefficient of about 8% (Table 4), which corresponds to a membrane rejection around 92%.

Lipids and nitrogen compounds were preferentially retained by NF membranes, probably through steric hindrance and non-electrostatic membrane–solute interactions, the main mechanisms responsible for the retention of uncharged molecules in nanofiltration membranes [30,31].

Relative to minerals, chloride and sodium are predominant due to their preferential permeation through UF membranes, as expected when negligible fouling problems occur. The distribution of ions between retentates and permeates in nanofiltration can result both from steric hindrance and electrostatic interactions between ions and surface charge, based on the Donnan exclusion mechanism [31]. Since NF membranes used in this work have an isoelectric point, pH_i = 4.2, this means that, at the pH of our samples, they carried a negative charge [30]. Then, the counter-ions, especially calcium and magnesium due to their higher density charge, were adsorbed at the membrane surface by electrostatic interactions, and the co-ions, such as chloride, were mainly repulsed by the membrane surface to satisfy the electroneutrality condition. The chloride ion even had a negative rejection during the concentration process by NF, probably because of its higher density charge.

In NF/DF stages, the concentration of monovalent ions clearly decreased due to their removal into the permeate streams. The removal efficiencies of salts were calculated based on their concentrations in CNF and CDNF3 (Table 6) and using Equation (4). Calcium and magnesium were preferentially retained (with negative removals); phosphates were slightly removed (circa 2.5%); potassium, sodium and chloride had removal efficiencies of 44%, 54% and 78%, respectively, after the three-stage DF process. It is possible that the high retention of calcium and magnesium contributed to a small reduction in the surface charge of the membrane, and therefore, to a lower removal of the chloride anion present in the retentates.

The predominant components of NF permeate are water, chloride and sodium, and thus, they can be used in cheese dairy plants as washing waters for cheese or during cheese processing.

4. Conclusions

The fractionation of goat cheese whey using the sequential membrane processes proposed in this study allowed a good separation between protein and lactose fractions through ultrafiltration followed by diafiltration. The recovery of lactose by the nanofiltration of permeates contributes to minimizing the environmental impact of this co-product of goat cheese manufacture and, at the same time, allows for possible applications of the separated fractions.

Author Contributions: A.M. and D.A. developed the experimental work. A.M., E.D. and C.P. contributed to the Results and Discussion. All authors have read and agreed to the published version of the manuscript.

Funding: This research was funded by partnership PORTUGAL2020-PDR, within the scope of the PDR2020-101-030768 project: LACTIES-Innovation, Eco-Efficiency and Safety in Dairy Industry.

Data Availability Statement: All the data of this study are presented along the text.

Acknowledgments: The authors would like to thank to project LACTIES, to Polytechnic Institute of Beja and to LEAF—Linking Landscape, Environment, Agriculture, and Food, Instituto Superior de Agronomia, University of Lisbon, Tapada da Ajuda, 1349-017 Lisboa, Portugal.

Conflicts of Interest: The authors declare no conflict of interest.

References

1. Guimarães, P.M.R.; Teixeira, J.A.; Domingues, L. Fermentation of lactose to bio-ethanol by yeasts as part of integrated solutions for the valorisation of cheese whey. *Biotechnol. Adv.* **2010**, *28*, 375–384. [CrossRef] [PubMed]
2. Hernández-Ledesma, B.; Ramos, M.; Gómez-Ruiz, J.Á. Bioactive components of ovine and caprine cheese whey. *Small Rumin. Res.* **2011**, *101*, 196–204. [CrossRef]
3. Maduko, C.; Park, Y. Production of infant formula analogs by membrane fractionation of caprine milk: Effect of temperature treatment on membrane performance. *J. Food Nutr. Sci.* **2011**, *2*, 1097–1104. [CrossRef]

4. Park, Y.W.; Juárez, M.; Ramos, M.; Haenlein, G.F.W. Physico-chemical characteristics of goat and sheep milk. *Small Rumin. Res.* **2007**, *68*, 88–113. [CrossRef]
5. Yangilar, F. As a potentially functional food: Goat's milk and products. *J. Food. Nutr. Res.* **2013**, *4*, 68–81.
6. Baldasso, C.; Barros, T.C.; Tessaro, I.C. Concentration and purification of whey proteins by ultrafiltration. *Desalination* **2011**, *278*, 381–386. [CrossRef]
7. Mollea, C.; Marmo, L.; Bosco, F. Valorisation of Cheese Whey, a By-Product from the Dairy Industry. *Food Ind.* **2013**, 549–588. [CrossRef]
8. Bordenave-Juchereau, S.; Almeida, B.; Piot, J.-M.; Sanier, F. Effect of protein concentration, pH, lactose content and pasteurization on thermal gelation of acid caprine whey protein concentrates. *J. Dairy Res.* **2005**, *72*, 34–38. [CrossRef] [PubMed]
9. Sanmartín, B.; Díaz, O.; Rodríguez-Turienzo, L.; Cobos, A. Composition of caprine whey protein concentrates produced by membrane technology after clarification of cheese whey. *Small Rumin. Res.* **2012**, *105*, 186–192. [CrossRef]
10. Palatnik, D.R.; Victoria, M.; Porcel, O.; Gonz, U.; Zaritzky, N.; Campderr, M.E. Recovery of caprine whey protein and its application in a food protein formulation. *LWT Food Sci. Technol.* **2015**, *63*, 331–338. [CrossRef]
11. Walstra, P.; Geurts, T.J.; Noomen, A.; Jellema, A.; Van Boekel, M.A.J.S. *Ciencia de la Leche y Tecnologia de los Productos Lácteos*; Editorial, A., Ed.; Acribia: Zaragoza, Spain, 2001.
12. Kovacs, Z.; Czermak, P. Diafiltration. In *Encyclopedia of Membrane Science and Technology*; Hoek, E.M.V., Tarabara, V.V., Eds.; Wiley: Hoboken, NJ, USA, 2013; Volume 3.
13. Cuartas-Uribe, B.; Alcaina-Miranda, M.I.; Soriano-Costa, J.A.; Mendoza-Roca, J.A.; Iborra-Clar, M.I.; Lora-Garcia, J. A study of the separation of lactose from whey ultrafiltration permeate using nanofiltration. *Desalination* **2009**, *241*, 244–255. [CrossRef]
14. Butylina, S.; Luque, S.; Nyström, M. Fractionation of whey-derived peptides using a combination of ultrafiltration and nanofiltration. *J. Membr. Sci.* **2006**, *280*, 418–426. [CrossRef]
15. Macedo, A.; Ochando-Pulido, J.M.; Fragoso, R.; Duarte, E. The use and performance of nanofiltration membranes for agro-industrial effluents purification. In *Nanofiltration*; Akhyar Farrukh, M., Ed.; Intechopen Limited: London, UK, 2018; pp. 65–84. [CrossRef]
16. Macedo, A.; Monteiro, J.; Duarte, E. A contribution for the valorisation of sheep and goat cheese whey through nanofiltration. *Membranes* **2018**, *8*, 114. [CrossRef] [PubMed]
17. Rice, G.S.; Kentish, S.E.; O´Connor, A.J.; Barber, A.R.; Pihljamaki, A.; Nystrom, M.; Stevens, G.W. Analysis of separation and fouling behaviour during nanofiltration of dairy ultrafiltration permeates. *Desalination* **2009**, *236*, 23–29. [CrossRef]
18. Macedo, A.; Duarte, E.; Fragoso, R. Assessment of the performance of three ultrafiltration membranes for fractionation of ovine second cheese whey. *Int. Dairy J.* **2015**, *48*, 31–37. [CrossRef]
19. Macedo, A.; Duarte, E.; Pinho, M. The role of concentration polarization in ultrafiltration of ovine cheese whey. *J. Membr. Sci.* **2011**, *381*, 34–40. [CrossRef]
20. Gallant, R.W. *Physical Properties of Hydrocarbons*; Gulf Publishing Company: Houston, TX, USA, 1970; Volume 2.
21. Mulder, M. *Basic Principles of Membrane Technology*, 2nd ed.; Kluwer Academic Publishers: Dordrecht, The Netherlands, 1996.
22. Portuguese Standard PS 675. *Milk and Dry Milk. Determination of Lactose Concentration*; Diário da República III Series, No. 123; Association of Official Analytical Chemists, Official Methods of Analysis: Washington, DC, USA, 1986.
23. AOAC Official Methods of Analysis. Solids (Total) in milk, IDF-ISO-AOAC method I. In *Official Methods of Analysis of AOAC International*; Cunliffe, P., Ed.; AOAC Official Methods of Analysis: Gaithersburg, MD, USA, 1990; p. 807.
24. Portuguese Standard PS 1986. Milk. In *Determination of Crude Protein*; Diário da República, III Series, No. 142; Association of Official Analytical Chemists, Official Methods of Analysis: Washington, DC, USA, 1991.
25. Gonçalves, M.L. Determination of sodium, potassium, calcium and magnesium in milk. In *Análise de Soluções*; Fundação Calouste Gulbenkian: Lisboa, Portugal, 1983.
26. Portuguese Standard PS 471. Milk. In *Determination of Chloride in Milk*; Diário da República; III Series, No. 236; Association of Official Analytical Chemists, Official Methods of Analysis: Washington, DC, USA, 1983.
27. American Public Health Association (APHA). Phosphate. In *Standard Methods for Examination of Water and Wastewater*, 22nd ed.; Rice, E.W., Clesceri, L.S., Baird, R.B., Eaton, A.D., Eds.; American Public Health Association: Washington, DC, USA, 2012.
28. Daufin, G.; René, F.; Aimar, P. *Les Separations par Membrane Dans les Procédés de l´Industrie Alimentaire*; Collection Sciences et Techniques Agroalimentaires: Paris, France, 1998.
29. Cheryan, M. *Ultrafiltration and Microfiltration Handbook*, 2nd ed.; CRC Press: Boca Raton, FL, USA, 1998.
30. Teixeira, M.R.; Rosa, M.J.; Nystrom, M. The role of Membrane charge on nanofiltration performance. *J. Membr. Sci.* **2005**, *265*, 160–166. [CrossRef]
31. Van der Bruggen, B.; Schaep, J.; Wilms, D.; Vandecasteele, C. Influence of molecular size, polarity and charge on the retention of organic molecules by nanofiltration. *J. Membr. Sci.* **1999**, *156*, 29–41. [CrossRef]

Review

Mixed-Matrix Membrane Fabrication for Water Treatment

Tawsif Siddique [†], Naba K. Dutta *[ ] and Namita Roy Choudhury *

Chemical and Environmental Engineering, School of Engineering, RMIT University, Melbourne, VIC 3000, Australia; s3642366@student.rmit.edu.au or tawsif@uctc.edu.bd
* Correspondence: naba.dutta@rmit.edu.au (N.K.D.); namita.choudhury@rmit.edu.au (N.R.C.)
† Currently on leave from Department of Mechanical Engineering, University of Creative Technology-Chittagong (UCTC), Chattogram 4212, Bangladesh.

Abstract: In recent years, technology for the fabrication of mixed-matrix membranes has received significant research interest due to the widespread use of mixed-matrix membranes (MMMs) for various separation processes, as well as biomedical applications. MMMs possess a wide range of properties, including selectivity, good permeability of desired liquid or gas, antifouling behavior, and desired mechanical strength, which makes them preferable for research nowadays. However, these properties of MMMs are due to their tailored and designed structure, which is possible due to a fabrication process with controlled fabrication parameters and a choice of appropriate materials, such as a polymer matrix with dispersed nanoparticulates based on a typical application. Therefore, several conventional fabrication methods such as a phase-inversion process, interfacial polymerization, co-casting, coating, electrospinning, etc., have been implemented for MMM preparation, and there is a drive for continuous modification of advanced, easy, and economic MMM fabrication technology for industrial-, small-, and bulk-scale production. This review focuses on different MMM fabrication processes and the importance of various parameter controls and membrane efficiency, as well as tackling membrane fouling with the use of nanomaterials in MMMs. Finally, future challenges and outlooks are highlighted.

Keywords: membrane; mixed-matrix membranes; MMMs; fabrication; membrane fouling; nanomaterials; phase-inversion process; interfacial polymerization; electrospinning

Citation: Siddique, T.; Dutta, N.K.; Choudhury, N.R. Mixed-Matrix Membrane Fabrication for Water Treatment. *Membranes* **2021**, *11*, 557. https://doi.org/10.3390/membranes 11080557

Academic Editors: Mohammad Peydayesh and Akihiko Tanioka

Received: 8 March 2021
Accepted: 16 July 2021
Published: 23 July 2021

Publisher's Note: MDPI stays neutral with regard to jurisdictional claims in published maps and institutional affiliations.

Copyright: © 2021 by the authors. Licensee MDPI, Basel, Switzerland. This article is an open access article distributed under the terms and conditions of the Creative Commons Attribution (CC BY) license (https://creativecommons.org/licenses/by/4.0/).

1. Introduction

Membranes can be described as films that act as selective barriers between two adjacent phases that allow the transportation of substances from one compartment to another [1]. Membranes play a vital role in separation technology, as well as in energy applications. Membranes are mostly polymer-based, which is adjusted by their synthesis process for the separation of specific substances, and results in efficient cost-effective separation technology with high performance. However, polymer-based membranes have some limitations due to their unavoidable built-in disadvantages, such as poor chemical and physical resilience.

Mixed-matrix membranes (MMMs) are an important class of organic–inorganic nanocomposite membranes with dispersed nanoparticles in polymeric films. Mixed-matrix membranes are based on either classical porous fillers such as zeolites, porous silica and carbon molecular sieves, or nonporous fillers such as graphene oxide, which has the ability to modify the free volume of a polymer by altering the molecular packing of the polymer chains in the membrane. The typical features of nanoparticles, such as stability, surface-area-to-volume ratio, surface charge, etc. [2], make them excellent candidates for inclusion in polymers for biomedical and environmental applications, including conventional water-treatment processes [3].

In the field of functional membranes, the use of a wide range of nanoparticles and the combination of them with other engineered novel materials gives great scope for engineering the shape and structure of the membranes with the desired performance. As a result, the use of mixed-matrix membranes (MMMs) is under development, in which nanoparticles are used as the filler materials in the polymeric matrix of MMMs [4] for applications such as water filtration, gas separation, fuel-cell application, and pervaporation [5–7]. MMMs have been developed substantially as per their applications, and new types of applications of MMMs also have been introduced in the past decade by incorporating inorganic nanomaterials such as metal oxides, including zinc oxide (ZnO) [8], titania (TiO_2) [9], iron oxides (Fe_2O_3, Fe_3O_4) [10], zeolite [5], silica [11], carbon nanotubes [4], graphene [12], graphene oxide (GO) [13], and metal–organic framework (MOF) [14] as fillers in the polymer matrix.

Currently, MMMs are fabricated using a wide range of fabrication processes based on the membrane materials and their applications. As the effective use of MMMs is increasing due to their various attractive properties, worldwide research on MMMs has experienced exponential growth, as indicated by the number of publications on MMMs in last 20 years (Figure 1).

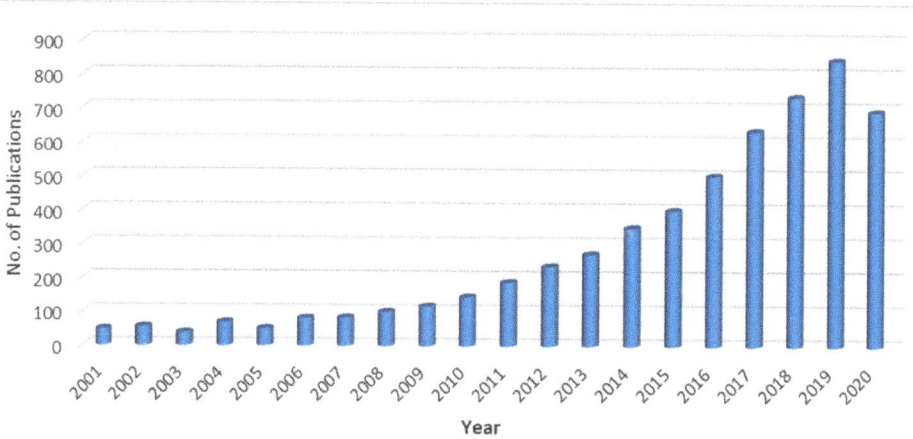

Figure 1. The number of publications each year since 2001 based on the keyword "Mixed matrix membrane" in the Web of Science database (data collected on 7 October 2020).

Due to the significant roles of various fabrication processes on MMMs' properties, the central focus of this review is the fabrication strategy of mixed-matrix membranes for water purification. We begin our discussion with the crucial issue of membrane fouling and ageing and the use of nanomaterials in membrane technology to address the issue, and then describe various fabrication strategies of MMMs, along with parameters that control membrane fabrication.

2. Membrane Fouling and Ageing: Major Challenges for Water-Separation Membranes

In general, membrane fouling occurs when undesirable particles, macromolecules, colloids, or salts are deposited on the surface of the membrane or inside the membrane's pores. Membrane fouling can be subdivided into a few categories such as inorganic fouling, organic fouling, and colloidal/biocolloidal fouling, based on the membranes' separation processes and the foulants' chemical properties [15,16].

Inorganic fouling occurs due to the higher concentration of inorganic salts, such as sulfates, carbonates of sodium, calcium, etc., mainly when their presence in the solvents is beyond the solubility limits and results in precipitation on the membrane surface or into the pores of the membranes [17]. Organic fouling occurs when irreversible and strong foulants like humic substances, proteins, and polysaccharides are deposited on membrane surfaces [18]. In the case of surface-water, brackish-water, and seawater treatment, the main organic foulant is natural organic matter (NOM) [19]. When the membrane fouling occurs due to the deposition of colloids and the suspension of the nanoparticles or microparticles, it is known as colloidal fouling. There are three types of colloids [20,21]: organic colloids such as natural organic matter, proteins, etc.; inorganic colloids such as SiO_2, iron oxides, and hydroxides of iron and heavy metals; and biocolloids such as viruses, bacteria, and other types of microorganisms. Biocolloid-induced membrane fouling is also called biofouling [22], and is caused by a range of bacteria like Aeromonas, Corynebacterium, Bacillus, Flavobacterium, Pseudomonas, and Arthrobacter, and also by fungi like Trichoderma, Penicillium, and other eukaryote microorganisms [23]. As a result, this foulant layer affects the permeate flux in two different ways [24,25]: first by creating an additional hydraulic resistance that results in low water flux and membrane permeability at a fixed applied pressure, which can be overcome by applying higher pressure; and also by the formation of a porous cake layer inside the unstirred cake layer, resulting in a higher concentration polarization, which leads to higher solute concentration on the membrane surface, as well as an increase in the osmotic pressure of the membrane surface and a decrease in the membrane flux.

Thus, it is widely recognized that the adherence of organic compounds and biocolloids to the surface of the membrane is the key parameter for the fouling of the membrane, and this adherence ability of the foulants is influenced by hydrogen bonding, London–van der Waals attractions, and hydrophobic and electrostatic interactions [18,26]. From the above discussion, it is evident that the inhibition or minimization of the fouling process might be possible by preventing the adhesion interactions between the membrane and the foulant. This could be possible through the development of MMMs with appropriate physiochemical properties, which could combine an efficient separation process with lower membrane fouling.

Additionally, various hydraulic cleaning procedures have been introduced for reversing or reducing membrane fouling [27]. Membrane backwashing with clean water is a common practice for foulant removal. After repeated filtration and backwash cycles, some materials are adsorbed on the membrane surface and need to be washed by a cleaning agent like hypochlorite for ultrafiltration membranes, as they cannot be removed otherwise [27]. Long-term exposure to foulants and cleaning agents has been reported to irreversibly change the performance and characteristics of membranes; these irreversible changes are defined as membrane ageing [28]. The characteristics of membranes are mainly chemical composition, pore size, etc.; and membrane performance factors are fouling rate, clean membrane resistance, etc. The main limitation is that complete full-scale ageing studies need many years of observation and cannot be controlled rigorously [29].

2.1. Effect of Membrane Surface Properties on Fouling and Ageing

The interactions between a membrane and foulants are determined by the membrane's surface properties such as hydrophobicity or hydrophilicity, surface charge, and surface roughness [30,31].

2.1.1. Hydrophilicity and Hydrophobicity of Membrane Surfaces

Usually, a membrane's hydrophilicity or hydrophobicity is evaluated with a wettability study using contact-angle measurement [32]. The commercial membranes are mostly fabricated from hydrophobic polymers with high thermal, chemical, and mechanical stability, including polysulfone (PSF), polyethersulfone (PES), polyvinylidenefluoride (PVDF), polyacrylonitrile (PAN), polypropylene (PP), polyethylene (PE), and polyamide (PA) [1]. These polymers exhibit a high contact angle, which leads to the adsorption of different solutes from the feed. It is established that a higher mass per unit area of hydrophobic solute is adsorbed by membranes with high contact angles than that by the membranes with a lower contact angle [33]. On the other hand, hydrophilic membranes attract fewer charged inorganic particles, microorganisms, and organic substances, and result in less fouling [34,35].

2.1.2. Surface Charge

In the case of charged foulants, membrane fouling can be controlled by the electrostatic charge of membranes. Membranes possessing the same charge as that of the foulants will reduce membrane fouling due to electrostatic repulsion occurring between the foulant and the membrane, which prevents foulant deposition on the membrane (Figure 2) [36,37]. Therefore, fouling can be reduced by incorporating ionizable functional groups on the surface of the membrane. For example, in protein filtration, when the protein is negatively charged at neutral pH, a negatively charged membrane surface could be a better choice [1]. Similarly, for organic compounds with a positive charge, the positively charged membrane surface is the solution for low membrane fouling [38]. So, low-fouling membranes could be fabricated and developed by considering the potential foulant's charge on the membrane surface and inside the membrane pores from feed streams.

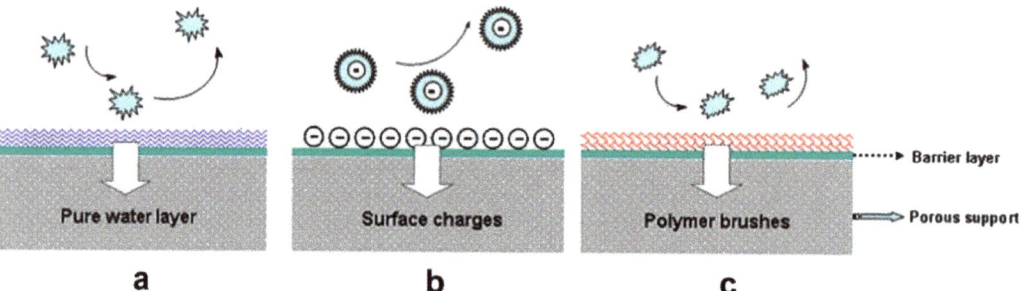

Figure 2. Schematic representation of various antifouling mechanisms with composite membranes: (**a**) thin layer of bounded water, (**b**) electrostatic repulsion, and (**c**) steric repulsion (adapted with permission from [39]).

2.1.3. Surface Roughness

Membrane fouling and surface roughness are strongly related to each other in nanofiltration (NF) and reverse osmosis (RO) membranes. Smooth and hydrophilic cellulose acetate (CA) RO membranes have less tendency toward colloidal fouling than hydrophobic and rough PA membranes [38]. Table 1 shows the relationship between surface roughness and relative fluxes for filtration of a sodium chloride solution containing silica particles with commercial NF (Osmonics HL, Dow-FilmTec NF-70) and RO (Trisep X-20, Hydranautics LFC-1) membranes. From the tabulated data of their flux and surface-roughness values, it is clearly visible that the flux decreased with the increase of surface roughness of the membrane during the filtration process. The increase in membrane surface roughness also led to an increase in the total surface area, resulting in more foulant attachment on the surface, and a ridge–valley structure also favoring the accumulation of foulants at the membrane surface. Using atomic force microscopy (AFM), Vrijenhoek et al. [40] showed that colloidal particles mostly accumulate in between the valleys of rough membrane surfaces, which results in valley clogging and causes lower flux and permeability than the membranes with smooth surfaces.

Table 1. Correlation between the surface roughness of commercial RO/NF membranes and their relative fluxes for the filtration of a 0.05 M NaCl solution containing 200 mg/L silica particles (0.10 μm); pH = 6.8. Flux-decline values determined for 10 L of permeate volume filtered *.

Membrane Type	Flux Decline, J/J$_o$, %	Average Roughness, nm	RMS Roughness, nm
Osmonics HL	13.9	10.1	12.8
Trisep X-20	38.3	33.4	41.6
Dow NF-70	46.9	43.3	56.5
HydranauticLFC-1	49.3	52.0	67.4

* Adapted with permission from [40].

Considering all the above points, it is clear that the top membrane layer is the key area to control the fouling process, so the main goal could be the surface modification of the membrane to develop a low-fouling composite membrane by introducing polymer brushes and charged groups on the membrane's surface, as well as hydrophilization and creating smooth surfaces, which would minimize the undesirable interactions between the foulants and the membrane surface for low or zero fouling of the membrane.

3. Mixed-Matrix Membrane Materials

3.1. Polymers

3.1.1. Glassy and Rubbery Polymers

In water-treatment processes, various polymers have been used in MMMs; some polymers employed are rubbery (e.g., polyethylene oxide) [41], but most are glassy (e.g., aromatic polyamides, cellulose acetate, and polysulfone). Classifying membranes for water-treatment processes as rubbery or glassy can be complex, since they are operated under hydrated conditions and can absorb substantial amounts of water (i.e., ~10–50 vol% water) [42–44].

Recently, ion and water transport in glassy hydrated polymers has been reported, and has become a topic of interest in the membrane field [45–49]. Xie et al. measured water and salt transport in a disulfonated poly (arylene ether sulfone) copolymer (i.e., BPS-32) [45]. BPS-32 was synthesized in the potassium counter-ion form (K) and acidified to the acid form (H), either in solid state or in solution, and subjected to various ion-exchange steps and thermal treatments. Due to its relatively high T_g (278 °C), the membrane remained glassy upon hydration, and therefore its processing history had a profound impact on its water and salt transport properties.

More recently, Chang et al. prepared two chemically similar copolymers, rubbery 2-hydroxyethyl acrylate-co-ethyl acrylate (HEA-co-EA) and glassy 2-hydroxyethyl methacrylate-co-methyl methacrylate (HEMA-co-MMA), to probe the impact of polymer backbone dynamics on ion and water transport properties [48,49]. Both had similar and relatively low water contents (~8% by mass). However, the rubbery membrane had salt permeability coefficients roughly 2–3 times higher than those of the glassy membrane. In a later study, Chang et al. reported water dynamics and tortuosity in the same membranes over several length scales [49]. Using pulsed-field gradient nuclear magnetic resonance (PFG nmR), they measured water diffusivity as a function of diffusion encoding time. The longer the diffusion encoding time, the greater the length scale over which diffusion was measured. Water-diffusion coefficients decreased with increasing encoding time, plateauing at long times as water-molecule diffusion became increasingly hindered by the polymer segmental obstructions on longer length scales. The long-duration plateau value of water diffusivity was regarded as equivalent to the value observed in measurements of bulk-transport properties [49]. Salt solubility and diffusivity were measured via equilibrium and kinetic desorption techniques, respectively. Equilibrium water solubility was also measured. Using the solution-diffusion model, water and salt permeabilities were calculated from these data. Water and salt diffusivity and permeability were lower in the glassy polymer than in the rubbery polymer. However, water/salt selectivity was enhanced in the glassy membranes, corroborating the enhanced size sieving observed in their earlier study [48]. This result was mainly attributed to enhanced diffusivity selectivity in the glassy polymer, since salt solubility was similar in both polymers.

3.1.2. Modification of Polymers

Chemical Cross-Linking

In many cases, membrane materials have reactive functional groups that can be linked through covalent bonds by applying a suitable cross-linker, which gives a remarkable scope of membrane fabrication using the chemical cross-linking process and for modifying polymers [50–59]. This chemical cross-linking method is used for a membrane's mechanical strength enhancement or swelling reduction, as well as the increase of a specific solutes' selectivity with better solvent permeability depending on the applications [60–62]. The cross-linking medium, the cross-linker's concentration and molecular structure, and the reaction time/temperature mainly influence the cross-linking degree, as well as the charge density, which can be confirmed by Fourier transform infrared spectroscopy (FTIR) [50,58,63]. A polyvinyl chloride (PVC) membrane has been cross-linked with an activated-carbon loaded 4,4′-oxidianiline to prepare the MMM for separation technology [60].

Chemical Grafting

Chemical grafting on a membrane surface can be performed by growing or grafting another polymer onto the surface. The hydrophilicity, selectivity, and antifouling property improve due to the grafted polymer. There are a few approaches to produce the active sites that can prompt the commencement of the graft polymerization; for example, plasma, UV, and ion-beam irradiation [64–66].

UV photo-grafting is performed on a polyimide membrane's active surface to modify it so it is suitable for wastewater-treatment applications. The outer active surface of a polysulfone UF hollow-fiber membrane was reported to be achieved by UV grafting, in which sodium p-styrene sulfonate (monomer), N,N′-methylene bis acrylamide (cross-linker), and 4-hydroxybenzophenone (photo-initiator) were used. Figure 3 shows a UV-photo-grafting setup in which the support layers of hollow fibers are wetted by water and immersed in a monomer solution. At that point, the fibers pass through two UV polychromatic lamps [67].

Graft polymerization of a methacrylic acid monomer was reported to contribute to membrane hydrophilicity and negatively charge the membrane surface, as it could eliminate the disrupting endocrine chemicals and active pharmaceutical compounds [68]. Furthermore, the introduction of a redox reaction at the initial stage of surface grafting also offered hydrophilicity, and the redox reaction could be achieved in aqueous media at room temperature without any external activation [66]. Additionally, the concentration of the monomer needed to be higher due to the slow reaction kinetics of the redox initiation [69]. Commercial polysulfone (PSF) has been grafted by poly(polyethylene glycol) methyl ether methacrylate (PEG) side chains to improve the interfacial interaction with zeolitic imidazolate framework-8 (ZIF-8) nanoparticles to prepare the desired MMMs [70].

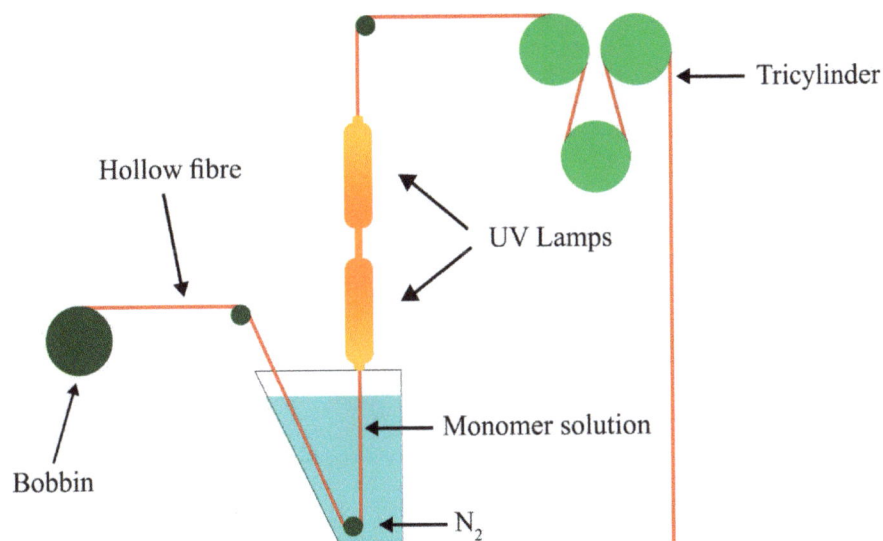

Figure 3. UV-photo-grafting setup for hollow-fiber membrane fabrication.

3.2. Nanoparticles (NPs)

Surface modifications of polymer membranes have led to various low-fouling membranes, and in some cases proved feasible for commercial purposes. However, the use of nanoparticles (NPs) in the membrane could be a better strategy for preparing low-fouling membranes in a simpler way with a long durability. The addition of a large variety of nanoparticles into the polymeric membrane has been extensively explored, leading to mitigation of membrane fouling with longer durability and high permeate flux [71,72]. The successful development of MMMs depends strongly on the polymer matrix selection, the inorganic filler, and the interfacial interaction between the two phases [73]. The selection of suitable types of inorganic filler and their surface modification dictates an MMM's overall performance. Various surface-modification strategies have been used to maximize the interfacial interactions. The superior permeability and selectivity of inorganic membranes with the processability of polymeric membranes are combined in MMMs to achieve synergistic performance, in which the rigid, porous-type inorganic NPs provide desirable properties, and the polymeric phase enables the ideal membrane formation, hence solving the issue of brittleness inherently obtained in the inorganic membranes [74].

3.2.1. Metal Oxides

Amongst various metal oxide nanoparticles, titanium dioxide (TiO_2) is very attractive due to features like ease of preparation, stability, and commercial availability, and membrane fouling could be significantly reduced by introducing TiO_2 into the polymer matrix of a membrane [9]. Additionally, the hydrophilicity and the free water fraction also increased with the deposition of TiO_2 nanoparticles on the polymer membrane surface. Studies on the effect of various sizes of TiO_2 nanoparticles in a hydrophobic polyvinylidenefluoride (PVDF) membrane revealed that the fouling activity of the PVDF membrane could be significantly improved using smaller nanoparticles [75,76], as this hydrophilic modification of PVDF membranes actually decreased the adsorption and deposition of hydrophobic organics on the membrane surface. For example, TiO_2 in polyvinyl acetate not only decreased the membrane-fouling activity, but also improved the thermal stability, which was determined by the increase in the glass transition temperature [77].

Silica nanoparticles also showed the same trend in polyester urethane and polyether urethane-based membranes [11]. Silica nanoparticles have shown performance enhancement of polydimethylsiloxane (PDMS) membranes in pervaporation [78], resulting in improved selectivity of the membranes in pervaporation as the polymer chains became more rigid, and the polymer-free volume was also decreased.

Zinc oxide (ZnO) is used as the filler material in membranes for photo-degradation of organic pollutants and dyes in water and wastewater, and provides antibacterial properties [8]. It has also good electrochemical activity [79].

3.2.2. Magnetic Nanoparticles

Nowadays, magnetic nanoparticles are considered as potential candidates for MMMs [80]. Iron-based magnetic nanoparticles have been studied for a vast number of environmental applications, as they also have the ability of bacterial inactivation [81,82]. Fe_3O_4 has been used as filler material in mixed-matrix membranes due to its attractive features for various applications such as oil–water separation [10], dye and magnetic-particle removal [83], etc.

3.2.3. Carbon-Based Nanoparticles

Carbon-based nanomaterials are also considered as an efficient family of filler materials for MMMs due to their improved chemical and mechanical properties and cost-effectiveness. Among them, graphene oxide (GO) has been explored extensively as a filler material in the polymer matrix for the fabrication of polymeric nanocomposite membranes [84–87]. GO is a two-dimensional material with one-atom thickness, resulting in ultrafast water transport across the GO nanocomposite membrane as it forms interconnected nanochannels [88]. The functional groups such as hydroxyl (—OH), carboxyl (—COOH), epoxide, and C=C on the GO surface offer excellent hydrophilic, antifouling, and antibacterial properties [12,89–91].

3.2.4. Zeolites

Mixed-matrix membranes with zeolite fillers have attracted attention due to their excellent advantages, such as high permeability and improved selectivity [92]. Zeolite–MMMs could be considered ideal for the purification industry, since they combine the properties of a polymeric matrix and zeolite inorganic fillers [93]. Nevertheless, only a few studies have been performed on zeolite–MMMs for water treatment; it was determined that the size of zeolite should be designed to match the expected polyimide active film thickness, thereby providing a preferential flow path through the nanochannels of zeolites [94,95]. Natural zeolite can readily form a suspension to coat the membrane as a support [96]. In another study by Damayanti and coworkers, zeolite-based membranes demonstrated excellent performance and high efficiency for removal of micro-pollutants for laundry-wastewater treatment [97]. Membrane performance was measured based on the flux and rejection values. They studied the superior ability of zeolite membrane to treat laundry wastewater as determined by turbidity measurements and phosphate removal as the two significant parameters. More importantly, another advantage of zeolite-based nanomembranes is that such membranes show an enhanced hydrophilicity when zeolites are used, since they are hydrophilic in nature, which in turn contributes to the enhanced removal of pollutants from wastewater. In addition, zeolite membranes showed improved separation performance and antifouling properties, and the structure and surface properties of the membrane's thin-film layers were modified [98,99].

3.2.5. Metal–Organic Frameworks (MOFs)

Metal–organic frameworks (MOFs) are a unique family of nanoparticles used with membranes for the enhancement of their separation performance, as well as in pervaporation to recover the bioalcohols [14]. MOFs decrease the ageing of the MMMs due to their good compatibility and interaction with the polymer matrix, which results in restrictions of chain mobility (one of the main causes of ageing) [100]. MOFs include ZIF-8 [101–103], HKUST-1 [103,104], and UiO-66 [100,105–107], mostly either as cast or modified [108]. MMMs with inorganic fillers or nanoparticles often have weak polymer–filler interfaces due to the lack of compatibility between the two components, which can create an adverse effect. MOFs containing organic functionality in their bridging ligands can potentially interact favorably with the organic functionality in polymers. However, the organic functionality does not completely eliminate this compatibility issue due to the rigid, crystalline nature of MOFs. Therefore, strategies to improve interfacial interactions, such as chemical and physical interactions, pre- and post-synthetic modifications to MOF ligands, chemically functionalizing the polymer, and employing cross-linking-type reactions to tether the MOF frameworks to the polymer, have been pursued [77–85].

Nanoparticles are also incorporated in membranes for pervaporation applications. As an example, for ethanol dehydration, phosphotungstic acid ($H_3PW_{12}O_{40}$) nanoparticles were added in a sodium alginate/poly(vinyl pyrrolidone) polymer blend [109]. Silica nanoparticles have shown performance enhancement of polydimethylsiloxane (PDMS) membranes in pervaporation [78], resulting in improved selectivity of the membranes in pervaporation as the polymer chains become more rigid, and the polymer-free volume was also decreased.

There is a large body of work using nanoparticles and their surface modification, leading to better properties in many fields, as the general trend of using nanoparticles is to improve and maintain the permeability of liquids and gases and to enhance the desired separation of the membranes. However, the mechanism behind these results was not studied extensively. It is noteworthy that the nanoparticles in the matrix influence the morphology and free volume of the membranes. So, the nanoparticles are used in the membrane for their performance enhancement, and the fabrication of mixed-matrix membranes with nanoparticles will be discussed in the following sections.

3.2.6. Loading or Addition of Nanoparticles in a Polymer Solution

MMMs are the combination of two phases: the polymer matrix and the filler material, such as NPs. Therefore, the mixing of NPs in the polymer matrix is an important part of MMM fabrication, as the homogeneous dispersion of NPs in polymer matrix needs to be ensured for good-quality membrane fabrication. To obtain this, preparation of a homogeneous solution of NPs and polymer is required, which can be done using one of the three established processes described below.

1. NPs are added to the solvent first and stirred for a predetermined time to prepare a well-dispersed solution, followed by the addition of a polymer in the dispersed solution [110–127].
2. The polymer is added to the solvent first and stirred for a specific time, and then the NPs are added to obtain the desired solution for MMM preparation [128–137].
3. The dispersed solution of NPs and the polymer solution are prepared separately in this process, and then the nanoparticle solution is added to the polymer solution [87,138–141].

Among these methods, the first and third methods are used for better distribution of inorganic particles because in a dilute suspension, the particles are prevented from agglomerating by a high shear rate during stirring, while the second method is commonly used for nanoparticle distribution in the polymer matrix [142].

4. Fabrication Processes of MMMs

Figure 4 shows the various membrane-fabrication processes that will be discussed in this review. The improvement in functional properties brought about by forming mixed-matrix membranes or nanocomposite membranes can be grouped in two categories: physical mixing and in situ synthesis [143]. The physical mixing method is very convenient to operate at a very low cost in large-scale production; as a consequence, it has been used extensively to fabricate nanocomposite MMMs. For any inorganic nanomaterials, the nanofillers and polymer dope typically are prepared independently and mixed using the solution, mechanical agitation, fusion, emulsion, etc. [144,145]. Inorganic particle deposition or direct coating onto the membrane surface could also be used to fabricate MMMs. Nonetheless, it is difficult to control the nanoparticles' distribution on or in the polymer matrix during MMM fabrication through the direct mixing method of polymers and nanofillers. The interfacial adhesion of nanoparticles with the polymer can lead to larger aggregates during mixing, thus noticeably diminishing the advantages of the nano dimensions. In addition, polymer degradation upon melt compounding and phase separation of nanoparticles from the polymer phase is sometimes detrimental. The uniform dispersion of nanoparticles on or in the polymer matrix can be achieved by adjusting different processing parameters like shear force, time, and temperature, etc. [146], and the use of dispersing agents could be a promising way of obtaining a well-dispersed membrane [147].

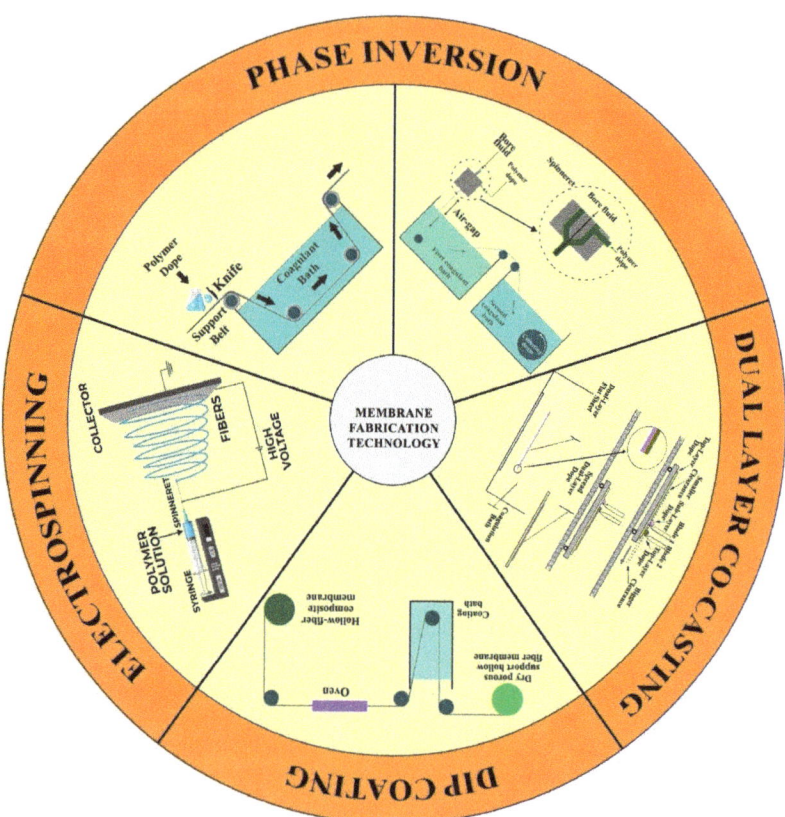

Figure 4. Various membrane-fabrication approaches.

The in situ synthesis process is also used for MMM fabrication, as some compounds like halides and sulfides can be easily and directly synthesized inside the polymer matrix. This in situ process has three categories, which are illustrated in Figure 5 in detail.

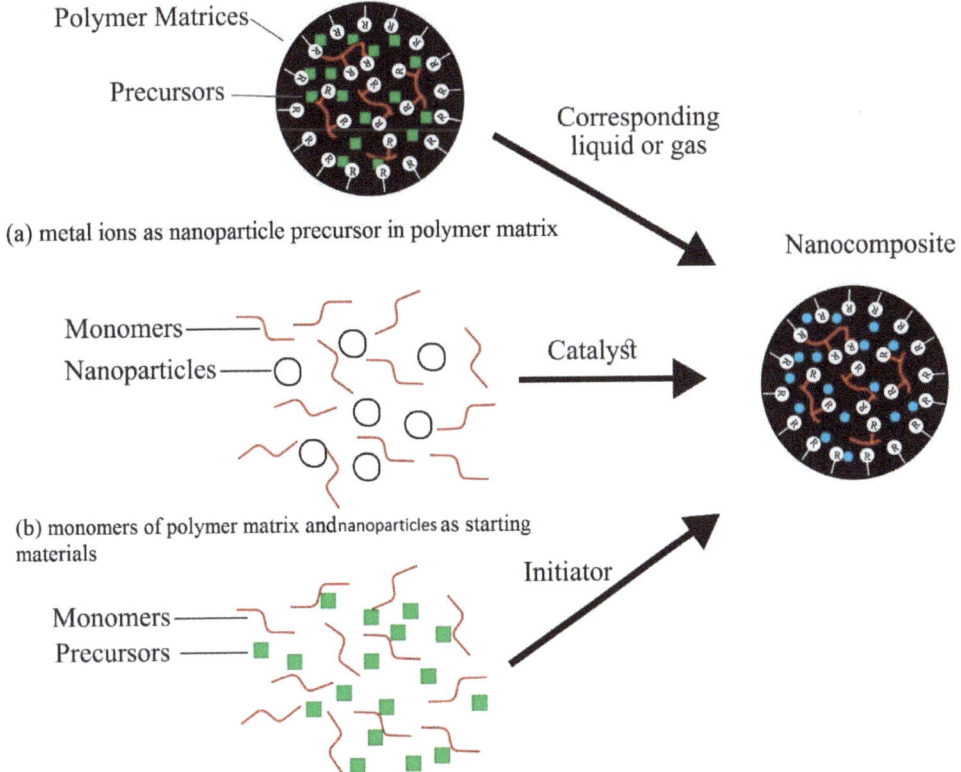

Figure 5. In situ synthesis process of mixed-matrix membranes.

(a) A precursor solution of metal ions and polymer is exposed to the appropriate liquid or gas, which results in the in situ synthesis of nanoparticles in or on the polymer matrix with a uniform distribution [148–150]. A sol–gel method has been developed based on this for fabricating polyimide-based MMMs, in which titanium alkoxide solution was used as the precursor solution of TiO_2 and modified by acetic acid [151].

(b) Another way is to start the synthesis with the solution of a monomer of the targeted polymer matrix and nanoparticles [152,153], in which polymerization takes place with the supplied desired catalyst at appropriate conditions just after the nanofillers dispersion into the monomer solution. This method allows the in situ nanocomposite synthesis of desired physical properties with a lower agglomeration tendency of the filler materials in the matrix.

(c) The other synthesis process is the combination of the above two, in which the precursor of desired nanoparticles and the monomers are dissolved in an appropriate solvent in the presence of an initiator for the in situ preparation of both the polymer and nanoparticles [154,155]. Based on this mechanism, a polyamide-based nanocomposite thin-film reverse-osmosis (TFN PA RO) membrane was synthesized from the dispersion of prepared zeolite in the trimesoyl chloride (TMC) solution [156].

Table 2 shows a list of membranes prepared according to various fabrication processes and their basic properties, and Table 3 compiles the merits and disadvantages of various fabrication processes. In the next section, we will discuss the main membrane-fabrication processes.

Table 2. Basic properties of some membranes with their fabrication process.

Membrane-Fabrication Process	System	Driving Force	Membrane Properties	References
Phase inversion	PS/PVP/MXene nanosheets	Solvent and non-solvent interaction (NMP vs. water)	Porosity—79.4% Pore size—29 nm	[157]
Phase inversion	Polyimide-GO	Solvent and non-solvent interaction (NMP vs. water) and solvent exchange (2-propanol)	Porosity—65.3% Pore size—0.69 nm Surface Zeta Potential—37.6 MV	[13]
Electrospinning	PVDF	Voltage difference	Porosity—88% Electrolyte uptake—440% Conductivity—1.88 mS cm^{-1}	[158]
Phase inversion	PVDF-PAN-SiO$_2$	Solubility parameter difference, solvent and non-solvent miscibility	Conductivity—3.32 mS cm^{-1} Electrochemical stability—5 V Electrolyte uptake—246.8% Porosity—78.7%	[159]
Graft polymerization	PMMA–g-PE	Grafting PMMA, results in large uptake of electrolyte	Electrolyte uptake—350% Electrochemical stability—5 V Conductivity—1.3 mS cm^{-1}	[160]
Electrospinning	Polyacrylonitrile/polyurethane	Voltage difference	Electrolyte uptake—776.1% Porosity—90.81% Conductivity—2.07 mS cm^{-1} Bulk resistance—1.2 Ω	[161]
Electrospinning	Poly(phthalazinone ether sulfone ketone)	Voltage difference	Electrolyte uptake—1210% Porosity—92% Conductivity—3.79 mS cm^{-1} Bulk resistance—1.2 Ω	[158]
Electrospinning	PS	Voltage difference	Fiber diameter—470 ± 150 nm Pore size—2.1 μm	[162]
Electrospinning and dip-coating	PEI/PVDF/x-PEGDA	Voltage difference for electrospun PEI/PVDF membrane and coating of x-PEGDA	Fracture Stress—12.1 MPa Pore size—2.56 μm Porosity—64.6% Electrolyte uptake—235.6% Conductivity—1.38 mS cm^{-1}	[163]
Electrospinning	PEI/PVDF	Voltage difference	Fracture Stress—6.6 MPa Pore size—3.11 μm Porosity—83.5% Electrolyte uptake—492.8% Conductivity—1.03 mS cm^{-1}	[163]
Electrospinning and coating	PE–PI–S	Voltage difference and coating	Porosity—60% Electrolyte uptake—400% Conductivity—1.34 mS cm^{-1}	[164]
Electrospinning	PVDF-HFP	Voltage difference	Porosity—70% Electrolyte uptake—247% Conductivity—3.2 mS cm^{-1}	[165]
Electrospinning	Trilayer (PVDF-HFP)/PVC/(PVDF-HFP)	Voltage difference	Porosity—62% Electrolyte uptake—230% Conductivity—1.58 mS cm^{-1}	[165]
Electrospinning	PVDF/SiO$_2$	Voltage difference	Porosity–85% Electrolyte uptake—646% Conductivity—7.47 mS cm^{-1}	[166]

Table 2. *Cont.*

Membrane-Fabrication Process	System	Driving Force	Membrane Properties	References
Electrospinning	Polyamic acid	Voltage difference	Pore size—800 nm Porosity—65.9% Electrolyte uptake—559%	[167]
Electrospinning	SiO_2/nylon 6,6	Voltage difference	Porosity—77% Electrolyte uptake—360% Conductivity—3.8 mS cm^{-1}	[168]
Electrospinning	PVDF-HFP/PEG/PEGDMA	Voltage difference	Electrolyte uptake—212% Porosity—71% Bulk resistance—0.94 Ω	[169]

Table 3. Merits and disadvantages of MMM fabrication processes.

MMM Fabrication Process	Merits	Disadvantages	References
Phase inversion	• Simple process • Economic	• Difficult to produce a pinhole-free membrane	[170–172]
Interfacial polymerization	• Produces a thin active layer with high flux of permeation and desired impurity rejection • Defect-free • Easy to scale up	• Mainly depends on the properties of the monomer, so appropriate monomer selection is a crucial point	[173]
Electrospinning	• Fabricates a membrane with high porosity, larger surface area, and outstanding pore interconnectivity. • A mechanically stable membrane can be fabricated	• Requires high voltage • Uses a solvent • Solubility issue for some polymers in a low-boiling-point solvent.	[174]

4.1. Phase Inversion Process

Phase-inversion is the most popular method to form an asymmetric polymer membrane, and was first developed by Loeb Sourirajan in 1963 [175]. It offers several advantages over other membrane-fabrication methods such as material selection flexibility and the capability of making membranes with different pore sizes (between 1 and 10,000 nm) by varying the process parameters, solvent, and membrane material. The phase-inversion process is also called the phase-separation process, in which a homogeneous polymer solution is separated into two different phases, polymer-rich and polymer-poor, leading to two different layers of the porous structure. The mechanism of phase inversion primarily involves controlled transformation of a polymer solution to a solid state through liquid–liquid demixing, as shown in the ternary phase diagram of a polymer–solvent–nonsolvent system (Figure 6). Thermally induced phase separation (TIPS) and non-solvent-induced phase separation (NIPS) are the two approaches for the separation of a polymer solution. In TIPS, the polymer and solvent are mixed at a high temperature followed by cooling, which results in phase separation, whereas NIPS is a three-component process in which a non-solvent is used with the polymer and the solvent, and the main phase change occurs via the immersion of the polymer solution into the non-solvent [176]. During this immersion, the non-solvent is absorbed by the polymer solution and the volatile solvent is evaporated. An electrolyte membrane of PVDF and PAN polymers in which SiO_2 was used as a nanofiller has been fabricated by phase inversion for lithium-ion batteries [159].

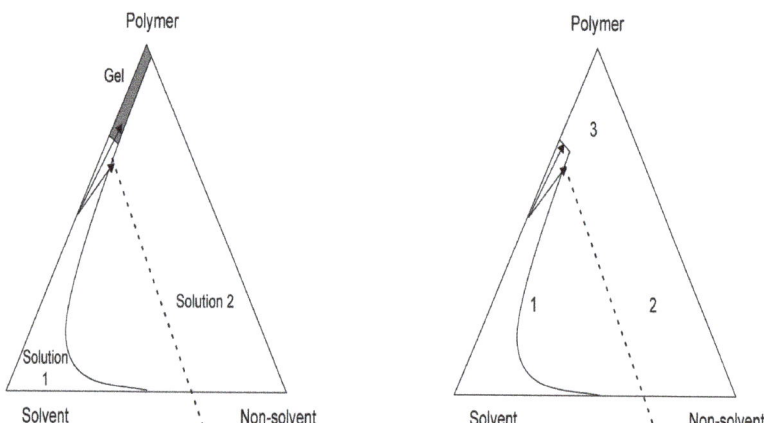

Figure 6. Phase diagram for the phase-inversion process.

Membrane Fabrication through Immersion Precipitation

In the immersion precipitation method, a coagulation bath and a casting knife are used (Figure 7). The prepared homogeneous polymer solution is poured over a non-woven supporting mat, and then the dope is spread to a pre-defined thickness by using the casting knife. Afterward, the membrane is dipped into the bath. Before dipping in the bath, the casted dope is exposed to an ambient environment. The membrane property can be adjusted by controlling the temperature of the coagulant bath, as well as the exposure time in that bath, and the condition of the ambient environment. Although mostly water is used for the coagulant solvent, other non-solvents can also be used.

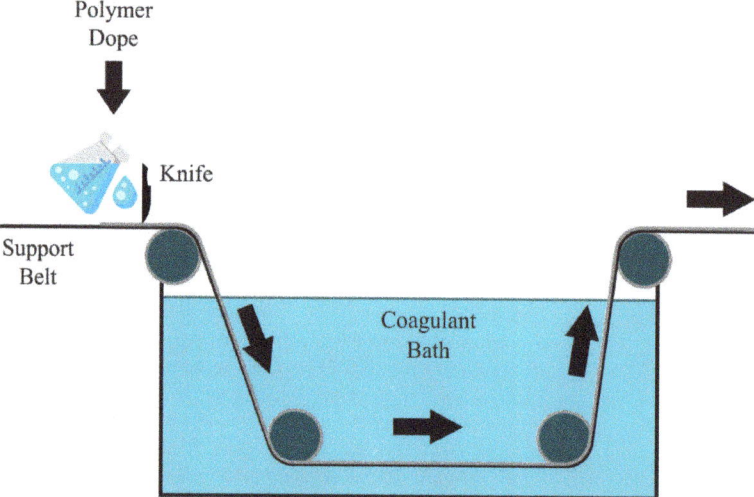

Figure 7. Illustration of membrane casting.

For hollow-fiber membranes [177,178], a bore fluid is required for hollow-fiber spinning as an internal coagulant. The process of hollow-membrane fabrication is complicated, as the phase separation occurs on both the inner and outer surfaces. A hollow-fiber fabrication process is illustrated in Figure 8. Extrusion of the bore fluid and the dope takes place simultaneously from the spinneret, and the pumps are used to control the flow rate. The developing fiber flows through an air gap and is finally immersed in the coagulant. A rotating drum is used to collect the final fiber at a constant speed, but it should be equal to or higher than the speed of free-falling fibers to avoid the coiling of the fiber. Finally, the solidified fiber is collected from the bath, followed by water soaking to remove the remaining solvent. Then the membrane is dried by freeze-drying or solvent exchange to avoid pore collapse during drying [177,178].

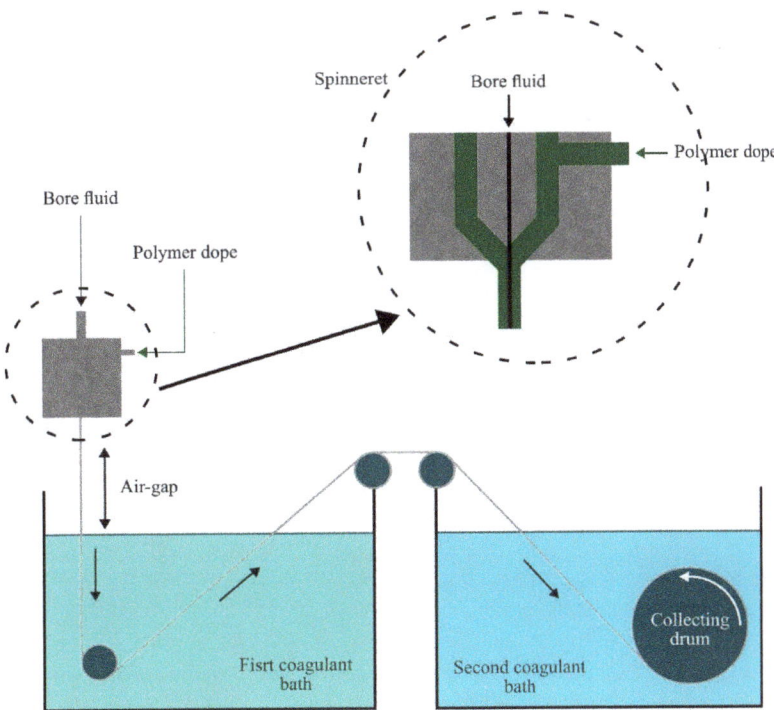

Figure 8. Spinning process of hollow-fiber membranes.

4.2. Interfacial Polymerization

Polyamide membrane development by interfacial polymerization has been recognized as the most regularly utilized method to form superior RO-like and NF-like active layers. Interfacial polymerization uses two exceptionally responsive monomers at the interface of two solvents that are immiscible with each other, one of which should be organic, and other of which should be inorganic/aqueous. There are two types of interfacial polymerization: (1) for drug delivery applications, micro/nanocapsules or micro/nanospheres are produced by dispersing one phase into another as tiny droplets using high-speed stirring [179]; and (2) the common process of introducing a continuous layer on a support, leading to a thin film [180,181].

A few types of monomers and prepolymers; for example, piperazine, N,N'-diaminopiperazine, and m-phenylenediamine for amine solution [182,183], and trimesoyl chloride, sebacoyl chloride, and iso-phthaloyl chloride for acyl halides solution [181] can be utilized for interfacial polymerization.

Mixed-matrix interfacial polymerization has been developed to insert nanoparticles throughout the polymer layer. The purpose is to improve the membrane's performance. Super-hydrophilic zeolite nanoparticles are utilized to improve the water permeability with high rejection of salts [156]. Aquaporin-based biomimetic membranes have been fabricated with a similar process, resulting in high separation performances [184].

4.3. Multilayer Polyelectrolyte Deposition

Polyelectrolyte is a polymer containing electrolyte(s) groups in its repeating units. Polyelectrolyte shows charge properties when it dissociates in an aqueous solution or water. The driving force of multi-layer polyelectrolyte deposition on the membrane surface is the electrostatic interaction between the oppositely charged molecules. Scheme 1 shows various polyanions and polycations used for layer-by-layer formation of a polyelectrolyte complex multilayer (PEM). Figure 9 shows such a process, in which it is clear that the deposition of an aqueous polyelectrolyte solution on a porous substrate in the desired sequence could be a facile method of membrane preparation [185].

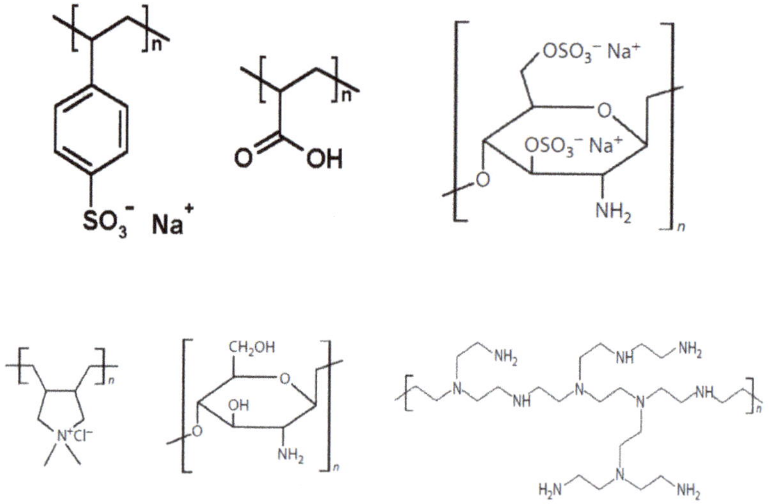

Scheme 1. Commonly used polyanions and polycations for the development of active–selective layer: **Top**: polyanions, from left to right: poly(styrene sulfonate) (PSS) sodium salt, poly(acrylic acid) (PAA), sulfated chitosan (S-Ch); **Bottom**: polycations, from left to right: poly(diallyldimethyl ammonium chloride) (PDADMAC); chitosan(Ch); polyethylenimine (PEI).

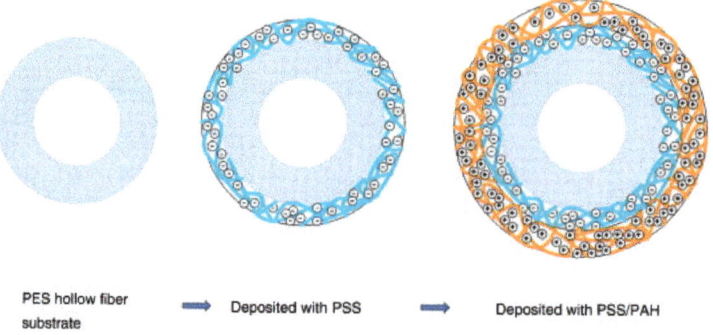

Figure 9. Schematic drawing of multi-layer polyelectrolyte deposition on the outer surface of a hollow-fiber membrane (adapted with permission from [185]).

Multi-layer polyelectrolyte deposition is easy and adaptable for membrane preparation with thinner thickness and containing specific desired layers for high selectivity of the desired content. The function and structure of the layers can be different for specific applications based on the charge density of the polyelectrolytes and their molecular structures. Polyelectrolyte membranes are used in different applications such as forward osmosis [185–187], NF [188], ion exchange [189], pervaporation [190,191], and gas separation [192,193].

Factors Affecting Multi-Layer Polyelectrolyte Deposition

The pH and ionic type of the polyelectrolyte multi-layer play a significant role in the development of a unique film of multi-layer polyelectrolytes. In the event that both permeable substrate and the polyelectrolyte are contrarily charged at a high pH and vice versa, the ideal pH utilized ought to be in the middle of the iso-electric point of the substrate and the polymer, as shown in Figure 10. Accordingly, inverse charges are conveyed by the substrate and the electrolyte [194]. An ultraviolet/ozone (UV/O_3)-cleaned permeable alumina membrane with surface pore measurement of 0.02 µm is becoming attractive as a substrate for polyelectrolyte layer deposition because of its positive charges [195]. Plasma-treated/hydrolyzed polyacrylonitrile and cellulose acetic acid derivatization are negatively charged [186,187]. Furthermore, PES is also appealing as a supporting material in spite of the fact that it is neutral. Hence, the connection of polyelectrolyte layer depends on hydrophobic cooperation [185].

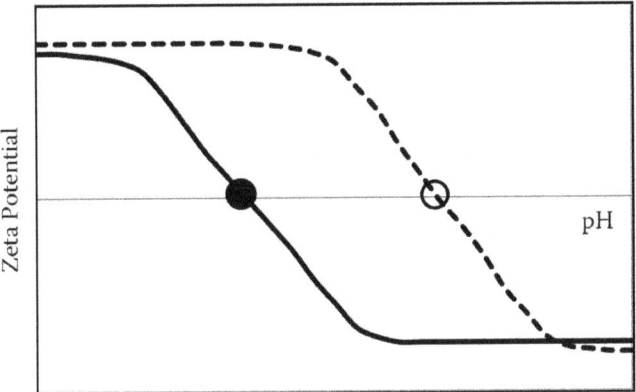

Figure 10. Plot of pH and zeta potential of polyelectrolyte (—) and substrate (—); ●, iso-electric point of polyelectrolyte; and ○, iso-electric point of substrate (adapted with permission from [196]).

Ionic strength of the polyelectrolyte solution can be expanded by including salts. At high ionic strength, the electrostatic repulsion of the polymer chain diminishes with the polymer coils becoming denser, with the deposited layer in collapsed form instead of a flat conformation. Subsequently, it builds the thickness of the individual layer [197]. In the climate of incredibly high salt concentration, just a limited quantity of polyelectrolyte can be absorbed by the substrate because of the opposition to the more modest charged particles from the salts [194]. To improve the density of the polyelectrolyte layers, cross-linking could be utilized to enhance the layers' stability. A cross-linking agent such as glutaraldehyde could be utilized in those cases [186,187]. Another parameter, the charge density of the polyelectrolytes, depends on the molecular structure and the degree of ionization of the polar groups. The charge density of the resultant multilayers, defined as the number of ionic groups per number of carbon atoms in the repeat unit of the polyelectrolyte complex/multi-layer [197,198] often guides the thickness. By adsorbing polyions from salt solutions of varying electrolyte concentrations, the layer thickness can be controlled over a wide range. In addition, consolidation of nanoparticles, such as silver on the active layer of a membrane, can also enhance the antifouling or antibacterial properties of the membrane. The layered structure of a multi-layer polyelectrolyte could improve the stability of the nanoparticles on the membrane surface [188].

4.4. Dual-Layer Co-Extrusion/Co-Casting

Improvement of composite dual-layer membranes is appealing, as beneficial properties of at least two polymeric materials can be consolidated for different applications. The material expense of the superior polymer can be decreased, and the polymer with extraordinary selectivity but poor mechanical strength can be reinforced, by consolidating them with an economical and strong polymer support layer [199,200].

Increasing uses of double-layer membranes include forward osmosis, gas separation, and NF membranes, which are made out of a thick selective layer supported by a porous polymer matrix [51,180,201–204]; and direct-contact membrane distillation, which requires an additional thin hydrophobic layer for wetting prevention and another hydrophilic layer for better water permeability [205].

Hollow-fiber [201,204] and flat-sheet [199,202] membranes are prepared by the dual-layer co-casting method on the basis of same principles, which are the casting of two different polymer solutions or a single-step co-extrusion. Synchronous development of the double-layer structure should be possible by utilizing a triple-orifice spinneret for hollow-fiber membranes, and a double-blade casting machine could be used to prepare the flat-sheet membrane by a co-casting process (Figure 11) [201,202].

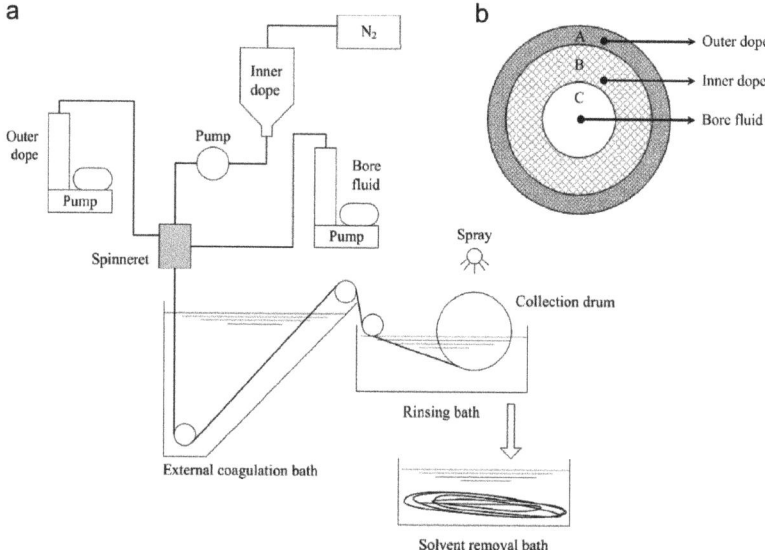

Figure 11. Cont.

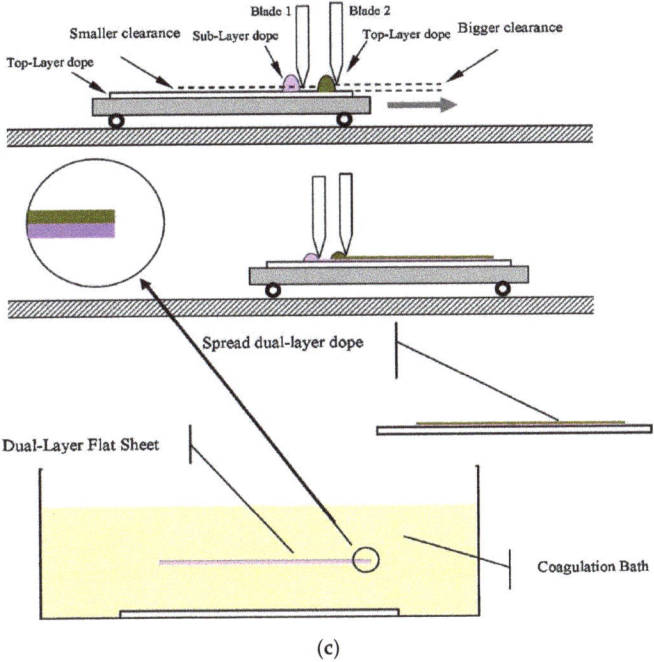

Figure 11. (**a**) Schematic diagram of a dual-layer hollow-fiber spinning process; (**b**) cross section of a triple-orifice spinneret (adapted with permission from [201]); and (**c**) fabrication process of a dual-layer flat-sheet membrane using a double-blade casting machine (adapted with permission from [202]).

Complexity arises while fabricating dual-layer membranes in either a hollow-fiber or flat-sheet configuration because of the involvement of many parameters controlling the thermodynamic properties and the energy of the phase change. This parameter control results in uniform cross-sectional morphology, as well as better lamination between the two layers of the synthesized membrane. The fabrication parameters can be divided into two categories: the chemistry of the polymer solution and the operating conditions. The chemistry of the polymer solution relies upon the polymer concentration and type, the solvent's affinity to the polymer or coagulant, and the concentration and variety of non-solvent additives (or pore formers) [201,202]. Working conditions incorporate an air gap for hollow-fiber spinning, the evaporation time for flat-sheet casting, the composition and temperature of the coagulant, the temperatures of the polymer solution, and the operating temperature [177,199,201,206,207].

4.5. Dip-Coating

In a dip-coating method, the membrane surface is coated by applying a polymer or organic materials. The polymer usually utilized as coating material should have some extraordinary properties; for example, it could be hydrophilic and negatively charged, and attach to the support layer easily. This group of polymers can be prepared by sulfonation; for example, sulfonated PES (SPES) and sulfonated poly(ether ether ketone) (SPEEK). The coating layer may upgrade the performance of the support layer; for example, by giving it a higher strength and better separation properties. Some basic properties should be taken into consideration while choosing the coating polymer; for example, the strength and stability of the polymer, layer-forming capabilities, easy solubility in solvents, cost, and cross-linking capability [208]. Three basic steps in the dip-coating process (Figure 12) are: (1) immersing a dry membrane in a coating solution, (2) permitting the coating material to interact with the substrate, and (3) drying the prepared membrane (Figure 12).

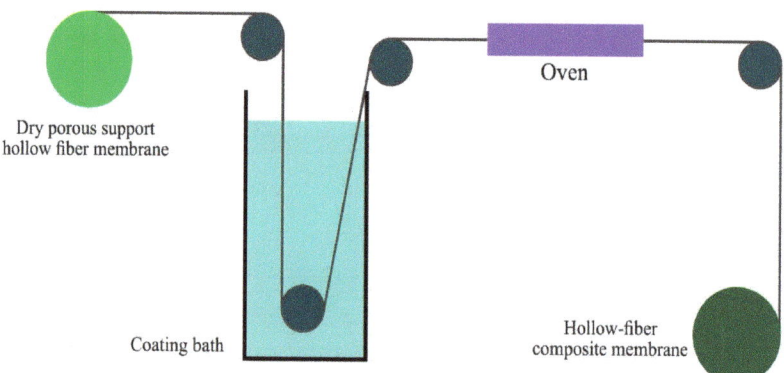

Figure 12. Hollow-fiber composite membrane fabrication by the dip-coating process.

SPES has been used as the selective layer of NF hollow-fiber membranes by dip coating due to its capacity of ion exchange (limit of 0.8 meq/g) and antifouling activities. SPES conveys negative charges due to the presence of a sulfonic acid group in the main chain. The significant disadvantage of this polymer is that it can swell in water easily. When the polymer is dried, the structure of the layer becomes brittle [199]. In addition, NF hollow-fiber membranes have been prepared using PES as the substrate, followed by the dip-coating of SPEEK as the selective layer. The thickness of the coating layer generally relies upon the viscosity of the coating solution, which is impacted by temperature, grouping of the solution, and added substances. At a lower concentration, the viscosity of the solution is low, and as a result, the coating solution will infiltrate to the substrate pores [209].

4.6. Electrospinning

Nanofibrous membranes are in high demand nowadays because of their scaffold structure, larger surface area, and interconnected porosity. Among different fabrication methods, electrospinning is attractive in developing nanofibrous membranes because of its scalability, simple design, and low cost [210,211]. Figure 13 shows a typical electrospinning setup.

Typically, the electrospinning system consists of a high-voltage power supply, syringe pump, syringe, needle, and a conductive collector where the fiber is gathered to make the membrane. Figure 13 represents a basic electrospinning system [212]. It can be classified as vertical and horizontal system based on the ordering of the spinneret. During the electrospinning process, the polymer solution is pumped at a suitable rate from the syringe to make small droplets at the tip of the spinneret. The voltage is supplied in the range of 1–50 kV from the high-voltage power supply, which results in charging of the droplet by the applied electric field, and eventually a solution jet is formed. The droplet is turned into a cone-shaped structure (Figure 14) to aim the solution jet toward the conductive collector. The threshold value of voltage causes the electrostatic force to overcome the surface tension of the droplets, which leads to the formation of jet from the cone's tip. However, an appropriate viscosity is required for a continuous jet of solution by avoiding the Rayleigh instability, which causes breakup into droplets [213]. This jet becomes thinner and dries before being deposited on the collector in fiber form [214].

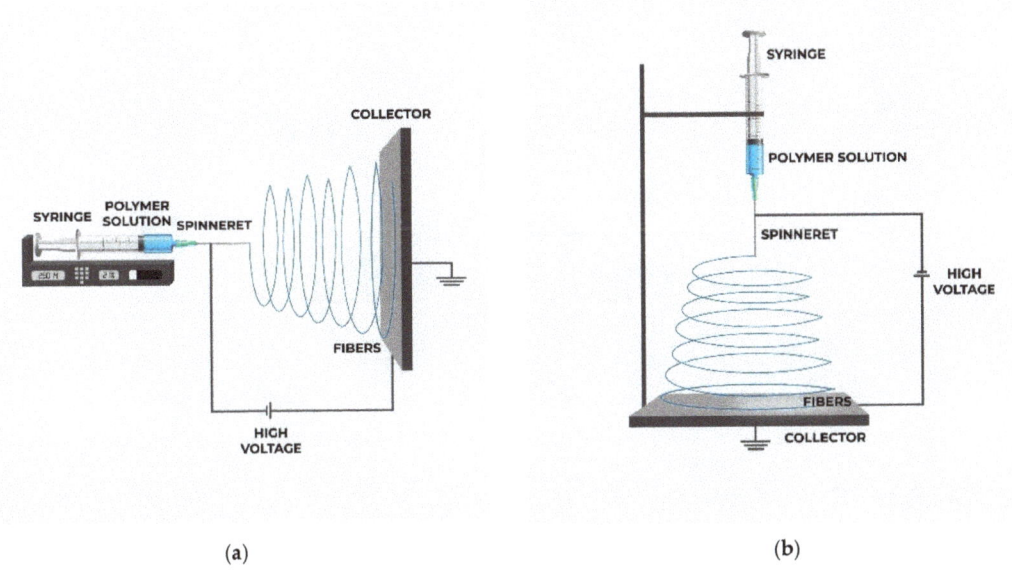

Figure 13. Schematic diagram of electrospinning setup: (**a**) horizontal; (**b**) vertical (adapted with permission from [212]).

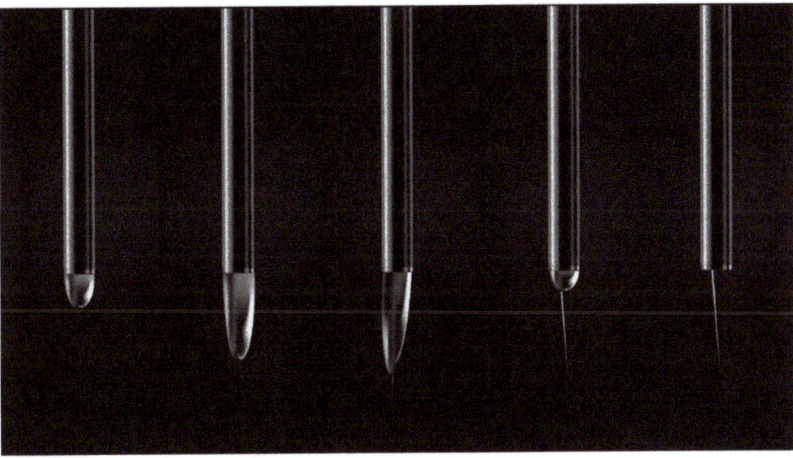

Figure 14. Formation of a Taylor cone with the increase of applied voltage.

In 1930, Formhals illustrated the principle of electrospinning first, though the first patent was obtained in United States earlier (1902) [215,216]. Nevertheless, the electrospinning process received attention after 1990, but it was recognized globally within a short time to prepare the polymer-based nanofibers of different diameters down to a few nanometers. In the last decade, the number of publications on electrospinning is notable (Figure 15).

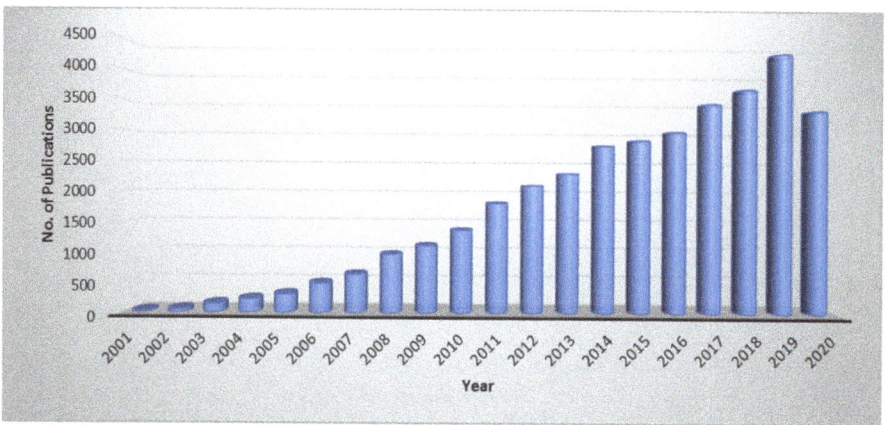

Figure 15. The number of publications each year since 2001 based on the keyword "Electrospinning" in the Web of Science database (data collected on 20 October 2020).

At present, the electrospinning process is more advanced than before, which allows a more controlled property by adjusting the process parameters. Eventually, the electrospinning method will become preferable in different fields of study, such as energy storage, separation and membrane technology, drug delivery, tissue engineering, and so on [217–220]. There is a drive to apply the electrospinning method in large-scale applications. Fortunately, Donaldson and Freudenberg [220] have successfully implemented electrospinning technology in making a filtration membrane.

The most interesting property of the electrospinning technique is the controllability of the fiber diameter by monitoring the variables such as solution concentration, loading of filler material, voltage, flow rate, temperature, and humidity [214]. A wide range of fiber diameters, from micron-sized to a few nanometers, can be achieved. Figure 16 shows a non-woven nanofibrous membrane of polyacrylonitrile [221].

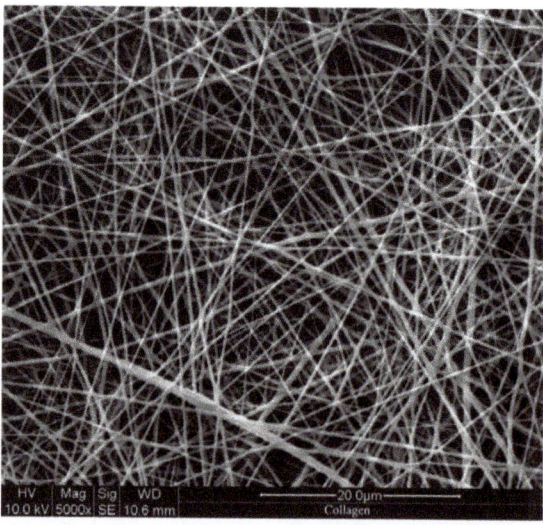

Figure 16. Scanning electron microscopy image of an electrospun polymer: a poly(acrylonitrile) non-woven nanofiber mat produced by electrospinning (adapted with permission from [221]).

The electrospinning method can be applied not only to polymers, but also to metals [222] and ceramics [223] for formation of micro- and nanofibers. However, polymers are mostly being studied, including mixed-matrix polymers containing polymer blends [224], drugs [225], and nanoparticles [226]. Although many polymers are being successfully electrospun into fiber, several polymer/solvent systems are very popular because of their suitable molecular weight, volatility, and conductivity of the solvent. This list includes polyamides [227], polyurethanes [228], polyester [229], poly(ethylene oxide) [230], polystyrene [231], poly(vinyl pyrrolidone) [232], poly(methylmethacrylate) [233], poly(vinyl alcohol) (PVA) [234], poly(lactic-co-glycolic acid) [235], polyacrylonitrile [236], and poly(caprolactone) [237], as well as bio-polymers such as chitosan [238], collagen [239], and gelatin [240].

In the next sections, the effect of properties of polymer solution and process parameter on the properties of electrospun membrane will be discussed. In addition, the roles of temperature and humidity are also mentioned.

4.6.1. Effect of Intrinsic Properties of Polymer Solutions

The properties of a polymer solution largely control the structure of the nanofiber. Currently, a large number of studies have reported the role of solution viscosity, surface tension, concentration, and conductivity on nanofiber fabrication [241–245]. In the next section, the effect of these parameters will be described.

Polymer Concentration and Solution Viscosity

Several research reports showed that the structure and morphology of the electrospun membrane largely depend on the solution viscosity and concentration [214,246–248]. Polymer concentration profoundly influences the surface tension and viscosity of a solution, which eventually controls the development of nanofibers. The low-viscosity solution results in bead-on-string fibers. On the other hand, with increasing viscosity, the shape of beads is changed from globular to a spindle-like structure, which leads to the formation of a uniform fiber [229,249]. However, high viscosity also increases the diameter of the nanofiber. Therefore, it is required to optimize the threshold value to obtain a preferable fiber structure.

Electrical Conductivity

The spinnability of a polymer largely depends on the electrical conductivity of the dope solution, as the rheological behavior largely depends on it [250,251]. The category of solvent and polymer and the concentration of ionizable salts determine the electrical conductivity of the polymer solution [249]. Usually, a highly conductive solution forms a finer fiber and a wide range of fiber-diameter distribution [217,218]. In addition, increased electrical conductivity can help to form a stable Taylor cone that leads to producing a dense scaffold structure [252]. The conductivity can be enhanced by adding ions in the dope solution. Moreover, due to a higher charge density, the smaller ion can create a stronger elongation force on the jet [217,253–255]. Electrical conductivity can also be enhanced by adding a suitable acid with a higher dielectric constant, such as formic acid [256,257].

Surface Tension

Surface tension of the dope solution is an important parameter in tailoring the nanofiber structure. It can be adjusted by adding surfactants [254,258–260]. Lower surface tension forms a stable jet, and consequently, a uniform woven structure is formed. However, a higher amount of surfactant can cause other defects, such as a clustered structure.

Solvent

Solvent plays an important role in determining the morphologies of a nanofibrous membrane. During fabrication of a nanofiber, the solvent is continuously evaporated. Therefore, solvents with different evaporation and solubility rates can change the final structure of the nanofibrous membrane [254,257]. It has been reported that a solvent with low solubility is suitable for electrospinning. Spinnability–solubility maps were used to select a suitable solvent for the polymer [261].

4.6.2. Effect of Electrospinning Process Parameters

The process parameters of the electrospinning technique such as flow rate, applied voltage and collector-to-spinneret distance play important roles in determining the quality of the electrospun membrane. In the following section, the effect of electrospinning parameters on the final product will be discussed.

Applied Voltage

The applied voltage determines the electrostatic force between the spinneret and the collector, and the charge density in the droplets [258]. The fiber diameter decreases with increasing voltage [252]. However, it may cause increased bead structure on the polymer net [262].

Electrode Distance

The distance between the spinneret and the collector defines the intensity of the electric field and the duration of the jet touching the collector. The distance should be enough to allow sufficient time for fiber elongation [263]. The fiber elongation and solvent evaporation can be decreased by decreasing the distance, which leads to formation of a thicker fiber [264]. However, reduced distance also helps to stabilize the solution jet [265], while an inappropriate distance causes formation of beads [263].

Solution Mass Flow Rate

The study of the impact of flow rate on the quality of nanofibers has not been studied extensively. However, Megelski et al. [231] noted that higher flow rates cause formation of thick nanofiber and beads. The fiber diameter is increased because of reduction of charge density of fiber jet [266]. A bead is formed as the unstable jet is formed by the removal of the higher solution from the tip [267].

Ambient Environment

The effect of temperature and humidity on the electrospinning process cannot be ruled out. A lower temperature decreases the evaporation rate of the solvent, and eventually fiber diameter is decreased, as there is more time to be elongated before solidification. On the other hand, at a higher temperature, the diameter of the fiber increases, as the solution jet solidifies faster [257,268]. Moreover, the relative humidity can also have an impact on the fiber properties. Higher humidity can form a finer membrane. On the other hand, a lower humidity increases the fiber diameter [252,257,258,262].

Although extensive research has been done to understand the effect, there is significant space for additional research to reach a better understanding of the possible cause of bead formation and control of the fiber diameter. More comprehensive study is required to control the solution properties in order to understand the effect on electrospinning.

5. Future Directions

Research on the fabrication process of mixed-matrix membranes is ongoing, as they have been found very useful in different applications. Among all the mixed-matrix membranes, nanofibrous-type MMMs are now more popular due to their properties and efficiency. Among all the spinning processes, electrospinning has some great features like high speed, capability, and low cost, resulting in a highly porous patterned nanofibrous polymer membrane [269,270]. The electrospinning process can fabricate a membrane with a larger specific area with smaller pores and fibers within a diameter of 10 to 1000 nm [271]. These unique properties of electrospun nanofibrous membranes make them desirable for a wide range of applications [272], such as SiO_2-incorporated electrospun SPEEK, which has been applied in a fuel cell [273]. Additionally, during the electrospinning process, it is easy to perform the ordering of the polymer, as well as the chain elongation. Considering all the mentioned characteristics of electrospinning, this process could be taken as the latest effective technology for the production of continuous, long-chain, mixed-matrix nanofibrous membranes using a combination of different polymers and nanomaterials for various applications on a large scale [246].

There is an opportunity for developing new technology combining 3D printing and electrospinning in the nanofiltration area, as has been done for biomedical applications (Figure 17). Recently, a 3D-printed mesh reinforcement on electrospun scaffolds was attempted, in which a poly (lactic acid) (PLA) mesh was 3D-printed into an electrospun poly(ε-caprolactone) (PCL) gelatin directly, resulting in better mechanical properties [274]. In Figure 18, the effect of 3D printing on the electrospun scaffold structure is clearly visible.

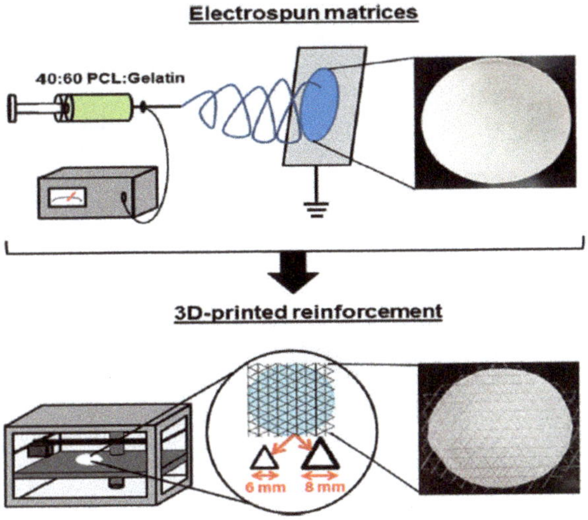

Figure 17. Fabrication of reinforced electrospun scaffolds. Electrospun scaffolds were produced from a 40:60 ratio of PCL:gelatin. The scaffolds were then placed in a 3D printer, and a PLA mesh was deposited onto one side of the scaffold. Two types of 3D-printed meshes were generated: one with a 6 mm distance between PLA struts, and the other with an 8 mm distance between struts (adapted with permission from [274]).

So, it can be concluded that this technique offers the same matrix-like structure with a higher mechanical strength of the electrospun membrane, and these modified and updated 3D-printed electrospun membranes could be used in a new range of membrane applications.

Nanocomposite materials; a combination of graphene, graphene oxides, or metal oxides such as ZnO, TiO_2, etc.; and magnetic nanoparticles, etc., could be better alternatives as filler materials in mixed-matrix membranes for various applications such as heavy metal removal, wastewater treatment, desalination, etc., as some previous research has shown that these types of nanocomposite particles excellently combine the properties that they exhibit individually [275,276]. The synthesis route of these nanocomposite particles is simple as well, using methods such as chemical mixing, chemical precipitation, sol–gel techniques, etc. [275–277].

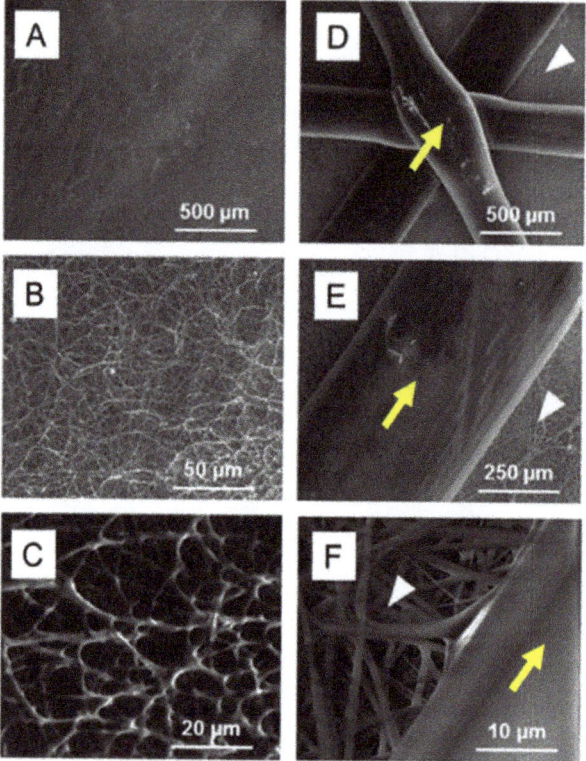

Figure 18. SEM images of reinforced electrospun scaffolds. (**A**–**C**) SEM images of the electrospun side of the reinforced scaffolds. The images show a uniform distribution of randomly oriented fibers. (**D**–**F**) SEM images of the 3D-printed side of the scaffolds. The high-magnification images (**F**) show that there is minimal damage to the electrospun fibers in the immediate vicinity of the 3D-printed PLA mesh. Yellow arrows depict the 3D-printed PLA. White arrowheads depict the PCL:gelatin scaffold (adapted with permission from [274]).

Finally, the demand from the end user based on the applications is the main driving force for obtaining a good market value and establishing a better position in the total global membrane market, including pharmaceutical and biomedical, water filtration and wastewater treatment, textile and metalworking industries, chemicals and petrochemicals, food and beverages, etc. So, the demand for such fabrication technology is also at its peak, and a cost-effective and easier fabrication technology is desirable for bulk and industrial production.

Author Contributions: Conceptualization, T.S. and N.R.C.; validation, T.S. and N.R.C.; writing—original draft preparation, T.S.; writing—review and editing, N.R.C. and N.K.D.; supervision, N.K.D. and N.R.C. All authors have read and agreed to the published version of the manuscript.

Funding: This research received no external funding.

Institutional Review Board Statement: Not applicable.

Informed Consent Statement: Not applicable.

Acknowledgments: The authors gratefully acknowledge the financial support of RMIT University for providing a PhD scholarship to Tawsif Siddique (T.S.) to carry out this work. T.S. acknowledges the support of Umma Habiba for his PhD work.

Conflicts of Interest: The authors declare no conflict of interest.

References

1. Ulbricht, M. Advanced functional polymer membranes. *Polymer* **2006**, *47*, 2217–2262. [CrossRef]
2. Narain, R. *Polymer Science and Nanotechnology: Fundamentals and Applications*; Elsevier: San Diego, CA, USA, 2020.
3. Matin, A.; Khan, Z.; Zaidi, S.; Boyce, M. Biofouling in reverse osmosis membranes for seawater desalination: Phenomena and prevention. *Desalination* **2011**, *281*, 1–16. [CrossRef]
4. Goh, P.; Ismail, A.; Ng, B. Carbon nanotubes for desalination: Performance evaluation and current hurdles. *Desalination* **2013**, *308*, 2–14. [CrossRef]
5. Baglio, V.; Arico, A.; Di Blasi, A.; Antonucci, P.; Nannetti, F.; Tricoli, V.; Antonucci, V. Zeolite-based composite membranes for high temperature direct methanol fuel cells. *J. Appl. Electrochem.* **2005**, *35*, 207–212. [CrossRef]
6. Hashemifard, S.; Ismail, A.; Matsuura, T. Mixed matrix membrane incorporated with large pore size halloysite nanotubes (HNTs) as filler for gas separation: Morphological diagram. *Chem. Eng. J.* **2011**, *172*, 581–590. [CrossRef]
7. Liu, G.; Xiangli, F.; Wei, W.; Liu, S.; Jin, W. Improved performance of PDMS/ceramic composite pervaporation membranes by ZSM-5 homogeneously dispersed in PDMS via a surface graft/coating approach. *Chem. Eng. J.* **2011**, *174*, 495–503. [CrossRef]
8. Kim, J.H.; Joshi, M.K.; Lee, J.; Park, C.H.; Kim, C.S. Polydopamine-assisted immobilization of hierarchical zinc oxide nanostructures on electrospun nanofibrous membrane for photocatalysis and antimicrobial activity. *J. Colloid Interface Sci.* **2018**, *513*, 566–574. [CrossRef]
9. Kwak, S.-Y.; Kim, S.H.; Kim, S.S. Hybrid Organic/Inorganic Reverse Osmosis (RO) Membrane for Bactericidal Anti-Fouling. 1. Preparation and Characterization of TiO2 Nanoparticle Self-Assembled Aromatic Polyamide Thin-Film-Composite (TFC) Membrane. *Environ. Sci. Technol.* **2001**, *35*, 2388–2394. [CrossRef]
10. Jiang, Z.; Tijing, L.D.; Amarjargal, A.; Park, C.H.; An, K.-J.; Shon, H.K.; Kim, C.S. Removal of oil from water using magnetic bicomponent composite nanofibers fabricated by electrospinning. *Compos. Part B Eng.* **2015**, *77*, 311–318. [CrossRef]
11. Hassanajili, S.; Masoudi, E.; Karimi, G.; Khademi, M. Mixed matrix membranes based on polyetherurethane and polyesterurethane containing silica nanoparticles for separation of CO_2/CH_4 gases. *Sep. Purif. Technol.* **2013**, *116*, 1–12. [CrossRef]
12. Power, A.; Chandra, S.; Chapman, J. Graphene, electrospun membranes and granular activated carbon for eliminating heavy metals, pesticides and bacteria in water and wastewater treatment processes. *Analyst* **2018**, *143*, 5629–5645.
13. Zaman, N.K.; Rohani, R.; Mohammad, A.W.; Isloor, A.M. Polyimide-graphene oxide nanofiltration membrane: Characterizations and application in enhanced high concentration salt removal. *Chem. Eng. Sci.* **2018**, *177*, 218–233. [CrossRef]
14. Liu, X.L.; Li, Y.S.; Zhu, G.Q.; Ban, Y.J.; Xu, L.Y.; Yang, W.S. An organophilic pervaporation membrane derived from metal–organic framework nanoparticles for efficient recovery of bio-alcohols. *Angew. Chem.* **2011**, *123*, 10824–10827. [CrossRef]
15. Kimura, K.; Hane, Y.; Watanabe, Y.; Amy, G.; Ohkuma, N. Irreversible membrane fouling during ultrafiltration of surface water. *Water Res.* **2004**, *38*, 3431–3441. [CrossRef]
16. Flemming, H.-C. Reverse osmosis membrane biofouling. *Exp. Therm. Fluid Sci.* **1997**, *14*, 382–391. [CrossRef]
17. Van de Lisdonk, C.; Van Paassen, J.; Schippers, J. Monitoring scaling in nanofiltration and reverse osmosis membrane systems. *Desalination* **2000**, *132*, 101–108. [CrossRef]
18. Escobar, I.C.; Hoek, E.M.; Gabelich, C.J.; DiGiano, F.A. Committee report: Recent advances and research needs in membrane fouling. *Am. Water Work. Assoc. J.* **2005**, *97*, 79.
19. Al-Amoudi, A.S.; Farooque, A.M. Performance restoration and autopsy of NF membranes used in seawater pretreatment. *Desalination* **2005**, *178*, 261–271. [CrossRef]
20. Buffle, J.; Leppard, G.G. Characterization of aquatic colloids and macromolecules. 1. Structure and behavior of colloidal material. *Environ. Sci. Technol.* **1995**, *29*, 2169–2175. [CrossRef]
21. Buffle, J.; Wilkinson, K.J.; Stoll, S.; Filella, M.; Zhang, J. A generalized description of aquatic colloidal interactions: The three-colloidal component approach. *Environ. Sci. Technol.* **1998**, *32*, 2887–2899. [CrossRef]
22. Flemming, H.-C.; Schaule, G.; Griebe, T.; Schmitt, J.; Tamachkiarowa, A. Biofouling—The Achilles heel of membrane processes. *Desalination* **1997**, *113*, 215–225. [CrossRef]
23. Baker, J.; Dudley, L. Biofouling in membrane systems—A review. *Desalination* **1998**, *118*, 81–89. [CrossRef]
24. Yuan, W.; Zydney, A.L. Humic acid fouling during ultrafiltration. *Environ. Sci. Technol.* **2000**, *34*, 5043–5050. [CrossRef]
25. Hoek, E.M.; Elimelech, M. Cake-enhanced concentration polarization: A new fouling mechanism for salt-rejecting membranes. *Environ. Sci. Technol.* **2003**, *37*, 5581–5588. [CrossRef] [PubMed]
26. Rana, D.; Kim, Y.; Matsuura, T.; Arafat, H.A. Development of antifouling thin-film-composite membranes for seawater desalination. *J. Membr. Sci.* **2011**, *367*, 110–118. [CrossRef]
27. Crittenden, J.C.; Trussell, R.R.; Hand, D.W.; Howe, K.J.; Tchobanoglous, G. *MWH's Water Treatment: Principles and Design*; John Wiley & Sons: Hoboken, NJ, USA, 2012.
28. Robinson, S.; Abdullah, S.Z.; Bérubé, P.; Le-Clech, P. Ageing of membranes for water treatment: Linking changes to performance. *J. Membr. Sci.* **2016**, *503*, 177–187. [CrossRef]
29. Robinson, S.J.; Bérubé, P.R. Seeking realistic membrane ageing at bench-scale. *J. Membr. Sci.* **2021**, *618*, 118606. [CrossRef]
30. Rana, D.; Matsuura, T. Surface modifications for antifouling membranes. *Chem. Rev.* **2010**, *110*, 2448–2471. [CrossRef] [PubMed]

31. Bellona, C.; Drewes, J.E.; Xu, P.; Amy, G. Factors affecting the rejection of organic solutes during NF/RO treatment—A literature review. *Water Res.* **2004**, *38*, 2795–2809. [CrossRef] [PubMed]
32. Tarboush, B.J.A.; Rana, D.; Matsuura, T.; Arafat, H.; Narbaitz, R. Preparation of thin-film-composite polyamide membranes for desalination using novel hydrophilic surface modifying macromolecules. *J. Membr. Sci.* **2008**, *325*, 166–175. [CrossRef]
33. Kimura, K.; Amy, G.; Drewes, J.; Watanabe, Y. Adsorption of hydrophobic compounds onto NF/RO membranes: An artifact leading to overestimation of rejection. *J. Membr. Sci.* **2003**, *221*, 89–101. [CrossRef]
34. Hilal, N.; Ogunbiyi, O.O.; Miles, N.J.; Nigmatullin, R. Methods employed for control of fouling in MF and UF membranes: A comprehensive review. *Sep. Sci. Technol.* **2005**, *40*, 1957–2005. [CrossRef]
35. Fane, A.; Fell, C. A review of fouling and fouling control in ultrafiltration. *Desalination* **1987**, *62*, 117–136. [CrossRef]
36. Van der Bruggen, B.; Mänttäri, M.; Nyström, M. Drawbacks of applying nanofiltration and how to avoid them: A review. *Sep. Purif. Technol.* **2008**, *63*, 251–263. [CrossRef]
37. Al-Amoudi, A.; Lovitt, R.W. Fouling strategies and the cleaning system of NF membranes and factors affecting cleaning efficiency. *J. Membr. Sci.* **2007**, *303*, 4–28. [CrossRef]
38. Kato, K.; Uchida, E.; Kang, E.-T.; Uyama, Y.; Ikada, Y. Polymer surface with graft chains. *Prog. Polym. Sci.* **2003**, *28*, 209–259. [CrossRef]
39. Kang, G.-D.; Cao, Y.-M. Development of antifouling reverse osmosis membranes for water treatment: A review. *Water Res.* **2012**, *46*, 584–600. [CrossRef] [PubMed]
40. Vrijenhoek, E.M.; Hong, S.; Elimelech, M. Influence of membrane surface properties on initial rate of colloidal fouling of reverse osmosis and nanofiltration membranes. *J. Membr. Sci.* **2001**, *188*, 115–128. [CrossRef]
41. Osborn, S.J.; Hassan, M.K.; Divoux, G.M.; Rhoades, D.W.; Mauritz, K.A. Glass transition temperature of perfluorosulfonic acid ionomers. *Macromolecules* **2007**, *40*, 3886–3890. [CrossRef]
42. Geise, G.M.; Lee, H.S.; Miller, D.J.; Freeman, B.D.; McGrath, J.E.; Paul, D.R. Water purification by membranes: The role of polymer science. *J. Polym. Sci. Part B Polym. Phys.* **2010**, *48*, 1685–1718. [CrossRef]
43. Baker, R.W. *Membrane Technology and Applications*; John Wiley & Sons: Hoboken, NJ, USA, 2012.
44. Merrick, M.M.; Sujanani, R.; Freeman, B.D. Glassy polymers: Historical findings, membrane applications, and unresolved questions regarding physical aging. *Polymer* **2020**, *211*, 123176. [CrossRef]
45. Xie, W.; Geise, G.M.; Freeman, B.D.; Lee, C.H.; McGrath, J.E. Influence of processing history on water and salt transport properties of disulfonated polysulfone random copolymers. *Polymer* **2012**, *53*, 1581–1592. [CrossRef]
46. Geise, G.; Freeman, B.; Paul, D. Characterization of a sulfonated pentablock copolymer for desalination applications. *Polymer* **2010**, *51*, 5815–5822. [CrossRef]
47. Xie, W.; Cook, J.; Park, H.B.; Freeman, B.D.; Lee, C.H.; McGrath, J.E. Fundamental salt and water transport properties in directly copolymerized disulfonated poly (arylene ether sulfone) random copolymers. *Polymer* **2011**, *52*, 2032–2043. [CrossRef]
48. Chang, K.; Xue, T.; Geise, G.M. Increasing salt size selectivity in low water content polymers via polymer backbone dynamics. *J. Membr. Sci.* **2018**, *552*, 43–50. [CrossRef]
49. Chang, K.; Korovich, A.; Xue, T.; Morris, W.A.; Madsen, L.A.; Geise, G.M. Influence of rubbery versus glassy backbone dynamics on multiscale transport in polymer membranes. *Macromolecules* **2018**, *51*, 9222–9233. [CrossRef]
50. Setiawan, L.; Wang, R.; Li, K.; Fane, A.G. Fabrication of novel poly (amide–imide) forward osmosis hollow fiber membranes with a positively charged nanofiltration-like selective layer. *J. Membr. Sci.* **2011**, *369*, 196–205. [CrossRef]
51. Setiawan, L.; Wang, R.; Tan, S.; Shi, L.; Fane, A.G. Fabrication of poly (amide-imide)-polyethersulfone dual layer hollow fiber membranes applied in forward osmosis by combined polyelectrolyte cross-linking and depositions. *Desalination* **2013**, *312*, 99–106. [CrossRef]
52. Liu, Y.; Chung, T.-S.; Wang, R.; Li, D.F.; Chng, M.L. Chemical cross-linking modification of polyimide/poly (ether sulfone) dual-layer hollow-fiber membranes for gas separation. *Ind. Eng. Chem. Res.* **2003**, *42*, 1190–1195. [CrossRef]
53. Stafie, N.; Stamatialis, D.; Wessling, M. Effect of PDMS cross-linking degree on the permeation performance of PAN/PDMS composite nanofiltration membranes. *Sep. Purif. Technol.* **2005**, *45*, 220–231. [CrossRef]
54. Le, N.L.; Wang, Y.; Chung, T.-S. Synthesis, cross-linking modifications of 6FDA-NDA/DABA polyimide membranes for ethanol dehydration via pervaporation. *J. Membr. Sci.* **2012**, *415*, 109–121. [CrossRef]
55. Papadimitriou, K.D.; Geormezi, M.; Neophytides, S.G.; Kallitsis, J.K. Covalent cross-linking in phosphoric acid of pyridine based aromatic polyethers bearing side double bonds for use in high temperature polymer electrolyte membrane fuelcells. *J. Membr. Sci.* **2013**, *433*, 1–9. [CrossRef]
56. Qiao, J.; Fu, J.; Liu, L.; Liu, Y.; Sheng, J. Highly stable hydroxyl anion conducting membranes poly (vinyl alcohol)/poly (acrylamide-co-diallyldimethylammonium chloride)(PVA/PAADDA) for alkaline fuel cells: Effect of cross-linking. *Int. J. Hydrogen Energy* **2012**, *37*, 4580–4589. [CrossRef]
57. Powell, C.E.; Duthie, X.J.; Kentish, S.E.; Qiao, G.G.; Stevens, G.W. Reversible diamine cross-linking of polyimide membranes. *J. Membr. Sci.* **2007**, *291*, 199–209. [CrossRef]
58. Ba, C.; Langer, J.; Economy, J. Chemical modification of P84 copolyimide membranes by polyethylenimine for nanofiltration. *J. Membr. Sci.* **2009**, *327*, 49–58. [CrossRef]
59. Huang, R.; Chen, G.; Yang, B.; Gao, C. Positively charged composite nanofiltration membrane from quaternized chitosan by toluene diisocyanate cross-linking. *Sep. Purif. Technol.* **2008**, *61*, 424–429. [CrossRef]

60. García, M.G.; Marchese, J.; Ochoa, N.A. High activated carbon loading mixed matrix membranes for gas separations. *J. Mater. Sci.* **2012**, *47*, 3064–3075. [CrossRef]
61. Asadi Tashvigh, A.; Luo, L.; Chung, T.-S.; Weber, M.; Maletzko, C. A novel ionically cross-linked sulfonated polyphenylsulfone (sPPSU) membrane for organic solvent nanofiltration (OSN). *J. Membr. Sci.* **2018**, *545*, 221–228. [CrossRef]
62. Davood Abadi Farahani, M.H.; Hua, D.; Chung, T.-S. Cross-linked mixed matrix membranes (MMMs) consisting of amine-functionalized multi-walled carbon nanotubes and P84 polyimide for organic solvent nanofiltration (OSN) with enhanced flux. *J. Membr. Sci.* **2018**, *548*, 319–331. [CrossRef]
63. Huang, R.; Chen, G.; Sun, M.; Gao, C. Preparation and characterization of quaterinized chitosan/poly (acrylonitrile) composite nanofiltration membrane from anhydride mixture cross-linking. *Sep. Purif. Technol.* **2008**, *58*, 393–399. [CrossRef]
64. Cohen, Y.; Lin, N.; Varin, K.J.; Chien, D.; Hicks, R.F. Membrane surface nanostructuring with terminally anchored polymer chains. In *Functional Nanostructured Materials and Membranes for Water Treatment*; John Wiley & Sons: Hoboken, NJ, USA, 2013; pp. 85–124.
65. Seman, M.A.; Khayet, M.; Ali, Z.B.; Hilal, N. Reduction of nanofiltration membrane fouling by UV-initiated graft polymerization technique. *J. Membr. Sci.* **2010**, *355*, 133–141. [CrossRef]
66. Bernstein, R.; Belfer, S.; Freger, V. Surface modification of dense membranes using radical graft polymerization enhanced by monomer filtration. *Langmuir* **2010**, *26*, 12358–12365. [CrossRef]
67. Akbari, A.; Desclaux, S.; Rouch, J.-C.; Remigy, J.-C. Application of nanofiltration hollow fibre membranes, developed by photografting, to treatment of anionic dye solutions. *J. Membr. Sci.* **2007**, *297*, 243–252. [CrossRef]
68. Kim, J.-H.; Park, P.-K.; Lee, C.-H.; Kwon, H.-H. Surface modification of nanofiltration membranes to improve the removal of organic micro-pollutants (EDCs and PhACs) in drinking water treatment: Graft polymerization and cross-linking followed by functional group substitution. *J. Membr. Sci.* **2008**, *321*, 190–198. [CrossRef]
69. Sarac, A.S. Redox polymerization. *Prog. Polym. Sci.* **1999**, *24*, 1149–1204. [CrossRef]
70. An, H.; Cho, K.Y.; Back, S.; Do, X.H.; Jeon, J.-D.; Lee, H.K.; Baek, K.-Y.; Lee, J.S. The significance of the interfacial interaction in mixed matrix membranes for enhanced propylene/propane separation performance and plasticization resistance. *Sep. Purif. Technol.* **2021**, *261*, 118279. [CrossRef]
71. Cortalezzi, M.M.; Rose, J.; Barron, A.R.; Wiesner, M.R. Characteristics of ultrafiltration ceramic membranes derived from alumoxane nanoparticles. *J. Membr. Sci.* **2002**, *205*, 33–43. [CrossRef]
72. Yan, L.; Li, Y.S.; Xiang, C.B.; Xianda, S. Effect of nano-sized Al2O3-particle addition on PVDF ultrafiltration membrane performance. *J. Membr. Sci.* **2006**, *276*, 162–167. [CrossRef]
73. Paul, M.; Jons, S.D. Chemistry and fabrication of polymeric nanofiltration membranes: A review. *Polymer* **2016**, *103*, 417–456. [CrossRef]
74. Mohammad, A.W.; Teow, Y.H.; Ang, W.L.; Chung, Y.T.; Oatley-Radcliffe, D.L.; Hilal, N. Nanofiltration membranes review: Recent advances and future prospects. *Desalination* **2015**, *356*, 226–254. [CrossRef]
75. Kim, S.H.; Kwak, S.-Y.; Sohn, B.-H.; Park, T.H. Design of TiO2 nanoparticle self-assembled aromatic polyamide thin-film-composite (TFC) membrane as an approach to solve biofouling problem. *J. Membr. Sci.* **2003**, *211*, 157–165. [CrossRef]
76. Bae, T.-H.; Tak, T.-M. Effect of TiO$_2$ nanoparticles on fouling mitigation of ultrafiltration membranes for activated sludge filtration. *J. Membr. Sci.* **2005**, *249*, 1–8. [CrossRef]
77. Ahmad, J.; Hågg, M.B. Polyvinyl acetate/titanium dioxide nanocomposite membranes for gas separation. *J. Membr. Sci.* **2013**, *445*, 200–210. [CrossRef]
78. Shirazi, Y.; Ghadimi, A.; Mohammadi, T. Recovery of alcohols from water using polydimethylsiloxane–silica nanocomposite membranes: Characterization and pervaporation performance. *J. Appl. Polym. Sci.* **2012**, *124*, 2871–2882. [CrossRef]
79. Kalpana, D.; Omkumar, K.; Kumar, S.S.; Renganathan, N. A novel high power symmetric ZnO/carbon aerogel composite electrode for electrochemical supercapacitor. *Electrochim. Acta* **2006**, *52*, 1309–1315. [CrossRef]
80. Tang, S.C.; Lo, I.M. Magnetic nanoparticles: Essential factors for sustainable environmental applications. *Water Res.* **2013**, *47*, 2613–2632. [CrossRef]
81. Auffan, M.; Achouak, W.; Rose, J.; Roncato, M.-A.; Chaneac, C.; Waite, D.T.; Masion, A.; Woicik, J.C.; Wiesner, M.R.; Bottero, J.-Y. Relation between the redox state of iron-based nanoparticles and their cytotoxicity toward *Escherichia coli*. *Environ. Sci. Technol.* **2008**, *42*, 6730–6735. [CrossRef]
82. Deng, C.-H.; Gong, J.-L.; Zeng, G.-M.; Niu, C.-G.; Niu, Q.-Y.; Zhang, W.; Liu, H.-Y. Inactivation performance and mechanism of *Escherichia coli* in aqueous system exposed to iron oxide loaded graphene nanocomposites. *J. Hazard. Mater.* **2014**, *276*, 66–76. [CrossRef]
83. Si, Y.; Ren, T.; Li, Y.; Ding, B.; Yu, J. Fabrication of magnetic polybenzoxazine-based carbon nanofibers with Fe$_3$O$_4$ inclusions with a hierarchical porous structure for water treatment. *Carbon* **2012**, *50*, 5176–5185. [CrossRef]
84. Liu, G.; Jin, W.; Xu, N. Graphene-based membranes. *Chem. Soc. Rev.* **2015**, *44*, 5016–5030. [CrossRef]
85. Ong, C.S.; Goh, P.; Lau, W.; Misdan, N.; Ismail, A.F. Nanomaterials for biofouling and scaling mitigation of thin film composite membrane: A review. *Desalination* **2016**, *393*, 2–15. [CrossRef]
86. Hebbar, R.S.; Isloor, A.M.; Asiri, A.M. Carbon nanotube-and graphene-based advanced membrane materials for desalination. *Environ. Chem. Lett.* **2017**, *15*, 643–671. [CrossRef]
87. Ganesh, B.; Isloor, A.M.; Ismail, A.F. Enhanced hydrophilicity and salt rejection study of graphene oxide-polysulfone mixed matrix membrane. *Desalination* **2013**, *313*, 199–207. [CrossRef]

88. Mi, B. Graphene oxide membranes for ionic and molecular sieving. *Science* **2014**, *343*, 740–742. [CrossRef] [PubMed]
89. Liu, S.; Hu, M.; Zeng, T.H.; Wu, R.; Jiang, R.; Wei, J.; Wang, L.; Kong, J.; Chen, Y. Lateral dimension-dependent antibacterial activity of graphene oxide sheets. *Langmuir* **2012**, *28*, 12364–12372. [CrossRef]
90. Lee, J.; Chae, H.-R.; Won, Y.J.; Lee, K.; Lee, C.-H.; Lee, H.H.; Kim, I.-C.; Lee, J.-M. Graphene oxide nanoplatelets composite membrane with hydrophilic and antifouling properties for wastewater treatment. *J. Membr. Sci.* **2013**, *448*, 223–230. [CrossRef]
91. Chae, H.-R.; Lee, J.; Lee, C.-H.; Kim, I.-C.; Park, P.-K. Graphene oxide-embedded thin-film composite reverse osmosis membrane with high flux, anti-biofouling and chlorine resistance. *J. Membr. Sci.* **2015**, *483*, 128–135. [CrossRef]
92. Ma, N.; Wei, J.; Liao, R.; Tang, C.Y. Zeolite-polyamide thin film nanocomposite membranes: Towards enhanced performance for forward osmosis. *J. Membr. Sci.* **2012**, *405*, 149–157. [CrossRef]
93. Maghami, M.; Abdelrasoul, A. Zeolites-mixed-matrix nanofiltration membranes for the next generation of water purification. In *Nanofiltration*; IntechOpen: London, UK, 2018.
94. Madhumala, M.; Satyasri, D.; Sankarshana, T.; Sridhar, S. Selective extraction of lactic acid from aqueous media through a hydrophobic H-Beta zeolite/PVDF mixed matrix membrane contactor. *Ind. Eng. Chem. Res.* **2014**, *53*, 17770–17781. [CrossRef]
95. Pechar, T.W.; Kim, S.; Vaughan, B.; Marand, E.; Baranauskas, V.; Riffle, J.; Jeong, H.K.; Tsapatsis, M. Preparation and characterization of a poly (imide siloxane) and zeolite L mixed matrix membrane. *J. Membr. Sci.* **2006**, *277*, 210–218. [CrossRef]
96. Dong, Y.; Chen, S.; Zhang, X.; Yang, J.; Liu, X.; Meng, G. Fabrication and characterization of low cost tubular mineral-based ceramic membranes for micro-filtration from natural zeolite. *J. Membr. Sci.* **2006**, *281*, 592–599. [CrossRef]
97. Damayanti, A.; Sari, T.K.; Afifah, A.; Sutikno, L.; Sunarno, E.; Soedjono, S. The performance operation of zeolite as membrane with using laundry waste water. *J. Membr. Sci. Technol.* **2016**, *6*, 148. [CrossRef]
98. Yurekli, Y. Removal of heavy metals in wastewater by using zeolite nano-particles impregnated polysulfone membranes. *J. Hazard. Mater.* **2016**, *309*, 53–64. [CrossRef] [PubMed]
99. Huang, H.; Qu, X.; Ji, X.; Gao, X.; Zhang, L.; Chen, H.; Hou, L. Acid and multivalent ion resistance of thin film nanocomposite RO membranes loaded with silicalite-1 nanozeolites. *J. Mater. Chem. A* **2013**, *1*, 11343–11349. [CrossRef]
100. Smith, S.J.; Ladewig, B.P.; Hill, A.J.; Lau, C.H.; Hill, M.R. Post-synthetic Ti exchanged UiO-66 metal-organic frameworks that deliver exceptional gas permeability in mixed matrix membranes. *Sci. Rep.* **2015**, *5*, 7823. [CrossRef] [PubMed]
101. Jusoh, N.; Yeong, Y.F.; Lau, K.K.; Shariff, A.M. Mixed matrix membranes comprising of ZIF-8 nanofillers for enhanced gas transport properties. *Procedia Eng.* **2016**, *148*, 1259–1265. [CrossRef]
102. Lee, H.; Park, S.C.; Roh, J.S.; Moon, G.H.; Shin, J.E.; Kang, Y.S.; Park, H.B. Metal–organic frameworks grown on a porous planar template with an exceptionally high surface area: Promising nanofiller platforms for CO_2 separation. *J. Mater. Chem. A* **2017**, *5*, 22500–22505. [CrossRef]
103. Fuoco, A.; Khdhayyer, M.R.; Attfield, M.P.; Esposito, E.; Jansen, J.C.; Budd, P.M. Synthesis and transport properties of novel MOF/PIM-1/MOF sandwich membranes for gas separation. *Membranes* **2017**, *7*, 7. [CrossRef]
104. Duan, C.; Jie, X.; Zhu, H.; Liu, D.; Peng, W.; Cao, Y. Gas-permeation performance of metal organic framework/polyimide mixed-matrix membranes and additional explanation from the particle size angle. *J. Appl. Polym. Sci.* **2018**, *135*, 45728. [CrossRef]
105. Tien-Binh, N.; Rodrigue, D.; Kaliaguine, S. In-situ cross interface linking of PIM-1 polymer and UiO-66-NH2 for outstanding gas separation and physical aging control. *J. Membr. Sci.* **2018**, *548*, 429–438. [CrossRef]
106. Wang, Z.; Ren, H.; Zhang, S.; Zhang, F.; Jin, J. Polymers of intrinsic microporosity/metal–organic framework hybrid membranes with improved interfacial interaction for high-performance CO_2 separation. *J. Mater. Chem. A* **2017**, *5*, 10968–10977. [CrossRef]
107. Ahmad, M.Z.; Navarro, M.; Lhotka, M.; Zornoza, B.; Téllez, C.; de Vos, W.M.; Benes, N.E.; Konnertz, N.M.; Visser, T.; Semino, R. Enhanced gas separation performance of 6FDA-DAM based mixed matrix membranes by incorporating MOF UiO-66 and its derivatives. *J. Membr. Sci.* **2018**, *558*, 64–77. [CrossRef]
108. Qian, Q.; Asinger, P.A.; Lee, M.J.; Han, G.; Mizrahi Rodriguez, K.; Lin, S.; Benedetti, F.M.; Wu, A.X.; Chi, W.S.; Smith, Z.P. MOF-Based Membranes for Gas Separations. *Chem. Rev.* **2020**, *120*, 8161–8266. [CrossRef] [PubMed]
109. Magalad, V.T.; Gokavi, G.S.; Ranganathaiah, C.; Burshe, M.H.; Han, C.; Dionysiou, D.D.; Nadagouda, M.N.; Aminabhavi, T.M. Polymeric blend nanocomposite membranes for ethanol dehydration—Effect of morphology and membrane–solvent interactions. *J. Membr. Sci.* **2013**, *430*, 321–329. [CrossRef]
110. Jiang, L.Y.; Chung, T.S.; Cao, C.; Huang, Z.; Kulprathipanja, S. Fundamental understanding of nano-sized zeolite distribution in the formation of the mixed matrix single-and dual-layer asymmetric hollow fiber membranes. *J. Membr. Sci.* **2005**, *252*, 89–100. [CrossRef]
111. Jiang, L.Y.; Chung, T.S.; Kulprathipanja, S. An investigation to revitalize the separation performance of hollow fibers with a thin mixed matrix composite skin for gas separation. *J. Membr. Sci.* **2006**, *276*, 113–125. [CrossRef]
112. Li, Y.; Chung, T.-S.; Huang, Z.; Kulprathipanja, S. Dual-layer polyethersulfone (PES)/BTDA-TDI/MDI co-polyimide (P84) hollow fiber membranes with a submicron PES–zeolite beta mixed matrix dense-selective layer for gas separation. *J. Membr. Sci.* **2006**, *277*, 28–37. [CrossRef]
113. Pechar, T.W.; Kim, S.; Vaughan, B.; Marand, E.; Tsapatsis, M.; Jeong, H.K.; Cornelius, C.J. Fabrication and characterization of polyimide–zeolite L mixed matrix membranes for gas separations. *J. Membr. Sci.* **2006**, *277*, 195–202. [CrossRef]
114. Xiao, Y.; Wang, K.Y.; Chung, T.-S.; Tan, J. Evolution of nano-particle distribution during the fabrication of mixed matrix TiO2-polyimide hollow fiber membranes. *Chem. Eng. Sci.* **2006**, *61*, 6228–6233. [CrossRef]

115. Jiang, L.Y.; Chung, T.S.; Kulprathipanja, S. Fabrication of mixed matrix hollow fibers with intimate polymer–zeolite interface for gas separation. *AIChE J.* **2006**, *52*, 2898–2908. [CrossRef]
116. Kusworo, T.D.; Ismail, A.F.; Mustafa, A.; Matsuura, T. Dependence of membrane morphology and performance on preparation conditions: The shear rate effect in membrane casting. *Sep. Purif. Technol.* **2008**, *61*, 249–257. [CrossRef]
117. Gorgojo, P.; Uriel, S.; Téllez, C.; Coronas, J. Development of mixed matrix membranes based on zeolite Nu-6 (2) for gas separation. *Microporous Mesoporous Mater.* **2008**, *115*, 85–92. [CrossRef]
118. Şen, D.; Kalıpçılar, H.; Yilmaz, L. Development of polycarbonate based zeolite 4A filled mixed matrix gas separation membranes. *J. Membr. Sci.* **2007**, *303*, 194–203. [CrossRef]
119. Widjojo, N.; Zhang, S.D.; Chung, T.S.; Liu, Y. Enhanced gas separation performance of dual-layer hollow fiber membranes via substructure resistance reduction using mixed matrix materials. *J. Membr. Sci.* **2007**, *306*, 147–158. [CrossRef]
120. Li, Y.; Chung, T.-S. Exploratory development of dual-layer carbon–zeolite nanocomposite hollow fiber membranes with high performance for oxygen enrichment and natural gas separation. *Microporous Mesoporous Mater.* **2008**, *113*, 315–324. [CrossRef]
121. Ismail, A.; Kusworo, T.; Mustafa, A. Enhanced gas permeation performance of polyethersulfone mixed matrix hollow fiber membranes using novel Dynasylan Ameo silane agent. *J. Membr. Sci.* **2008**, *319*, 306–312. [CrossRef]
122. Zhang, Y.; Musselman, I.H.; Ferraris, J.P.; Balkus, K.J., Jr. Gas permeability properties of Matrimid® membranes containing the metal-organic framework Cu–BPY–HFS. *J. Membr. Sci.* **2008**, *313*, 170–181. [CrossRef]
123. Li, Q.; Liu, Q.; Zhao, J.; Hua, Y.; Sun, J.; Duan, J.; Jin, W. High efficient water/ethanol separation by a mixed matrix membrane incorporating MOF filler with high water adsorption capacity. *J. Membr. Sci.* **2017**, *544*, 68–78. [CrossRef]
124. Wu, G.; Ma, J.; Wang, S.; Chai, H.; Guo, L.; Li, J.; Ostovan, A.; Guan, Y.; Chen, L. Cationic metal-organic framework based mixed-matrix membrane for extraction of phenoxy carboxylic acid (PCA) herbicides from water samples followed by UHPLC-MS/MS determination. *J. Hazard. Mater.* **2020**, *394*, 122556. [CrossRef] [PubMed]
125. El-Mehalmey, W.A.; Safwat, Y.; Bassyouni, M.; Alkordi, M.H. Strong Interplay between Polymer Surface Charge and MOF Cage Chemistry in Mixed-Matrix Membrane for Water Treatment Applications. *ACS Appl. Mater. Interfaces* **2020**, *12*, 27625–27631. [CrossRef]
126. De Guzman, M.R.; Andra, C.K.A.; Ang, M.B.M.Y.; Dizon, G.V.C.; Caparanga, A.R.; Huang, S.-H.; Lee, K.-R. Increased performance and antifouling of mixed-matrix membranes of cellulose acetate with hydrophilic nanoparticles of polydopamine-sulfobetaine methacrylate for oil-water separation. *J. Membr. Sci.* **2021**, *620*, 118881. [CrossRef]
127. Alkhouzaam, A.; Qiblawey, H. Novel polysulfone ultrafiltration membranes incorporating polydopamine functionalized graphene oxide with enhanced flux and fouling resistance. *J. Membr. Sci.* **2021**, *620*, 118900. [CrossRef]
128. Cong, H.; Radosz, M.; Towler, B.F.; Shen, Y. Polymer–inorganic nanocomposite membranes for gas separation. *Sep. Purif. Technol.* **2007**, *55*, 281–291. [CrossRef]
129. Kim, S.; Pechar, T.W.; Marand, E. Poly (imide siloxane) and carbon nanotube mixed matrix membranes for gas separation. *Desalination* **2006**, *192*, 330–339. [CrossRef]
130. Zhang, Y.; Li, H.; Lin, J.; Li, R.; Liang, X. Preparation and characterization of zirconium oxide particles filled acrylonitrile-methyl acrylate-sodium sulfonate acrylate copolymer hybrid membranes. *Desalination* **2006**, *192*, 198–206. [CrossRef]
131. Genne, I.; Kuypers, S.; Leysen, R. Effect of the addition of ZrO2 to polysulfone based UF membranes. *J. Membr. Sci.* **1996**, *113*, 343–350. [CrossRef]
132. Wara, N.M.; Francis, L.F.; Velamakanni, B.V. Addition of alumina to cellulose acetate membranes. *J. Membr. Sci.* **1995**, *104*, 43–49. [CrossRef]
133. Kim, S.; Chen, L.; Johnson, J.K.; Marand, E. Polysulfone and functionalized carbon nanotube mixed matrix membranes for gas separation: Theory and experiment. *J. Membr. Sci.* **2007**, *294*, 147–158. [CrossRef]
134. Ahn, J.; Chung, W.-J.; Pinnau, I.; Guiver, M.D. Polysulfone/silica nanoparticle mixed-matrix membranes for gas separation. *J. Membr. Sci.* **2008**, *314*, 123–133. [CrossRef]
135. Ciobanu, G.; Carja, G.; Ciobanu, O. Structure of mixed matrix membranes made with SAPO-5 zeolite in polyurethane matrix. *Microporous Mesoporous Mater.* **2008**, *115*, 61–66. [CrossRef]
136. Jamshidi Gohari, R.; Lau, W.J.; Matsuura, T.; Halakoo, E.; Ismail, A.F. Adsorptive removal of Pb(II) from aqueous solution by novel PES/HMO ultrafiltration mixed matrix membrane. *Sep. Purif. Technol.* **2013**, *120*, 59–68. [CrossRef]
137. Jamshidi Gohari, R.; Lau, W.J.; Matsuura, T.; Ismail, A.F. Fabrication and characterization of novel PES/Fe-Mn binary oxide UF mixed matrix membrane for adsorptive removal of As(III) from contaminated water solution. *Sep. Purif. Technol.* **2013**, *118*, 64–72. [CrossRef]
138. Husain, S.; Koros, W.J. Mixed matrix hollow fiber membranes made with modified HSSZ-13 zeolite in polyetherimide polymer matrix for gas separation. *J. Membr. Sci.* **2007**, *288*, 195–207. [CrossRef]
139. Rafizah, W.; Ismail, A. Effect of carbon molecular sieve sizing with poly (vinyl pyrrolidone) K-15 on carbon molecular sieve–polysulfone mixed matrix membrane. *J. Membr. Sci.* **2008**, *307*, 53–61. [CrossRef]
140. Kim, S.; Marand, E. High permeability nano-composite membranes based on mesoporous MCM-41 nanoparticles in a polysulfone matrix. *Microporous Mesoporous Mater.* **2008**, *114*, 129–136. [CrossRef]
141. Shu, L.; Xie, L.-H.; Meng, Y.; Liu, T.; Zhao, C.; Li, J.-R. A thin and high loading two-dimensional MOF nanosheet based mixed-matrix membrane for high permeance nanofiltration. *J. Membr. Sci.* **2020**, *603*, 118049. [CrossRef]

142. Aroon, M.; Ismail, A.; Matsuura, T.; Montazer-Rahmati, M. Performance studies of mixed matrix membranes for gas separation: A review. *Sep. Purif. Technol.* **2010**, *75*, 229–242. [CrossRef]
143. Chen, J.; Wang, G.; Zeng, X.; Zhao, H.; Cao, D.; Yun, J.; Tan, C.K. Toughening of polypropylene–ethylene copolymer with nanosized CaCO3 and styrene–butadiene–styrene. *J. Appl. Polym. Sci.* **2004**, *94*, 796–802. [CrossRef]
144. Zhang, Q.-X.; Yu, Z.-Z.; Xie, X.-L.; Mai, Y.-W. Crystallization and impact energy of polypropylene/$CaCO_3$ nanocomposites with nonionic modifier. *Polymer* **2004**, *45*, 5985–5994. [CrossRef]
145. Daming, W.; Qingyun, M.; Ying, L.; Yumei, D.; Weihong, C.; Hong, X.; Dongyun, R. In situ bubble-stretching dispersion mechanism for additives in polymers. *J. Polym. Sci. Part B Polym. Phys.* **2003**, *41*, 1051–1058. [CrossRef]
146. Zha, L.; Fang, Z. Polystyrene/$CaCO_3$ composites with different $CaCO_3$ radius and different nano-$CaCO_3$ content—structure and properties. *Polym. Compos.* **2010**, *31*, 1258–1264. [CrossRef]
147. Yu, Q.; Wu, P.; Xu, P.; Li, L.; Liu, T.; Zhao, L. Synthesis of cellulose/titanium dioxide hybrids in supercritical carbon dioxide. *Green Chem.* **2008**, *10*, 1061–1067. [CrossRef]
148. Kango, S.; Kalia, S.; Celli, A.; Njuguna, J.; Habibi, Y.; Kumar, R. Surface modification of inorganic nanoparticles for development of organic–inorganic nanocomposites—A review. *Prog. Polym. Sci.* **2013**, *38*, 1232–1261. [CrossRef]
149. Ahmad, S.; Ahmad, S.; Agnihotry, S. Synthesis and characterization of in situ prepared poly (methyl methacrylate) nanocomposites. *Bull. Mater. Sci.* **2007**, *30*, 31–35. [CrossRef]
150. Luo, Y.-B.; Li, W.-D.; Wang, X.-L.; Xu, D.-Y.; Wang, Y.-Z. Preparation and properties of nanocomposites based on poly (lactic acid) and functionalized TiO_2. *Acta Mater.* **2009**, *57*, 3182–3191. [CrossRef]
151. Tong, Y.; Li, Y.; Xie, F.; Ding, M. Preparation and characteristics of polyimide–TiO_2 nanocomposite film. *Polym. Int.* **2000**, *49*, 1543–1547. [CrossRef]
152. Tang, E.; Cheng, G.; Ma, X. Preparation of nano-ZnO/PMMA composite particles via grafting of the copolymer onto the surface of zinc oxide nanoparticles. *Powder Technol.* **2006**, *161*, 209–214. [CrossRef]
153. Wang, Z.; Lu, Y.; Liu, J.; Dang, Z.; Zhang, L.; Wang, W. Preparation of nano-zinc oxide/EPDM composites with both good thermal conductivity and mechanical properties. *J. Appl. Polym. Sci.* **2011**, *119*, 1144–1155. [CrossRef]
154. Utracki, L.; Sepehr, M.; Boccaleri, E. Synthetic, layered nanoparticles for polymeric nanocomposites (PNCs). *Polym. Adv. Technol.* **2007**, *18*, 1–37. [CrossRef]
155. Li, X.; Wang, D.; Cheng, G.; Luo, Q.; An, J.; Wang, Y. Preparation of polyaniline-modified TiO_2 nanoparticles and their photocatalytic activity under visible light illumination. *Appl. Catal. B Environ.* **2008**, *81*, 267–273. [CrossRef]
156. Jeong, B.-H.; Hoek, E.M.; Yan, Y.; Subramani, A.; Huang, X.; Hurwitz, G.; Ghosh, A.K.; Jawor, A. Interfacial polymerization of thin film nanocomposites: A new concept for reverse osmosis membranes. *J. Membr. Sci.* **2007**, *294*, 1–7. [CrossRef]
157. Shen, Z.; Chen, W.; Xu, H.; Yang, W.; Kong, Q.; Wang, A.; Ding, M.; Shang, J. Fabrication of a Novel Antifouling Polysulfone Membrane with in Situ Embedment of Mxene Nanosheets. *Int. J. Environ. Res. Public Health* **2019**, *16*, 4659. [CrossRef] [PubMed]
158. Qi, W.; Lu, C.; Chen, P.; Han, L.; Yu, Q.; Xu, R. Electrochemical performances and thermal properties of electrospun Poly (phthalazinone ether sulfone ketone) membrane for lithium-ion battery. *Mater. Lett.* **2012**, *66*, 239–241. [CrossRef]
159. Liu, L.; Wang, Z.; Zhao, Z.; Zhao, Y.; Li, F.; Yang, L. PVDF/PAN/SiO_2 polymer electrolyte membrane prepared by combination of phase inversion and chemical reaction method for lithium ion batteries. *J. Solid State Electrochem.* **2016**, *20*, 699–712. [CrossRef]
160. Gwon, S.-J.; Choi, J.-H.; Sohn, J.-Y.; Ihm, Y.-E.; Nho, Y.-C. Preparation of a new micro-porous poly (methyl methacrylate)-grafted polyethylene separator for high performance Li secondary battery. *Nucl. Instrum. Methods Phys. Res. Sect. B Beam Interact. Mater. At.* **2009**, *267*, 3309–3313. [CrossRef]
161. Zainab, G.; Wang, X.; Yu, J.; Zhai, Y.; Babar, A.A.; Xiao, K.; Ding, B. Electrospun polyacrylonitrile/polyurethane composite nanofibrous separator with electrochemical performance for high power lithium ion batteries. *Mater. Chem. Phys.* **2016**, *182*, 308–314. [CrossRef]
162. Gopal, R.; Kaur, S.; Feng, C.Y.; Chan, C.; Ramakrishna, S.; Tabe, S.; Matsuura, T. Electrospun nanofibrous polysulfone membranes as pre-filters: Particulate removal. *J. Membr. Sci.* **2007**, *289*, 210–219. [CrossRef]
163. Zhai, Y.; Xiao, K.; Yu, J.; Ding, B. Closely packed x-poly (ethylene glycol diacrylate) coated polyetherimide/poly (vinylidene fluoride) fiber separators for lithium ion batteries with enhanced thermostability and improved electrolyte wettability. *J. Power Sources* **2016**, *325*, 292–300. [CrossRef]
164. Shi, C.; Zhang, P.; Huang, S.; He, X.; Yang, P.; Wu, D.; Sun, D.; Zhao, J. Functional separator consisted of polyimide nonwoven fabrics and polyethylene coating layer for lithium-ion batteries. *J. Power Sources* **2015**, *298*, 158–165. [CrossRef]
165. Angulakshmi, N.; Stephan, A.M. Electrospun trilayer polymeric membranes as separator for lithium–ion batteries. *Electrochim. Acta* **2014**, *127*, 167–172. [CrossRef]
166. Zhang, F.; Ma, X.; Cao, C.; Li, J.; Zhu, Y. Poly (vinylidene fluoride)/SiO_2 composite membranes prepared by electrospinning and their excellent properties for nonwoven separators for lithium-ion batteries. *J. Power Sources* **2014**, *251*, 423–431. [CrossRef]
167. Wang, Q.; Song, W.-L.; Wang, L.; Song, Y.; Shi, Q.; Fan, L.-Z. Electrospun polyimide-based fiber membranes as polymer electrolytes for lithium-ion batteries. *Electrochim. Acta* **2014**, *132*, 538–544. [CrossRef]
168. Yanilmaz, M.; Dirican, M.; Zhang, X. Evaluation of electrospun SiO_2/nylon 6, 6 nanofiber membranes as a thermally-stable separator for lithium-ion batteries. *Electrochim. Acta* **2014**, *133*, 501–508. [CrossRef]

169. Kimura, N.; Sakumoto, T.; Mori, Y.; Wei, K.; Kim, B.-S.; Song, K.-H.; Kim, I.-S. Fabrication and characterization of reinforced electrospun poly (vinylidene fluoride-co-hexafluoropropylene) nanofiber membranes. *Compos. Sci. Technol.* **2014**, *92*, 120–125. [CrossRef]
170. Valappil, R.S.K.; Ghasem, N.; Al-Marzouqi, M. Current and future trends in polymer membrane-based gas separation technology: A comprehensive review. *J. Ind. Eng. Chem.* **2021**, *98*, 103–129. [CrossRef]
171. Sanders, D.F.; Smith, Z.P.; Guo, R.; Robeson, L.M.; McGrath, J.E.; Paul, D.R.; Freeman, B.D. Energy-efficient polymeric gas separation membranes for a sustainable future: A review. *Polymer* **2013**, *54*, 4729–4761. [CrossRef]
172. Jung, J.T.; Kim, J.F.; Wang, H.H.; Di Nicolo, E.; Drioli, E.; Lee, Y.M. Understanding the non-solvent induced phase separation (NIPS) effect during the fabrication of microporous PVDF membranes via thermally induced phase separation (TIPS). *J. Membr. Sci.* **2016**, *514*, 250–263. [CrossRef]
173. Sridhar, S.; Smitha, B.; Mayor, S.; Prathab, B.; Aminabhavi, T. Gas permeation properties of polyamide membrane prepared by interfacial polymerization. *J. Mater. Sci.* **2007**, *42*, 9392–9401. [CrossRef]
174. Li, Z.; Wang, C. *One-Dimensional Nanostructures: Electrospinning Technique and Unique Nanofibers*; Springer: Berlin/Heidelberg, Germany, 2013.
175. Loeb, S.; Sourirajan, S. *Sea Water Demineralization by Means of an Osmotic Membrane*; ACS Publications: Washington, DC, USA, 1962.
176. Kajitvichyanukul, P.; Hung, Y.-T.; Wang, L.K. Membrane technologies for oil–water separation. In *Membrane and Desalination Technologies*; Springer: Berlin/Heidelberg, Germany, 2011; pp. 639–668.
177. Jiang, L.; Chung, T.-S.; Li, D.F.; Cao, C.; Kulprathipanja, S. Fabrication of Matrimid/polyethersulfone dual-layer hollow fiber membranes for gas separation. *J. Membr. Sci.* **2004**, *240*, 91–103. [CrossRef]
178. Teoh, M.M.; Chung, T.-S.; Yeo, Y.S. Dual-layer PVDF/PTFE composite hollow fibers with a thin macrovoid-free selective layer for water production via membrane distillation. *Chem. Eng. J.* **2011**, *171*, 684–691. [CrossRef]
179. Gaudin, F.; Sintes-Zydowicz, N. Correlation between the polymerization kinetics and the chemical structure of poly (urethane-urea) nanocapsule membrane obtained by interfacial step polymerization in miniemulsion. *Colloids Surf. A Physicochem. Eng. Asp.* **2012**, *415*, 328–342. [CrossRef]
180. Chou, S.; Wang, R.; Shi, L.; She, Q.; Tang, C.; Fane, A.G. Thin-film composite hollow fiber membranes for pressure retarded osmosis (PRO) process with high power density. *J. Membr. Sci.* **2012**, *389*, 25–33. [CrossRef]
181. Veríssimo, S.; Peinemann, K.-V.; Bordado, J. Thin-film composite hollow fiber membranes: An optimized manufacturing method. *J. Membr. Sci.* **2005**, *264*, 48–55. [CrossRef]
182. Veríssimo, S.; Peinemann, K.-V.; Bordado, J. New composite hollow fiber membrane for nanofiltration. *Desalination* **2005**, *184*, 1–11. [CrossRef]
183. Wei, J.; Qiu, C.; Tang, C.Y.; Wang, R.; Fane, A.G. Synthesis and characterization of flat-sheet thin film composite forward osmosis membranes. *J. Membr. Sci.* **2011**, *372*, 292–302. [CrossRef]
184. Zhao, Y.; Qiu, C.; Li, X.; Vararattanavech, A.; Shen, W.; Torres, J.; Helix-Nielsen, C.; Wang, R.; Hu, X.; Fane, A.G. Synthesis of robust and high-performance aquaporin-based biomimetic membranes by interfacial polymerization-membrane preparation and RO performance characterization. *J. Membr. Sci.* **2012**, *423*, 422–428. [CrossRef]
185. Liu, C.; Fang, W.; Chou, S.; Shi, L.; Fane, A.G.; Wang, R. Fabrication of layer-by-layer assembled FO hollow fiber membranes and their performances using low concentration draw solutions. *Desalination* **2013**, *308*, 147–153. [CrossRef]
186. Qi, S.; Li, W.; Zhao, Y.; Ma, N.; Wei, J.; Chin, T.W.; Tang, C.Y. Influence of the properties of layer-by-layer active layers on forward osmosis performance. *J. Membr. Sci.* **2012**, *423*, 536–542. [CrossRef]
187. Qiu, C.; Qi, S.; Tang, C.Y. Synthesis of high flux forward osmosis membranes by chemically crosslinked layer-by-layer polyelectrolytes. *J. Membr. Sci.* **2011**, *381*, 74–80. [CrossRef]
188. Liu, X.; Qi, S.; Li, Y.; Yang, L.; Cao, B.; Tang, C.Y. Synthesis and characterization of novel antibacterial silver nanocomposite nanofiltration and forward osmosis membranes based on layer-by-layer assembly. *Water Res.* **2013**, *47*, 3081–3092. [CrossRef]
189. Liu, X.; Dotzauer, D.M.; Bruening, M.L. Ion-exchange membranes prepared using layer-by-layer polyelectrolyte deposition. *J. Membr. Sci.* **2010**, *354*, 198–205. [CrossRef]
190. Zhang, P.; Qian, J.; Yang, Y.; An, Q.; Liu, X.; Gui, Z. Polyelectrolyte layer-by-layer self-assembly enhanced by electric field and their multilayer membranes for separating isopropanol–water mixtures. *J. Membr. Sci.* **2008**, *320*, 73–77. [CrossRef]
191. Zhang, G.; Gao, X.; Ji, S.; Liu, Z. One-step dynamic assembly of polyelectrolyte complex membranes. *Mater. Sci. Eng. C* **2009**, *29*, 1877–1884. [CrossRef]
192. van Ackern, F.; Krasemann, L.; Tieke, B. Ultrathin membranes for gas separation and pervaporation prepared upon electrostatic self-assembly of polyelectrolytes. *Thin Solid Films* **1998**, *327*, 762–766. [CrossRef]
193. Krasemann, L.; Tieke, B. Composite membranes with ultrathin separation layer prepared by self-assembly of polyelectrolytes. *Mater. Sci. Eng. C* **1999**, *8*, 513–518. [CrossRef]
194. Van Tassel, P.R. Polyelectrolyte adsorption and layer-by-layer assembly: Electrochemical control. *Curr. Opin. Colloid Interface Sci.* **2012**, *17*, 106–113. [CrossRef]
195. Miller, M.D.; Bruening, M.L. Controlling the nanofiltration properties of multilayer polyelectrolyte membranes through variation of film composition. *Langmuir* **2004**, *20*, 11545–11551. [CrossRef]

196. Loh, C.H.; Liao, Y.; Setiawan, L.; Wang, R. Fabrication of Polymeric and Composite Membranes. In *Membrane Fabrication*; CRC Press: Boca Raton, FL, USA, 2015; pp. 511–568. [CrossRef]
197. Krasemann, L.; Tieke, B. Selective ion transport across self-assembled alternating multilayers of cationic and anionic polyelectrolytes. *Langmuir* **2000**, *16*, 287–290. [CrossRef]
198. Ouyang, L.; Malaisamy, R.; Bruening, M.L. Multilayer polyelectrolyte films as nanofiltration membranes for separating monovalent and divalent cations. *J. Membr. Sci.* **2008**, *310*, 76–84. [CrossRef]
199. He, T.; Mulder, M.; Strathmann, H.; Wessling, M. Preparation of composite hollow fiber membranes: Co-extrusion of hydrophilic coatings onto porous hydrophobic support structures. *J. Membr. Sci.* **2002**, *207*, 143–156. [CrossRef]
200. Liu, R.X.; Qiao, X.Y.; Chung, T.-S. Dual-layer P84/polyethersulfone hollow fibers for pervaporation dehydration of isopropanol. *J. Membr. Sci.* **2007**, *294*, 103–114. [CrossRef]
201. Setiawan, L.; Shi, L.; Krantz, W.B.; Wang, R. Explorations of delamination and irregular structure in poly (amide-imide)-polyethersulfone dual layer hollow fiber membranes. *J. Membr. Sci.* **2012**, *423*, 73–84. [CrossRef]
202. Hashemifard, S.; Ismail, A.; Matsuura, T. Co-casting technique for fabricating dual-layer flat sheet membranes for gas separation. *J. Membr. Sci.* **2011**, *375*, 258–267. [CrossRef]
203. Ding, X.; Cao, Y.; Zhao, H.; Wang, L.; Yuan, Q. Fabrication of high performance Matrimid/polysulfone dual-layer hollow fiber membranes for O_2/N_2 separation. *J. Membr. Sci.* **2008**, *323*, 352–361. [CrossRef]
204. Sun, S.P.; Wang, K.Y.; Peng, N.; Hatton, T.A.; Chung, T.-S. Novel polyamide-imide/cellulose acetate dual-layer hollow fiber membranes for nanofiltration. *J. Membr. Sci.* **2010**, *363*, 232–242. [CrossRef]
205. Bonyadi, S.; Chung, T.S. Flux enhancement in membrane distillation by fabrication of dual layer hydrophilic–hydrophobic hollow fiber membranes. *J. Membr. Sci.* **2007**, *306*, 134–146. [CrossRef]
206. Li, D.; Chung, T.-S.; Wang, R. Morphological aspects and structure control of dual-layer asymmetric hollow fiber membranes formed by a simultaneous co-extrusion approach. *J. Membr. Sci.* **2004**, *243*, 155–175. [CrossRef]
207. Widjojo, N.; Chung, T.S.; Krantz, W.B. A morphological and structural study of Ultem/P84 copolyimide dual-layer hollow fiber membranes with delamination-free morphology. *J. Membr. Sci.* **2007**, *294*, 132–146. [CrossRef]
208. Schaefer, A.; Fane, A.G.; Waite, T.D. *Nanofiltration: Principles and Applications*; Elsevier: Amsterdam, The Netherlands, 2005.
209. He, T.; Frank, M.; Mulder, M.; Wessling, M. Preparation and characterization of nanofiltration membranes by coating polyethersulfone hollow fibers with sulfonated poly (ether ether ketone)(SPEEK). *J. Membr. Sci.* **2008**, *307*, 62–72. [CrossRef]
210. Iwamoto, S.; Nakagaito, A.; Yano, H. Nano-fibrillation of pulp fibers for the processing of transparent nanocomposites. *Appl. Phys. A* **2007**, *89*, 461–466. [CrossRef]
211. Lin, Y.; Yao, Y.; Yang, X.; Wei, N.; Li, X.; Gong, P.; Li, R.; Wu, D. Preparation of poly (ether sulfone) nanofibers by gas-jet/electrospinning. *J. Appl. Polym. Sci.* **2008**, *107*, 909–917. [CrossRef]
212. Siddique, T.; Dutta, N.K.; Roy Choudhury, N. Nanofiltration for Arsenic Removal: Challenges, Recent Developments, and Perspectives. *Nanomaterials* **2020**, *10*, 1323. [CrossRef]
213. Yarin, A.L.; Koombhongse, S.; Reneker, D.H. Taylor cone and jetting from liquid droplets in electrospinning of nanofibers. *J. Appl. Phys.* **2001**, *90*, 4836–4846. [CrossRef]
214. Deitzel, J.M.; Kleinmeyer, J.; Harris, D.; Tan, N.B. The effect of processing variables on the morphology of electrospun nanofibers and textiles. *Polymer* **2001**, *42*, 261–272. [CrossRef]
215. Formhals, A. Process and Apparatus for Preparing Artificial Threads. US Patent 1975504, 2 October 1934.
216. Morton, W.J. Method of Dispersing Fluids. US Patents 705691A, 29 July 1902.
217. Liao, Y.; Wang, R.; Tian, M.; Qiu, C.; Fane, A.G. Fabrication of polyvinylidene fluoride (PVDF) nanofiber membranes by electro-spinning for direct contact membrane distillation. *J. Membr. Sci.* **2013**, *425*, 30–39. [CrossRef]
218. Sill, T.J.; Von Recum, H.A. Electrospinning: Applications in drug delivery and tissue engineering. *Biomaterials* **2008**, *29*, 1989–2006. [CrossRef]
219. Chen, C.; Liu, K.; Wang, H.; Liu, W.; Zhang, H. Morphology and performances of electrospun polyethylene glycol/poly (dl-lactide) phase change ultrafine fibers for thermal energy storage. *Solar Energy Mater. Sol. cells* **2013**, *117*, 372–381. [CrossRef]
220. Ramakrishna, S.; Fujihara, K.; Teo, W.-E.; Yong, T.; Ma, Z.; Ramaseshan, R. Electrospun nanofibers: Solving global issues. *Mater. Today* **2006**, *9*, 40–50. [CrossRef]
221. Liao, S.; Li, B.; Ma, Z.; Wei, H.; Chan, C.; Ramakrishna, S. Biomimetic electrospun nanofibers for tissue regeneration. *Biomed. Mater.* **2006**, *1*, R45. [CrossRef] [PubMed]
222. Wu, H.; Hu, L.; Rowell, M.W.; Kong, D.; Cha, J.J.; McDonough, J.R.; Zhu, J.; Yang, Y.; McGehee, M.D.; Cui, Y. Electrospun metal nanofiber webs as high-performance transparent electrode. *Nano Lett.* **2010**, *10*, 4242–4248. [CrossRef]
223. Sigmund, W.; Yuh, J.; Park, H.; Maneeratana, V.; Pyrgiotakis, G.; Daga, A.; Taylor, J.; Nino, J.C. Processing and structure relationships in electrospinning of ceramic fiber systems. *J. Am. Ceram. Soc.* **2006**, *89*, 395–407. [CrossRef]
224. Bognitzki, M.; Frese, T.; Steinhart, M.; Greiner, A.; Wendorff, J.H.; Schaper, A.; Hellwig, M. Preparation of fibers with nanoscaled morphologies: Electrospinning of polymer blends. *Polym. Eng. Sci.* **2001**, *41*, 982–989. [CrossRef]
225. Katti, D.S.; Robinson, K.W.; Ko, F.K.; Laurencin, C.T. Bioresorbable nanofiber-based systems for wound healing and drug delivery: Optimization of fabrication parameters. *J. Biomed. Mater. Res. Part B Appl. Biomater. Off. J. Soc. Biomater. Jpn. Soc. Biomater. Aust. Soc. Biomater. Korean Soc. Biomater.* **2004**, *70*, 286–296. [CrossRef]

226. Son, W.K.; Youk, J.H.; Park, W.H. Antimicrobial cellulose acetate nanofibers containing silver nanoparticles. *Carbohydr. Polym.* **2006**, *65*, 430–434. [CrossRef]
227. Tsai, P.P.; Schreuder-Gibson, H.; Gibson, P. Different electrostatic methods for making electret filters. *J. Electrost.* **2002**, *54*, 333–341. [CrossRef]
228. Schreuder-Gibson, H.; Gibson, P.; Senecal, K.; Sennett, M.; Walker, J. Protective textile materials based on electrospun nanofibers. *J. Adv. Mater.* **2002**, *34*, 44–55.
229. Reneker, D.H.; Chun, I. Nanometre diameter fibres of polymer, produced by electrospinning. *Nanotechnology* **1996**, *7*, 216. [CrossRef]
230. Son, W.K.; Youk, J.H.; Lee, T.S.; Park, W.H. The effects of solution properties and polyelectrolyte on electrospinning of ultrafine poly (ethylene oxide) fibers. *Polymer* **2004**, *45*, 2959–2966. [CrossRef]
231. Megelski, S.; Stephens, J.S.; Chase, D.B.; Rabolt, J.F. Micro-and nanostructured surface morphology on electrospun polymer fibers. *Macromolecules* **2002**, *35*, 8456–8466. [CrossRef]
232. Yu, D.G.; Zhang, X.F.; Shen, X.X.; Brandford-White, C.; Zhu, L.M. Ultrafine ibuprofen-loaded polyvinylpyrrolidone fiber mats using electrospinning. *Polym. Int.* **2009**, *58*, 1010–1013. [CrossRef]
233. Gupta, P.; Elkins, C.; Long, T.E.; Wilkes, G.L. Electrospinning of linear homopolymers of poly (methyl methacrylate): Exploring relationships between fiber formation, viscosity, molecular weight and concentration in a good solvent. *Polymer* **2005**, *46*, 4799–4810. [CrossRef]
234. Lee, J.S.; Choi, K.H.; Ghim, H.D.; Kim, S.S.; Chun, D.H.; Kim, H.Y.; Lyoo, W.S. Role of molecular weight of atactic poly (vinyl alcohol)(PVA) in the structure and properties of PVA nanofabric prepared by electrospinning. *J. Appl. Polym. Sci.* **2004**, *93*, 1638–1646. [CrossRef]
235. Shin, H.J.; Lee, C.H.; Cho, I.H.; Kim, Y.-J.; Lee, Y.-J.; Kim, I.A.; Park, K.-D.; Yui, N.; Shin, J.-W. Electrospun PLGA nanofiber scaffolds for articular cartilage reconstruction: Mechanical stability, degradation and cellular responses under mechanical stimulation in vitro. *J. Biomater. Sci. Polym. Ed.* **2006**, *17*, 103–119. [CrossRef] [PubMed]
236. Gupta, P.; Wilkes, G.L. Some investigations on the fiber formation by utilizing a side-by-side bicomponent electrospinning approach. *Polymer* **2003**, *44*, 6353–6359. [CrossRef]
237. Yoshimoto, H.; Shin, Y.; Terai, H.; Vacanti, J. A biodegradable nanofiber scaffold by electrospinning and its potential for bone tissue engineering. *Biomaterials* **2003**, *24*, 2077–2082. [CrossRef]
238. Bhattarai, N.; Edmondson, D.; Veiseh, O.; Matsen, F.A.; Zhang, M. Electrospun chitosan-based nanofibers and their cellular compatibility. *Biomaterials* **2005**, *26*, 6176–6184. [CrossRef]
239. Matthews, J.A.; Wnek, G.E.; Simpson, D.G.; Bowlin, G.L. Electrospinning of collagen nanofibers. *Biomacromolecules* **2002**, *3*, 232–238. [CrossRef]
240. Huang, Z.-M.; Zhang, Y.; Ramakrishna, S.; Lim, C. Electrospinning and mechanical characterization of gelatin nanofibers. *Polymer* **2004**, *45*, 5361–5368. [CrossRef]
241. Sun, B.; Long, Y.; Zhang, H.; Li, M.; Duvail, J.; Jiang, X.; Yin, H. Advances in three-dimensional nanofibrous macrostructures via electrospinning. *Prog. Polym. Sci.* **2014**, *39*, 862–890. [CrossRef]
242. Agarwal, S.; Greiner, A.; Wendorff, J.H. Functional materials by electrospinning of polymers. *Prog. Polym. Sci.* **2013**, *38*, 963–991. [CrossRef]
243. Wang, X.; Ding, B.; Sun, G.; Wang, M.; Yu, J. Electro-spinning/netting: A strategy for the fabrication of three-dimensional polymer nano-fiber/nets. *Prog. Mater. Sci.* **2013**, *58*, 1173–1243. [CrossRef]
244. Li, D.; McCann, J.T.; Xia, Y.; Marquez, M. Electrospinning: A simple and versatile technique for producing ceramic nanofibers and nanotubes. *J. Am. Ceram. Soc.* **2006**, *89*, 1861–1869. [CrossRef]
245. Li, D.; Xia, Y. Electrospinning of nanofibers: Reinventing the wheel? *Adv. Mater.* **2004**, *16*, 1151–1170. [CrossRef]
246. Huang, Z.-M.; Zhang, Y.-Z.; Kotaki, M.; Ramakrishna, S. A review on polymer nanofibers by electrospinning and their applications in nanocomposites. *Compos. Sci. Technol.* **2003**, *63*, 2223–2253. [CrossRef]
247. Ryu, Y.J.; Kim, H.Y.; Lee, K.H.; Park, H.C.; Lee, D.R. Transport properties of electrospun nylon 6 nonwoven mats. *Eur. Polym. J.* **2003**, *39*, 1883–1889. [CrossRef]
248. Lee, K.; Kim, H.; Bang, H.; Jung, Y.; Lee, S. The change of bead morphology formed on electrospun polystyrene fibers. *Polymer* **2003**, *44*, 4029–4034. [CrossRef]
249. Bhardwaj, N.; Kundu, S.C. Electrospinning: A fascinating fiber fabrication technique. *Biotechnol. Adv.* **2010**, *28*, 325–347. [CrossRef]
250. Subbiah, T.; Bhat, G.S.; Tock, R.W.; Parameswaran, S.; Ramkumar, S.S. Electrospinning of nanofibers. *J. Appl. Polym. Sci.* **2005**, *96*, 557–569. [CrossRef]
251. Rutledge, G.C.; Fridrikh, S.V. Formation of fibers by electrospinning. *Adv. Drug Deliv. Rev.* **2007**, *59*, 1384–1391. [CrossRef]
252. Wang, X.; Ding, B.; Yu, J.; Si, Y.; Yang, S.; Sun, G. Electro-netting: Fabrication of two-dimensional nano-nets for highly sensitive trimethylamine sensing. *Nanoscale* **2011**, *3*, 911–915. [CrossRef]
253. Barakat, N.A.; Kanjwal, M.A.; Sheikh, F.A.; Kim, H.Y. Spider-net within the N6, PVA and PU electrospun nanofiber mats using salt addition: Novel strategy in the electrospinning process. *Polymer* **2009**, *50*, 4389–4396. [CrossRef]
254. Yang, S.; Wang, X.; Ding, B.; Yu, J.; Qian, J.; Sun, G. Controllable fabrication of soap-bubble-like structured polyacrylic acid nano-nets via electro-netting. *Nanoscale* **2011**, *3*, 564–568. [CrossRef] [PubMed]

255. Zong, X.; Kim, K.; Fang, D.; Ran, S.; Hsiao, B.S.; Chu, B. Structure and process relationship of electrospun bioabsorbable nanofiber membranes. *Polymer* **2002**, *43*, 4403–4412. [CrossRef]
256. Sun, S.P.; Wang, K.Y.; Rajarathnam, D.; Hatton, T.A.; Chung, T.S. Polyamide-imide nanofiltration hollow fiber membranes with elongation-induced nano-pore evolution. *AIChE J.* **2010**, *56*, 1481–1494. [CrossRef]
257. Wang, X.; Ding, B.; Yu, J.; Yang, J. Large-scale fabrication of two-dimensional spider-web-like gelatin nano-nets via electro-netting. *Colloids Surf. B Biointerfaces* **2011**, *86*, 345–352. [CrossRef] [PubMed]
258. Hu, J.; Wang, X.; Ding, B.; Lin, J.; Yu, J.; Sun, G. One-step Electro-spinning/netting Technique for Controllably Preparing Polyurethane Nano-fiber/net. *Macromol. Rapid Commun.* **2011**, *32*, 1729–1734. [CrossRef]
259. Talwar, S.; Krishnan, A.S.; Hinestroza, J.P.; Pourdeyhimi, B.; Khan, S.A. Electrospun nanofibers with associative polymer−surfactant systems. *Macromolecules* **2010**, *43*, 7650–7656. [CrossRef]
260. Lin, T.; Wang, H.; Wang, H.; Wang, X. The charge effect of cationic surfactants on the elimination of fibre beads in the electrospinning of polystyrene. *Nanotechnology* **2004**, *15*, 1375. [CrossRef]
261. Luo, C.; Nangrejo, M.; Edirisinghe, M. A novel method of selecting solvents for polymer electrospinning. *Polymer* **2010**, *51*, 1654–1662. [CrossRef]
262. Ding, B.; Li, C.; Miyauchi, Y.; Kuwaki, O.; Shiratori, S. Formation of novel 2D polymer nanowebs via electrospinning. *Nanotechnology* **2006**, *17*, 3685. [CrossRef]
263. Zhang, X.; Reagan, M.R.; Kaplan, D.L. Electrospun silk biomaterial scaffolds for regenerative medicine. *Adv. Drug Deliv. Rev.* **2009**, *61*, 988–1006. [CrossRef] [PubMed]
264. Buchko, C.J.; Chen, L.C.; Shen, Y.; Martin, D.C. Processing and microstructural characterization of porous biocompatible protein polymer thin films. *Polymer* **1999**, *40*, 7397–7407. [CrossRef]
265. Sun, D.; Chang, C.; Li, S.; Lin, L. Near-field electrospinning. *Nano Lett.* **2006**, *6*, 839–842. [CrossRef] [PubMed]
266. Theron, S.; Zussman, E.; Yarin, A. Experimental investigation of the governing parameters in the electrospinning of polymer solutions. *Polymer* **2004**, *45*, 2017–2030. [CrossRef]
267. Haghi, A. *Electrospinning of Nanofibers in Textiles*; CRC Press: Boca Raton, FL, USA, 2011.
268. Wang, N.; Wang, X.; Ding, B.; Yu, J.; Sun, G. Tunable fabrication of three-dimensional polyamide-66 nano-fiber/nets for high efficiency fine particulate filtration. *J. Mater. Chem.* **2012**, *22*, 1445–1452. [CrossRef]
269. Zucchelli, A.; Fabiani, D.; Gualandi, C.; Focarete, M. An innovative and versatile approach to design highly porous, patterned, nanofibrous polymeric materials. *J. Mater. Sci.* **2009**, *44*, 4969–4975. [CrossRef]
270. Zhang, D.; Karki, A.B.; Rutman, D.; Young, D.P.; Wang, A.; Cocke, D.; Ho, T.H.; Guo, Z. Electrospun polyacrylonitrile nanocomposite fibers reinforced with Fe3O4 nanoparticles: Fabrication and property analysis. *Polymer* **2009**, *50*, 4189–4198. [CrossRef]
271. Frenot, A.; Chronakis, I.S. Polymer nanofibers assembled by electrospinning. *Curr. Opin. colloid Interface Sci.* **2003**, *8*, 64–75. [CrossRef]
272. Fang, J.; Wang, X.; Lin, T. Functional applications of electrospun nanofibers. *Nanofibers Prod. Prop. Funct. Appl.* **2011**, *14*, 287–302.
273. Lee, C.; Jo, S.M.; Choi, J.; Baek, K.-Y.; Truong, Y.B.; Kyratzis, I.L.; Shul, Y.-G. SiO$_2$/sulfonated poly ether ether ketone (SPEEK) composite nanofiber mat supported proton exchange membranes for fuel cells. *J. Mater. Sci.* **2013**, *48*, 3665–3671. [CrossRef]
274. Pensa, N.W.; Curry, A.S.; Bonvallet, P.P.; Bellis, N.F.; Rettig, K.M.; Reddy, M.S.; Eberhardt, A.W.; Bellis, S.L. 3D printed mesh reinforcements enhance the mechanical properties of electrospun scaffolds. *Biomater. Res.* **2019**, *23*, 1–7. [CrossRef] [PubMed]
275. Chang, Y.-N.; Ou, X.-M.; Zeng, G.-M.; Gong, J.-L.; Deng, C.-H.; Jiang, Y.; Liang, J.; Yuan, G.-Q.; Liu, H.-Y.; He, X. Synthesis of magnetic graphene oxide–TiO$_2$ and their antibacterial properties under solar irradiation. *Appl. Surf. Sci.* **2015**, *343*, 1–10. [CrossRef]
276. El-Shafai, N.; El-Khouly, M.E.; El-Kemary, M.; Ramadan, M.; Eldesoukey, I.; Masoud, M. Graphene oxide decorated with zinc oxide nanoflower, silver and titanium dioxide nanoparticles: Fabrication, characterization, DNA interaction, and antibacterial activity. *RSC Adv.* **2019**, *9*, 3704–3714. [CrossRef]
277. Lee, K.S.; Park, C.W.; Lee, S.J.; Kim, J.-D. Hierarchical zinc oxide/graphene oxide composites for energy storage devices. *J. Alloys Compd.* **2018**, *739*, 522–528. [CrossRef]

Article

Impact of Pre-Ozonation during Nanofiltration of MBR Effluent

Zoulkifli Amadou-Yacouba, Julie Mendret *, Geoffroy Lesage, François Zaviska and Stephan Brosillon

IEM (Institut Européen des Membranes), UMR 5635 (CNRS-ENSCM-UM2), Université de Montpellier, 34095 Montpellier, France; zoulkifli.amadou-yacouba@umontpellier.fr (Z.A.-Y.); geoffroy.lesage@umontpellier.fr (G.L.); francois.zaviska@umontpellier.fr (F.Z.); stephan.brosillon@umontpellier.fr (S.B.)
* Correspondence: julie.mendret@umontpellier.fr; Tel.: +33-467-144-624

Abstract: This study aimed to investigate the impact of real MBR effluent pre-ozonation on nanofiltration performances. Nanofiltration experiments were separately run with non-ozonated real MBR effluent, ozonated real MBR effluent and synthetic ionic solution mimicking the ionic composition of the real MBR effluent. The specific UV absorbance and the chemical oxygen demand were monitored during ozonation of real effluent, and the mineralization rate was calculated through the quantitative analysis of dissolved organic carbon. The membrane structure was characterized using SEM on virgin and fouled membrane surfaces and after different cleaning steps. The results confirm the low effect of the ozonation process in terms of organic carbon mineralization. However, the chemical oxygen demand and the specific UV absorbance were decreased by 50% after ozonation, demonstrating the efficiency of ozonation in degrading a specific part of the organic matter fraction. A benefic effect of pre-ozonation was observed, as it limits both fouling and flux decrease. This study shows that the partial mineralization of dissolved and colloidal organic matter by ozonation could have a positive effect on inorganic scaling and decrease severe NF membrane fouling.

Keywords: wastewater reuse; organic matter; ozonation; nanofiltration; membrane fouling

Citation: Amadou-Yacouba, Z.; Mendret, J.; Lesage, G.; Zaviska, F.; Brosillon, S. Impact of Pre-Ozonation during Nanofiltration of MBR Effluent. *Membranes* **2022**, *12*, 341. https://doi.org/10.3390/membranes12030341

Academic Editor: Mohammad Peydayesh

Received: 1 March 2022
Accepted: 16 March 2022
Published: 18 March 2022

Publisher's Note: MDPI stays neutral with regard to jurisdictional claims in published maps and institutional affiliations.

Copyright: © 2022 by the authors. Licensee MDPI, Basel, Switzerland. This article is an open access article distributed under the terms and conditions of the Creative Commons Attribution (CC BY) license (https://creativecommons.org/licenses/by/4.0/).

1. Introduction

A promising solution to the challenge of water shortage is to consider urban and industrial wastewaters no longer as wastes but more as renewable resources of water, nutrients and energy. One of the most challenging limiting factors to wastewater reuse is the widespread occurrence of micropollutants in different environmental compartments. To overcome this issue, membrane processes have been demonstrated to remove well micropollutants [1,2]. Among the numerous available membrane processes, nanofiltration is widely recognized for the compromise it offers in terms of selectivity and flux permeability [3,4].

Nonetheless, the fouling propensity remains a very big challenge for a widespread usage of this process [5]. In fact, the nanofiltration membranes seem particularly vulnerable to severe fouling, which constitutes their main drawback. Their propensity to both organic and inorganic fouling was demonstrated by numerous previous studies [6,7]. The reduction of membrane permeability due to fouling causes a substantial increase in operational and maintenance costs and a decrease in effluent quality and membrane lifetime.

As a solution to fouling challenges, numerous authors have investigated the impacts of different types of pretreatment processes such as advanced oxidation. Ozonation, as pretreatment to mitigate the fouling propensity in NF process, is one of the most promising technologies [8–10]. Former studies have pointed out the increase in permeate flux, but very few studies have focused on fouling dynamics in the combined process of ozonation/nanofiltration [11,12]. In particular, there is a lack of data about the specific roles of inorganic salts and organic matter during pre-ozonation of real wastewater before nanofiltration.

The aim of this study was to analyze the NF process applied to a real MBR secondary effluent. Specifically, it consisted in identifying the role of organic and inorganic matters in fouling mechanisms, monitoring the degradation rate of organic matter by an ozonation process and investigating the impact of pre-ozonation on performances of subsequent NF processes. In particular, the effect of the mineralization rate of organic matter on fouling mechanisms in NF was identified.

2. Materials and Methods

2.1. Matrix Used for the Study

2.1.1. Real Secondary Effluent Matrix

An effluent from a full-scale domestic WWTP equipped with MBR, located close to Montpellier, France, was used as real matrix. The plant was designed to treat 13,000 m^3/d of domestic wastewater. The MBR was equipped with KUBOTA Submerged Membrane Unit (SMU RW400) (KUBOTA, London, England flat-sheet microporous membranes made of chlorinated polyethylene (total surface of 16,240 m^2), with an average pore size of 0.2 µm. The characteristics of the MBR permeate are presented in Table 1. SUVA254 is the specific ultraviolet absorbance. The MBR effluent was immediately stored at nearly 4 °C after sampling in order to limit the variation of the composition and re-warmed at room temperature (20 °C ± 1 °C) before conducting the experiments.

Table 1. Characteristics of real MBR effluent ($n = 5$).

Parameters	Unit	Average	Minimum	Maximum
pH		7.40	7.10	7.80
Electric conductivity	µS/cm	3300	2460	3940
TOC	mgC/L	6.70	5.50	8.60
COD	mg O$_2$/L	19.10	13.60	23.00
Absorbance at 254 nm		0.14	0.13	0.16
SUVA$_{254}$	L/mg/m	2.1	1.9	2.4
TSS	mg/L	2.50	2.30	2.70

2.1.2. Synthetic Ionic Solution Matrix Composition

In order to deeply investigate the impact of organic and ionic matter on fouling mechanisms, it was chosen to conduct experiments with matrix free of organic matter. Therefore, a synthetic ionic solution was prepared in ultra-pure water imitating the ionic composition of the real MBR effluent (Table 2).

Table 2. Ionic composition of real MBR effluent and synthetic ionic solution.

Parameters	Unit	Real MBR Effluent	Synthetic Ionic Solution	Diffusion Coefficient k (m^2/s)
Ammonium NH$_4^+$	mg/L	1.80	2.00	0.51
Bromide Br$^-$	mg/L	1.20	0.00	1.46
Calcium Ca^{2+}	mg/L	134.70	130.70	0.58
Chloride Cl$^-$	mg/L	602.10	640.30	1.47
Hydrogen carbonate HCO$_3^-$	mg/L	254.00	290.50	1.00
Magnesium Mg^{2+}	mg/L	48.50	47.50	0.51
Nitrate NO$_3^-$	mg/L	9.00	7.30	1.38
Nitrite NO$_2^-$	mg/L	7.70	0.00	1.39
Orthophosphate PO$_4^{3-}$	mg/L	10.00	9.60	0.44
Potassium K$^+$	mg/L	34.10	30.40	1.42
Sodium Na$^+$	mg/L	321.90	324.10	0.96
Sulfate SO$_4^{2-}$	mg/L	153.70	101.40	0.78

To prepare the solution with the aforementioned ions, different salts were used in the following concentrations (Table 3).

Table 3. Salts used to prepare the synthetic ionic solution.

Compounds	Concentration (mg/L)
NaCl	400
$CaCl_2, 2H_2O$	477
$MgCl_2, 6H_2O$	400
$Na_2HPO_4, 2H_2O$	18
Na_2SO_4	150
$NaHCO_3$	400
KCl	60
$NaNO_3$	10
NH_4Cl	2

2.2. Nanofiltration Experiments

2.2.1. Membrane Selection and Characterization

The membrane used for this study is an NF-90 polyamide membrane from DOW Filmtec. It is considered as a "tight" NF membrane with an estimated MWCO around 150 Da. Before experiments, each membrane was firstly soaked in ultrapure water to remove preservative agent and then compacted at 18 bars for at least one hour or until stability of the permeate flux was reached. Thereafter, the membranes were fully characterized in terms of pure water permeability and sodium chloride rejection, with values corresponding to 8.4 ± 1.0 L h^{-1} m^{-2} bar^{-1} and $88 \pm 4\%$, respectively.

2.2.2. Cross-Flow Nanofiltration Unit and Experimental Protocol

The filtration experiments were carried out with 140 cm^2 flat-sheet membrane samples in an Osmonics Sepa CF II cell (Sterlitech Corp., Auburn, WA, USA). The Sepa cell was fed by a pump Hydra-Cell, Wanner Engineering, Inc, Minneapolis, MN, USA) with the solution from a 16 L feed vessel (Figure 1). The wastewater temperature was kept constant (20 ± 1 °C) using a cryothermostat (F32, Julabo, Seelbach, Germany). The bench-scale NF experiments were performed at a cross-flow velocity (vT) of 0.5 m s^{-1} with a medium foulant spacer, 47 Mil (1.194 mm). The transmembrane pressure (TMP) was set constant at 10 bars using a micrometric pressure control valve located on the retentate outlet. The membrane performances were monitored throughout the filtration experiment at ~0%, 15%, 40% and 60% until reaching 80% of water recovery (or the maximum water recovery rate reachable in case of earlier severe fouling). The flux was recorded throughout the experiment by measuring the permeate weight every 60 s. Retentate and permeate samples were collected for physico-chemical analysis. The volume of the collected sample for different analyses was considered in the apparent rejection determination. Considering that the NF system is made of stainless steel and all the tubing is in Teflon, it was assumed that compounds (organic and inorganic matter) adsorption was exclusively occurring on membrane material.

To evaluate the impact of organic and inorganic matters on membrane fouling mechanisms, three types of NF experiments were run: (1) non-ozonated real MBR effluent, (2) synthetic ionic solution and (3) ozonated real MBR effluent.

2.2.3. Membrane Fouling Propensity Test

After each filtration experiment, the NF unit was cleaned first by ultrapure water cleaning, then recirculating caustic soda (NaOH, 2%) for 6 h and finally recirculating acid solution (HNO$_3$, 2%) for 6 h. After each base and acid cleaning, the system was fully rinsed with deionized water until a conductivity of 50 µS cm^{-1} and a neutral pH were reached in the NF permeate. Membrane fouling was characterized according to the flux recovery after effluent filtration and after different cleaning steps. Reversible fouling was estimated immediately after ultrapure water cleaning by comparison with water flux before the filtration, at the beginning of the experiment. Then, the irreversible fouling was determined using chemical cleaning. Two types of irreversible fouling were distinguished: organic irreversible fouling evaluated by the determination of flux recovery after NaOH cleaning and inorganic irreversible fouling (scaling) determined after acid cleaning (HCl). Flux was measured after these cleaning steps and compared to the initial flux so as to estimate the flux recovery proportion of each type of cleaning.

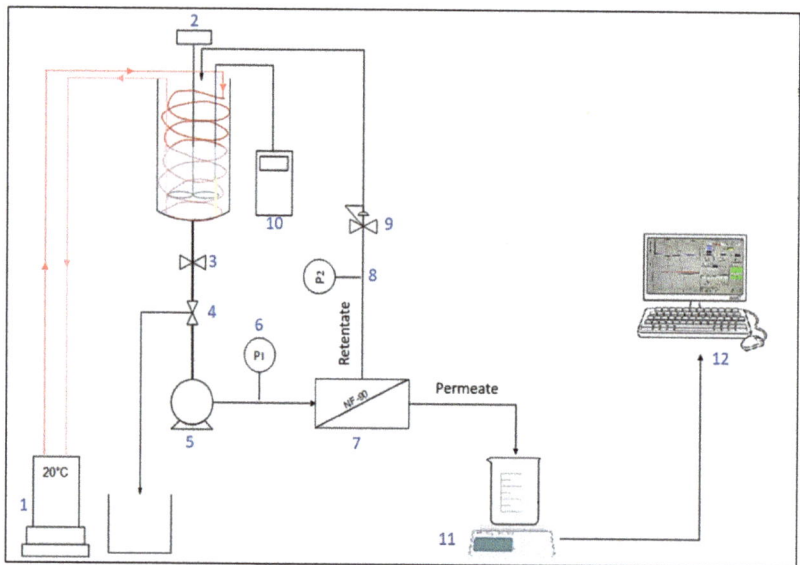

Figure 1. Experimental setup of nanofiltration bench-scale pilot. (**1**) Cryothermostat. (**2**) Mechanical stirrer. (**3**) Tank isolation valve. (**4**) Valve for sampling. (**5**) Pump. (**6**) And. (**8**) Pressure sensors. (**7**) Filtration unit. (**9**) Pressure control valve. (**10**) Conductivity meter. (**11**) Precision scale. (**12**) Data processing.

Membrane surface morphology, for virgin and fouled membranes and after each cleaning step, were characterized with a Scanning Electron Microscope (SEM, Hitachi Table top Microscope S-4800) interfaced with an Energy-Dispersive X-ray (EDX) spectroscopy system (Thermo-Fisher, Waltham, MA, USA). Membrane samples were coated with a thin layer of gold before SEM analysis. EDX measurements were performed at different locations on the membrane surface, in order to obtain a comprehensive elemental composition. SEM micrographs were obtained at an accelerating voltage of 2 kV and magnification of 25,000.

2.2.4. Osmotic Pressure

The difference in osmotic pressure ($\Delta\pi$) between feed and permeate sides of the membrane was calculated using Equation (1) [13]:

$$\Delta\pi = \pi_{feed} - \pi_{perm} \tag{1}$$

with π feed representing osmotic pressure in the feed side and π perm representing osmotic pressure in permeate side.

The NF removal was high, and the ions concentrations (and consequently the induced osmotic pressure) at permeate side were negligible compared to that of feed side.

For each ion, the osmotic pressure is given by Equation (2) [13]:

$$\pi = C \cdot R \cdot T \tag{2}$$

For all the identified ions, Equation (3) enables estimation of π [13]:

$$\pi = R \cdot T \sum_{i=1}^{n} C_i \qquad (3)$$

with:
- R: gas constant (= 8.314 J/mol K);
- T: temperature of solution (°K);
- C: concentration of ion (mol/m^3);
- n: number of ions in the solution.

2.2.5. Concentration Polarization

Due to concentration polarization, the osmotic pressure is not homogeneous in feed solution. In fact, the ions concentration and the induced osmotic pressure (π) are more important at membrane surface (π_{memb}) than in the bulk solution (π_{bulk}). These values are linked by the relation given in Equation (4) [13]:

$$\pi_{memb} = \pi_{bulk} \cdot e^{\frac{Jp}{k}} \qquad (4)$$

with:
- k: diffusion coefficient (m^2/s);
- Jp: flux (m^3/s/m^2).

2.3. Bench-Scale Ozonation System Setup

Experiments were performed in a glass stirred batch reactor (Vreactor = 3 L) where the liquid solution is maintained at room temperature (20 °C) using a cryothermostat (Figure 2). The ozone was continuously produced from a lab-grade pure oxygen tank by an ozone generator (BMT 803 N). Before diffusion in the reactor, the ozone was diluted with oxygen at a gas flow of 60 L h^{-1} and introduced through a porous diffuser at the bottom of the reactor. The gas ozone concentration ([O$_3$]gas,in) was monitored after dehumidification by an ozone gas analyzer (BMT 964). The impact of pre-ozonation on NF process was investigated for 30 min reaction contact time, and the dissolved ozone dose (TOD) was determined using indigo method [14].

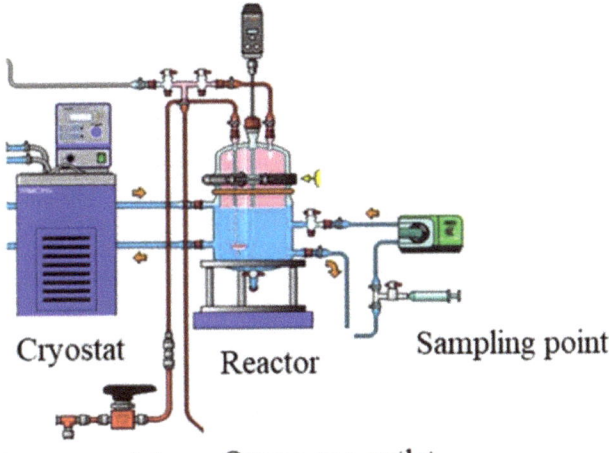

Figure 2. Experimental setup of ozonation bench-scale pilot.

The desired oxygen/ozone ratio was determined using two electro-valves connected to the monitoring software. The ozone dissolution rate was increased in the solution us-ing an agitator (400 rpm). The experiment consisted in applying an ozone gas concen-tration of 5 gO_3/Nm^3 to determine the transferred ozone dose through Equation (5).

$$TOD = \frac{(C_{ge} - C_{gs}) * Q_g * t}{V_{reactor}} \quad (5)$$

with:

TOD: transferred ozone dose (gO_3/m^3);
C_{ge}: gas-phase ozone inlet concentration (g/Nm^3);
C_{gs}: gas-phase ozone outlet concentration (g/Nm^3);
Q_g: gas flow (m^3/h);
t: reaction time (h);
$V_{reactor}$: reactor volume (m^3).

Finally, the specific ozone dose [O_3]$_{specific}$ was calculated with Equation (6):

$$[O_3]specific = TOD/TOC \quad (6)$$

with:

TOC: total organic carbon (gC/m^3)

2.4. Chemical Analysis

2.4.1. Ionic Chromatography

The concentrations of ionic compounds were determined in all samples by ionic chromatography:

- Anionic compounds concentrations were determined with an ICS 1000 system (Thermo-Fisher, Waltham, MA, USA) equipped with a Dionex AS19 column fed by an eluent flow rate of 1 mL·min^{-1}. A KOH eluent was used as mobile phase through the following gradient: 10 mM for 10 min, then 45 mM for 20 min and 10 mM for 10 min.
- Cationic compounds concentrations were determined with an ICS 900 system (Thermofisher Dionex, France) equipped with a Dionex CS12A column fed by 20 mM methanesulfonic acid at a flow rate of 1 mL·min^{-1}.

2.4.2. Global Indicators for Pollution Monitoring: TOC, UV254 and SUVA Analysis

The specific UV absorbance (SUVA254) corresponds to the ratio of UV absorbance at wavelength of 254 nm, measured in a 1 cm quartz cuvette using a UV–vis spectrophotometer (UV-2401PC, Shimadzu, Kyoto, Japan) and TOC value [15]. TOC analysis was performed using a TOC-VCSN Shimadzu analyzer (Shimadzu Japan).

2.4.3. Scanning Electron Microscopy (SEM)

A Hitachi Microscope (Hitachi S4800 SEM) was used to inspect surfaces of the virgin and pre-fouled membranes. Small pieces were cut from the surfaces of membranes (post-mortem analysis). Before analysis, the samples were dried in desiccator until measurement in order to remove residual moisture and then metalized with platinum. The surfaces of fouled and virgin membrane were magnified 5000–15,000 times.

3. Results

3.1. Flux Evolution and Fouling Mechanisms during Nanofiltration

One of the criteria to evaluate NF efficiency is the evolution of the permeate flux with the time of filtration. The recovery rate (Y) was calculated corresponding to the ratio between the extracted permeate volume and initial feed volume. In order to compare the flux evolution for different experiments, the relative flux corresponding to the ratio between the flux at any time (J) and the initial flux (J0) was considered. Figure 3 presents the normalized flux (J/J0) during nanofiltration of real MBR effluent under a TMP of 10 bars.

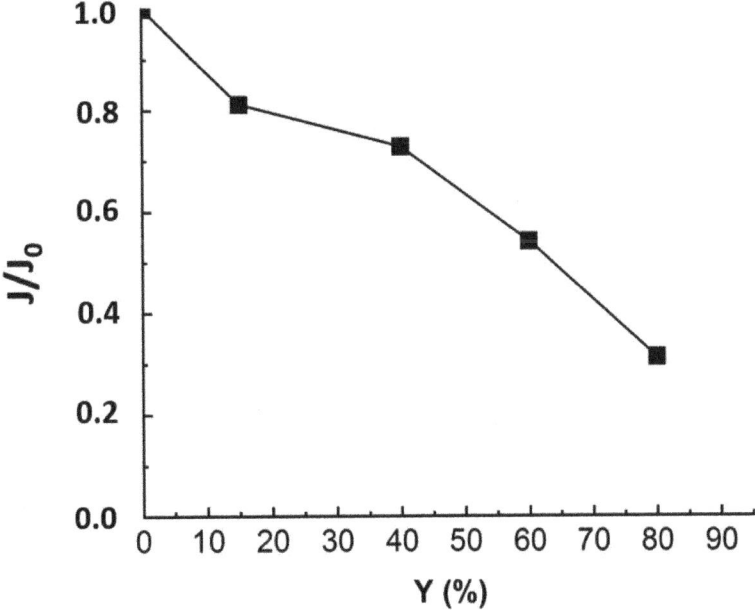

Figure 3. Flux evolution in NF experiment applied to MBR real effluent matrix: TMP = 10 bars, T° = 20 °C, J_0 = 53 L·m^{-2}·h^{-1}, duration of the experiment = 24 h.

Figure 3 revealed a drop of almost 70% in the initial flux value when reaching the maximum conversion rate of 80%. As established in previous studies, the main fouling mechanism during MBR effluent filtration by NF is organic fouling [11]. According to some authors, the reason that could explain the flux drop is that the organic matters, particularly those with higher MW and hydrophobicity, corresponding to humic-like substances, deposited into the pores and onto the membrane [16–18]. The deposited organics enhance gel layer formation, which was related to the rapid flux decline at the first stage. Then, the slower flux decrease could come from gel layer compaction and interactions between inorganic salts and organic matter deposited on the membrane surface [19]. For instance, Lin et al. have studied the roles of organic, inorganic and biological fouling along with NF applied to raw effluent. The authors noticed that organic/inorganic binary fouling became dominant, contributing up to 39.7% of flux decline due to metal/organic complexation [7]. The third stage, corresponding to a more pronounced flux drop, could come from concentration polarization [5,17]. Nonetheless, to establish a clear distinction between the impacts of organic and inorganic contributions to flux decline, it is required to run NF experiments with OM-free matrix.

3.2. Influence of Ionic Matrix during Nanofiltration

To evaluate the impact of organic matter on fouling, experiments were run with synthetic ionic solution (SIS) mimicking the ionic composition of the real MBR effluent. The flux was monitored along with permeate recovery rate and is presented in Figure 4 with that of real MBR effluent matrix.

The Figure 4 revealed a decline of 75% in permeate flux at 60% of recovery for SIS solution. The occurrence of the severe fouling may be linked to an inner fouling caused by ionic compounds. In fact, as the organic matter playing the role of competitor in ions adsorption is no longer present in solution, the ions are free to adsorb onto membrane surfaces and enhance membrane fouling while diffusing through membrane pores. Thus, it was not possible to reach such high conversion rates as with real effluent (Y = 60% instead of 80%). Teixeira and Rosa have studied the impact of the water inorganic matrix on the permeate flux and the natural organic matter (NOM) removal by nanofiltration [20]. They noticed a decrease in flux in the presence of calcium. According to the authors, the flux and rejection decreased further in the presence of 1 mM Ca^{2+}, which reduced the membrane negative charge and sieving effects and increased chemical interactions. In fact, in the present study, all the detected ionic composition was mimicked by a synthetic ionic solution free of OM that could compete with the membrane in adsorbing the inorganic and mitigate the inorganic fouling.

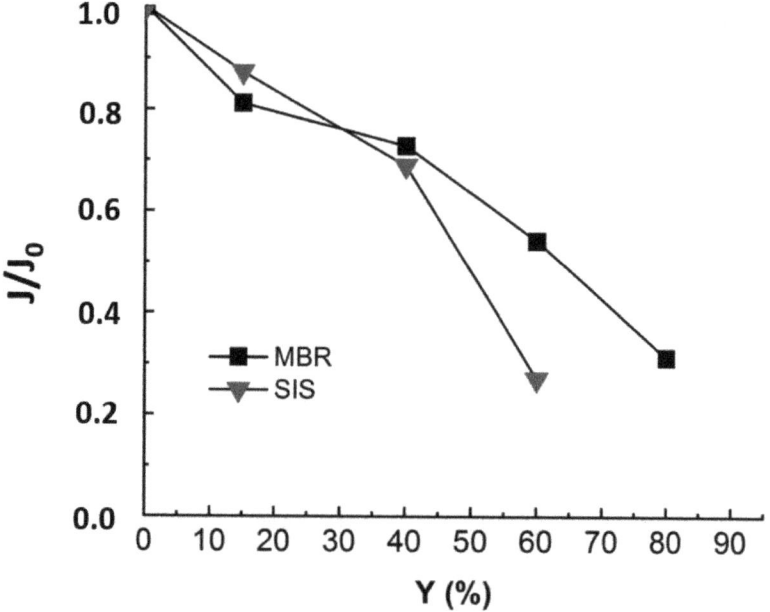

Figure 4. Flux evolution during NF experiment applied to MBR real effluent (duration of the experiment = 24 h) and synthetic ionic solution matrixes SIS (duration of the experiment = 18 h). TMP = 10 bars, T° = 20 °C, J_0-SIS = 64 L·m^{-2}·h^{-1}, J_0-MBR = 53 L·m^{-2}·h^{-1}.

3.2.1. Influence of the Osmotic Pressure

During the nanofiltration experiments, the ionic compounds became more and more concentrated and induced an osmotic pressure, which is supposed to increase with permeate recovery. The osmotic pressure constitutes a resistance to physical pressure and should be overcome in order to get permeate flux through the membrane. The differential osmotic pressure between retentate and permeate streams was calculated for real MBR effluent and synthetic ionic solution and compared in Figure 5.

The monitoring of the osmotic pressure revealed that it increases with permeate recovery rate from around 1 bar at the beginning to 2.2 bars at 60% of recovery rate for SIS solution and up to 4 bars for the MBR effluent matrix. The Figure 5 clearly displays a similarity in the evolution of osmotic pressure for both real MBR effluent and the synthetic ionic solution mimicking the MBR ionic composition, even though the permeate flux drastically dropped in the case of SIS much earlier than in the MBR effluent case (Figure 4). This result confirms the suspected inner fouling due to inorganic scaling. As the solution is free of OM, which would adsorb the ions, they are free to interact with each other and with the membrane, enhancing the scaling [21,22].

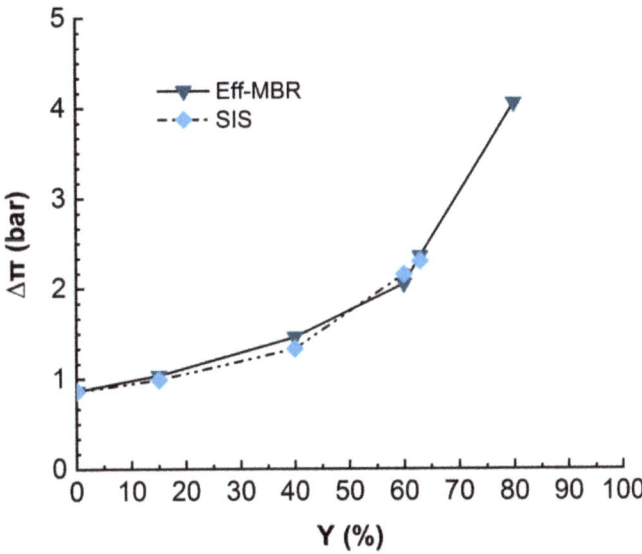

Figure 5. Evolution of differential osmotic pressure in NF for real MBR effluent (duration of the experiment = 24 h) and SIS (duration of the experiment = 18 h). TMP = 10 bars, T° = 20 °C.

3.2.2. Characterization of Membrane Fouling

At the end of each experiment, the membrane goes through different cleaning steps beginning with ultrapure water, followed by basic and acid-based cleanings, respectively. Scanning Electron Microscopy (SEM) analysis was applied to samples from membrane used for both real MBR effluent and SIS. Samples of virgin and fouled membrane and membrane after the different cleaning steps were used, and the results are presented in Figure 6.

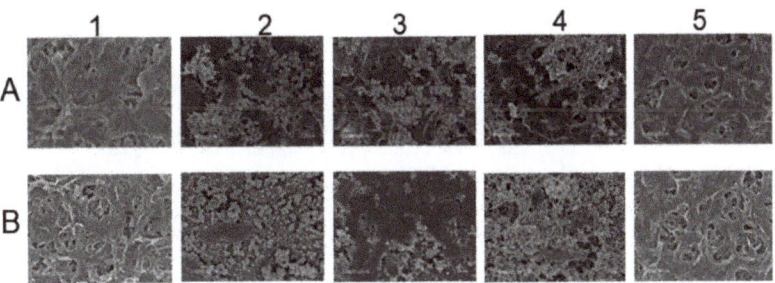

Figure 6. SEM of membrane surfaces at different states: (**A**) For membrane used in MBR effluent experiment, (**B**) Membrane used in SIS experiment. 1. Virgin membrane, 2. Fouled after experiment, 3. UPW-cleaned membrane, 4. Base-cleaned and 5. Acid-cleaned.

Figure 6 visually illustrates the membrane surface state throughout the different steps. A mixture of inorganic and dissolved organic matter can be noticed on the fouled membrane used with the real MBR effluent (2.A), while the membrane fouled with SIS (2.B) displays disaggregated inorganic compounds only. The subsequent cleaning methods helped to identify the type of fouling that occurred during these experiments through foulants characterization [16,23]. In fact, for the membrane fouled by real MBR effluent, while the ultrapure-water-based cleaning likely removed part of the fouling matter (3.A), the sodium hydroxide cleaning significantly removed it, except for some inorganics (4.A) that were totally removed by hydrogen chloride acid washing (5.A). For the SIS-fouled membrane, on the other hand, the ultrapure-water-based cleaning was able to remove part of scaling (3.B). The sodium hydroxide cleaning was not able to remove the inorganics on the membrane surface (4.B). Only the acid cleaning totally recovered the fouled membrane surface to almost virgin state (5.B).

3.3. Influence of Pre-Ozonation during Nanofiltration of Real MBR Effluent

3.3.1. Monitoring of Organic Matter

To evaluate the impact of organic matter and its degradation by ozone on the performances of nanofiltration process, the mineralization rate of organic matter was monitored during ozonation process, and the results are given in Figure 7.

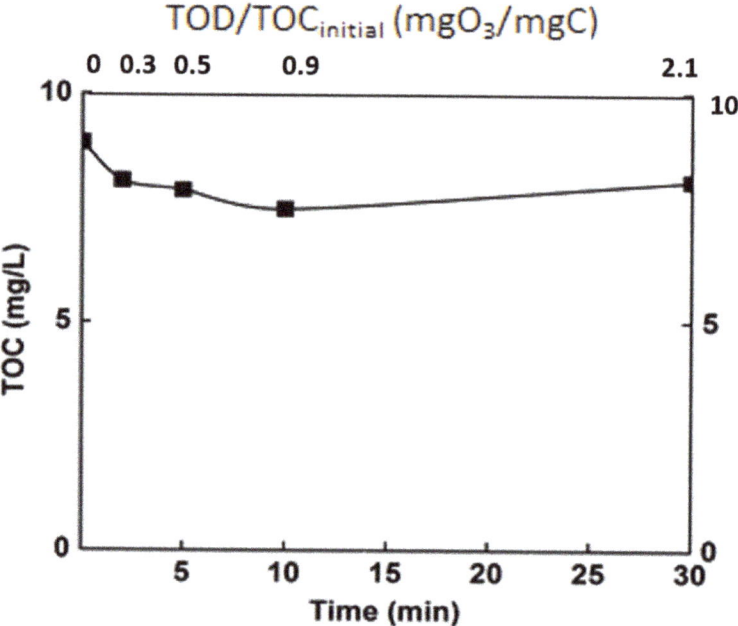

Figure 7. Monitoring of the mineralization rate of the real matrix during ozonation. T° = 20 °C, Vreactor = 3 L, Vstir = 400 rpm, [O_3]gas = 5 gO_3/Nm^3.

Ozonation, as revealed by some previous studies, is not sufficient to completely degrade organic matter [8]. This is confirmed by the current study, in which only a mineralization of 15% was achieved after 30 min (TOC around 8 mg/L). Even though the mineralization rate was relatively low, the ozonation engendered an important change in the organic matter. Indeed, even if the mineralization of the organic matter was moderate, chemical changes occur, and the efficiency of ozonation in terms of modification of organic matter structure was monitored through some common parameters. The chemical oxygen demand and the specific UV absorbance (SUVA254) are some of these indicators (Figure 8).

Both of the two parameters indicate the efficiency of ozonation process in oxidizing organic matter. The COD decreased from 33 mgO_2/L to 23 mgO_2/L after 3 min reaction time and to less than 20 mgO_2/L at 30 min of reaction time (Figure 8a), corresponding to the introduction of oxygen in the chemical structure of the organic matter. This level of mineralization was already observed by Gong et al. and Justo et al. [15,24]. In addition, the ozonation decreased the SUVA by half after 30 min reaction time; this indicates the opening of the double bond mainly in the aromatic group. This parameter is a good indicator of the change in the chemical structure of the organic matter [25,26] (Figure 8b).

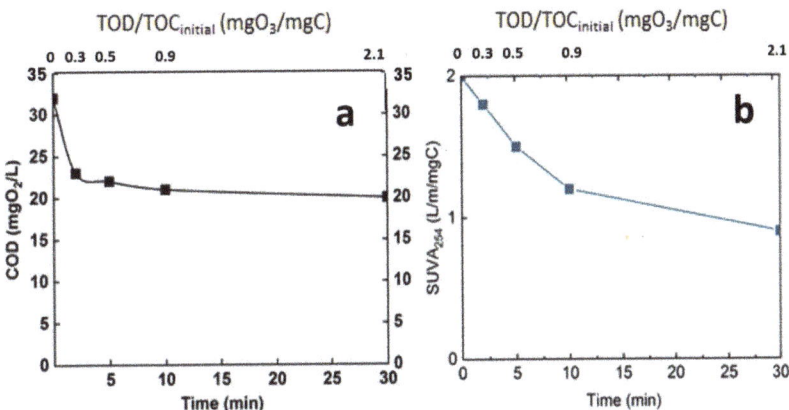

Figure 8. Evolution of global parameters of the real matrix during ozonation. (**a**): COD, (**b**): SUVA254, T° = 20 °C, Vreactor = 3 L, Vstir = 400 rpm, [O$_3$]gas = 5 gO$_3$/Nm^3.

3.3.2. Nanofiltration of Ozonated Real MBR Effluent

During the NF experiment applied to the ozonated real MBR effluent, the flux evolution was monitored, and the relative flux is displayed in Figure 9 in comparison with non-ozonated real MBR and SIS matrix.

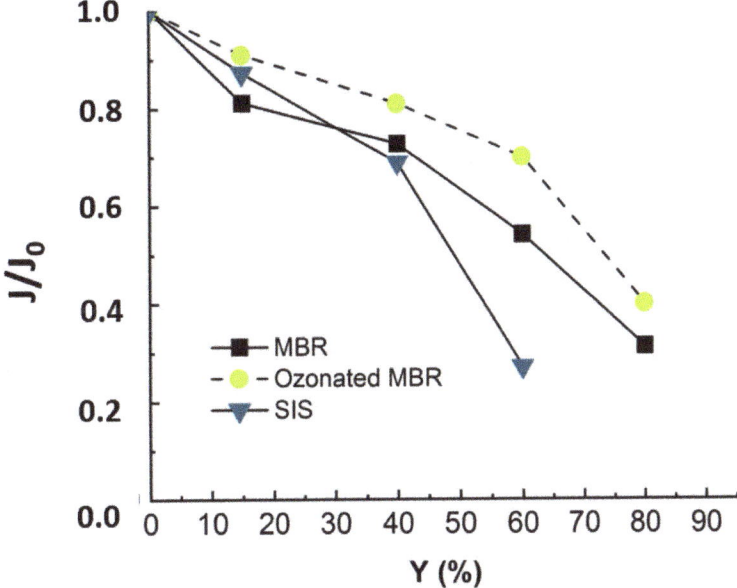

Figure 9. Flux evolution in NF experiment applied to synthetic ionic solution (duration of the experiment = 18 h) and non-ozonated (duration of the experiment = 24 h) and ozonated MBR real effluent matrixes (duration of the experiment = 24 h). Ozonation reaction time = 30 min, T° = 20 °C, Vreactor = 3 L, Vstir = 400 rpm, [O$_3$]gas = 5 gO$_3$/Nm^3 TMP = 10 bars, J$_0$-MBR = 53 L·m^{-2}·h^{-1}, J$_0$-MBR + O$_3$ = 54 L·m^{-2}·h^{-1}, J$_0$-SIS = 64 L·m^{-2}·h^{-1}.

Figure 9 reveals that when the nanofiltration experiment is run with ozonated real MBR effluent, the drop in flux trends is slightly slower, as around 10% of flux was re-covered by pre-ozonation. Even though the ozonation is not efficient in terms of mineralization, it changes the structure of the organic matter [12]. According to the cited authors, pre-ozonation increases the hydrophilic fraction and anionic charge of organics and alters their size distribution [1,12]. In fact, the gel layer (coming from organic and inorganic complexation) was demonstrated to be responsible for membrane fouling. Therefore, the ozonation, by degrading part of this gel layer, leads to improve nanofiltration conditions by reducing the fouling celerity [11].

When the NF experiment was applied to non-ozonated real MBR effluent, the fouling essentially came from complexation of organic and inorganic matter [11]. When the nanofiltration experiment was run with SIS free of organic matter, the drop in flux trends was much more severe and occurred earlier (Figure 9). This demonstrated that during NF experiments, the propensity to inorganic fouling is much higher than that of organic fouling. After ozonating the real MBR, the trend of permeate flux in NF is improved due to delayed fouling, as the ozonation was insufficient to totally mineralize the effluent organic matter. These results demonstrate that the main drawback of the ozonation process, which is its limited mineralization rate, rather constitutes an advantage for a subsequent NF process: the residual organic matter prevents a severe inorganic fouling by competing with the membrane for adsorption of inorganics. For instance, Li et al. have studied the operational optimization and membrane fouling analysis of nanofiltration in municipal wastewater advanced treatment [16]. One of the main conclusions they came to is that inorganic fouling was mitigated because the inorganics were assumed to adsorb on the effluent organic matter.

3.3.3. Cleaning and Nanofiltration Performances Recovery

Two other parameters used to characterize the fouling that occurred during NF experiments are the type of cleaning and the rate of flux that it allowed to be recovered. The values of flux at the beginning and the end of the experiments are recapitulated for all matrixes in Table 4.

Table 4. Values of permeate flux at the beginning and end of studied NF experiments.

Matrixes Used for NF Experiments	Unit	Flux at the Beginning of the Filtration Experiment	Flux at the End of the Filtration Experiment
Non-ozonated MBR effluent	$L \cdot m^{-2} \cdot h^{-1}$	53	17
SIS	$L \cdot m^{-2} \cdot h^{-1}$	64	17
Ozonated MBR effluent	$L \cdot m^{-2} \cdot h^{-1}$	54	21

Hence, after each nanofiltration experiment, ultrapure water was used to clean the membrane, and the permeability was measured. A basic cleaning using NaOH (0.1 N) and acid cleaning using HCl (0.1 N) were successively performed as well. It consisted in imbibing the membrane in the cleaning solution for 6 h for both chemical solutions. The membrane permeability recovery rates were determined for the three studied matrixes and are presented in Figure 10.

According to Figure 9, the SIS induced more severe fouling than the real MBR effluent. Then, ultrapure water cleaning enabled 53%, 69% and 15% flux recovery for MBR effluent, ozonated MBR effluent and SIS solution, respectively. Fouling corresponding to both MBR effluent and ozonated MBR effluent thus seem easier to remove, which is consistent with Figure 9. Sodium-hydroxide-based cleanings allowed non-negligible permeate flux recovery in the cases of real MBR effluent (35%) and ozonated MBR effluent (18%) and a significant flux recovery of SIS-fouled membrane (60%), for which chemicals are needed. The acid-based cleaning allowed the most important recovery for SIS solution, which is consistent with the inorganic nature of fouling in this case. According to Li et al., the water flushing samples after nanofiltration of wastewater were essentially composed of low MW with high intensity, which is typically related to humic substances, indicating that the humic substances could be removed easily by physical cleaning, which is in accordance with the present result during ultrapure water cleaning [16]. In the case of severe inorganic fouling, the acid-based cleaning is required for flux recovery [17,23,27].

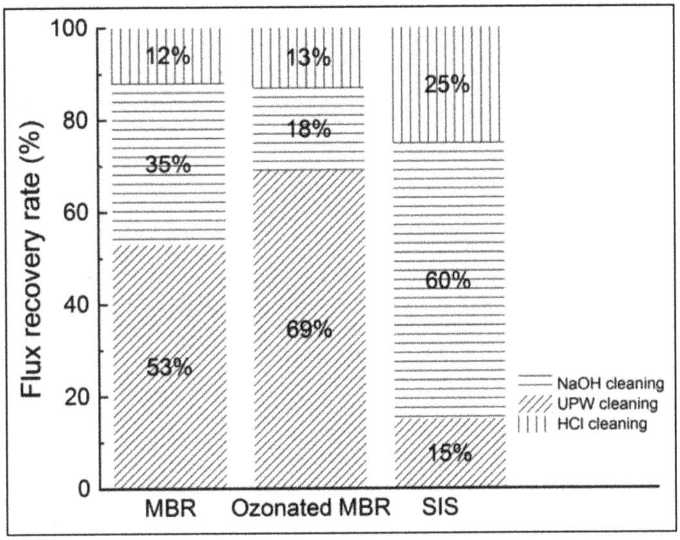

Figure 10. Flux recovery after pure water, NaOH and HCl cleanings in NF experiments applied to non-ozonated and ozonated real MBR effluent and synthetic ionic solution.

4. Conclusions

This research aimed to evaluate the impact of pre-ozonation on fouling propensity in nanofiltration. The fouling was mainly due to organics and inorganics complexation forming a gel layer (70% drop in flux at 80% of permeate recovery). When the NF experiment was run with an organics-free synthetic ionic solution, the fouling was more severe because of the high propensity of NF to inorganic fouling (75% drop in flux at 60% of permeate recovery). When the ozonated real MBR effluent was used for NF experiment, not only was the fouling delayed (62% drop in flux at 80% of permeate recovery), but the flux recovery was improved as well by a mere water cleaning. Therefore, pre-ozonating the effluent presents two advantages: it allows economical use of chemicals needed for chemical cleaning, and it contributes to improving the membrane lifetime by delaying chemical cleaning.

The SEM analysis confirmed that the acid cleaning was the most efficient to recover a virgin membrane state, even though the ultrapure water and basic cleanings can allow recovering an important part of flux, depending on the type of fouling linked to the nature of matrix used for the experiment.

The results demonstrate that the low mineralization rate of ozonation process is of high value to preventing a severe inorganic fouling. It mitigates the organic fouling by degrading partially and modifying the molecular structures of organic matter, which improves its hydrophilicity. On the other hand, the remaining organic matter, which resulted from the partial mineralization, prevented the membrane from a severe fouling, as a total mineralization would lead to occurrence of inorganic scaling. To sum up, ozonation might be the best AOP to couple with an NF process for better organic and inorganic fouling mitigation for wastewater reuse.

Author Contributions: Conceptualization, J.M. and G.L.; methodology, Z.A.-Y.; validation, S.B.; formal analysis, F.Z.; investigation, Z.A.-Y.; resources, J.M.; data curation, Z.A.-Y.; writing—original draft preparation, Z.A.-Y.; writing—review and editing, J.M., G.L., S.B.; visualization, Z.A.-Y.; supervision, S.B.; project administration, J.M.; funding acquisition, J.M. All authors have read and agreed to the published version of the manuscript.

Funding: This research was funded by ANR, grant number SAWARE ANR-16-CE04-0002-01.

Institutional Review Board Statement: Not applicable.

Informed Consent Statement: Not applicable.

Conflicts of Interest: The authors declare no conflict of interest.

Abbreviations

AOP	Advanced Oxidation Process
COD	Chemical Oxygen Demand (gO_2/m^3)
Da	Dalton
DCOM	Dissolved and Colloidal Organic Matter
DWW	Domestic Wastewater
EDX	Energy-Dispersive X-ray
MBR	Membrane Bioreactor
MWCO	Molecular Weight Cut-Off
NF	Nanofiltration
NOM	Natural Organic Matter
OM	Organic Matter
$[O_3]gas$	Applied gas ozone concentration (gO_3/Nm^3)
SEM	Scanning Electron Microscopy
SIS	Synthetic Ionic Solution
T°	Temperature
TMP	Transmembrane Pressure (bar)
TOC	Total Organic Carbon
TOD	Transferred Ozone Dose
TSS	Total Suspended Solid (mg/L)
UPW	Ultrapure Water
UV254	UV absorbance at 254 nm of wavelength
v	Cross-flow velocity (m/s)
Vreactor	Volume of reactor (m3)
Vstir	Stirring velocity (rpm)
WWTP	Wastewater Treatment Plant
Y	Permeate recovery rate (%)

References

1. Alturki, A.A.; Tadkaew, N.; McDonald, J.A.; Khan, S.J.; Price, W.E.; Nghiem, L.D. Combining MBR and NF/RO membrane filtration for the removal of trace organics in indirect potable water reuse applications. *J. Membr. Sci.* **2010**, *365*, 206–215. [CrossRef]
2. Rizzo, L.; Gernjak, W.; Krzeminski, P.; Malato, S.; McArdell, C.S.; Perez, J.A.S. Best available technologies and treatment trains to address current challenges in urban wastewater reuse for irrigation of crops in EU countries. *Sci. Total Environ.* **2010**, *710*, 136312. [CrossRef] [PubMed]
3. Bellona, C.; Heil, D.; Yu, C.; Fu, P.; Drewes, J.E. The pros and cons of using nanofiltration in lieu of reverse osmosis for indirect potable reuse applications. *Sep. Purif. Technol.* **2012**, *85*, 69–76. [CrossRef]
4. Khanzada, N.K.; Farid, M.U.; Kharraz, J.A.; Choi, J.; Tang, C.Y.; Nghiem, L.D. Removal of organic micropollutants using advanced membrane-based water and wastewater treatment: A review. *J. Membr. Sci.* **2020**, *598*, 117672. [CrossRef]
5. Azaïs, A.; Mendret, J.; Petit, E.; Brosillon, S. Evidence of solute-solute interactions and cake enhanced concentration polarization during removal of pharmaceuticals from urban wastewater by nanofiltration. *Water Res.* **2016**, *104*, 156–167. [CrossRef]
6. Jarusutthirak, C.; Amy, G. Role of Soluble Microbial Products (SMP) in Membrane Fouling and Flux Decline. *Environ. Sci. Technol.* **2016**, *40*, 969–974. [CrossRef]
7. Lin, W.; Li, M.; Xiao, K.; Huang, X. The role shifting of organic, inorganic and biological foulants along different positions of a two-stage nanofiltration process. *J. Membr. Sci.* **2020**, *602*, 117979. [CrossRef]
8. Byun, S.; Taurozzi, J.S.; Tarabara, V.V. Ozonation as a pretreatment for nanofiltration: Effect of oxidation pathway on the permeate flux. *Sep. Purif. Technol.* **2015**, *149*, 174–182. [CrossRef]

9. Park, M.; Anumol, T.; Simon, J.; Zraick, F.; Snyder, S.A. Pre-ozonation for high recovery of nanofiltration (NF) membrane system: Membrane fouling reduction and trace organic compound attenuation. *J. Membr. Sci.* **2017**, *523*, 255–263. [CrossRef]
10. Mansas, C.; Atfane-Karfane, L.; Petit, E.; Mendret, J.; Brosillon, S.; Ayral, A. Functionalized ceramic nanofilter for wastewater treatment by coupling membrane separation and catalytic ozonation. *J. Environ. Chem. Eng.* **2020**, *8*, 104043. [CrossRef]
11. Vatankhah, H.; Murray, C.C.; Brannum, J.W.; Vanneste, J.; Bellona, C. Effect of pre-ozonation on nanofiltration membrane fouling during water reuse applications. *Sep. Purif. Technol.* **2018**, *205*, 203–211. [CrossRef]
12. Yu, W.; Liu, T.; Crawshaw, J.; Liu, T.; Graham, N. Ultrafiltration and nanofiltration membrane fouling by natural organic matter: Mechanisms and mitigation by pre-ozonation and pH. *Water Res.* **2018**, *139*, 353–362. [CrossRef]
13. Johnson, D.; Hashaikeh, R.; Hilal, N. Basic principles of osmosis and osmotic pressure. In *Osmosis Engineering*; Elsevier: Amsterdam, The Netherlands, 2021; Volume 1, pp. 1–15.
14. Bader, H.; Hoigné, J. Determination of ozone in water by the indigo method. *Water Res.* **1981**, *15*, 449–456. [CrossRef]
15. Justo, A.; González, O.; Aceña, J.; Pérez, S.; Barceló, D.; Sans, C. Pharmaceuticals and organic pollution mitigation in reclamation osmosis brines by UV/H_2O_2 and ozone. *J. Hazard. Mater.* **2013**, *263*, 268–274. [CrossRef] [PubMed]
16. Mänttäri, M.; Puro, L.; Nuortila-Jokinen, J.; Nyström, M. Fouling effects of polysaccharides and humic acid in nanofiltration. *J. Membr. Sci.* **2000**, *165*, 1–17. [CrossRef]
17. Li, K.; Wang, J.; Liu, J.; Wei, Y.; Chen, M. Advanced treatment of municipal wastewater by nanofiltration: Operational optimization and membrane fouling analysis. *J. Environ. Sci.* **2016**, *43*, 106–117. [CrossRef]
18. Lan, Y.; Groenen-Serrano, K.; Coetsier, C.; Causserand, C. Nanofiltration performances after membrane bioreactor for hospital wastewater treatment: Fouling mechanisms and the quantitative link between stable fluxes and the water matrix. *Water Res.* **2018**, *146*, 77–87. [CrossRef]
19. Fersi, C.; Gzara, L.; Dhahbi, M. Flux decline study for textile wastewater treatment by membrane processes. *Desalination* **2009**, *244*, 321–332. [CrossRef]
20. Teixeira, M.R.; Rosa, M.J. The impact of the water background inorganic matrix on the natural organic matter removal by nanofiltration. *J. Membr. Sci.* **2006**, *279*, 513–520. [CrossRef]
21. Anwar, N.; Rahaman, M.S. Membrane desalination processes for water recovery from pre-treated brewery wastewater: Performance and fouling. *Sep. Purif. Technol.* **2020**, *252*, 117420. [CrossRef]
22. Song, W.; Lee, L.Y.; Liu, E.; Shi, X.; Ong, S.L.; Ng, H.Y. Spatial variation of fouling behavior in high recovery nanofiltration for industrial reverse osmosis brine treatment towards zero liquid discharge. *J. Membr. Sci.* **2020**, *609*, 118185. [CrossRef]
23. Chon, K.; Sarp, S.; Lee, S.; Lee, J.-H.; Lopez-Ramirez, J.A.; Cho, J. Evaluation of a membrane bioreactor and nanofiltration for municipal wastewater reclamation: Trace contaminant control and fouling mitigation. *Desalination* **2011**, *272*, 128–134. [CrossRef]
24. Gong, J.; Liu, Y.; Sun, X. O_3 and UV/O_3 oxidation of organic constituents of biotreated municipal wastewater. *Water Res.* **2008**, *42*, 1238–1244. [CrossRef] [PubMed]
25. Weishaar, J.L.; Aiken, G.R.; Bergamaschi, B.A.; Fram, M.S.; Fujii, R.; Mopper, K. Evaluation of Specific Ultraviolet Absorbance as an Indicator of the Chemical Composition and Reactivity of Dissolved Organic Carbon. *Environ. Sci. Technol.* **2003**, *37*, 4702–4708. [CrossRef] [PubMed]
26. Bahr, C.; Schumacher, J.; Ernst, M.; Luck, F.; Heinzmann, B.; Jekel, M. SUVA as control parameter for the effective ozonation of organic pollutants in secondary effluent. *Water Sci. Technol.* **2007**, *55*, 267–274. [CrossRef]
27. Kim, Y.; Li, S.; Ghaffour, N. Evaluation of different cleaning strategies for different types of forward osmosis membrane fouling and scaling. *J. Membr. Sci.* **2020**, *596*, 117731. [CrossRef]

Article

Life Cycle Assessment of Hybrid Nanofiltration Desalination Plants in the Persian Gulf

Benyamin Bordbar [1], Arash Khosravi [1,*], Ali Ahmadi Orkomi [2] and Mohammad Peydayesh [3,*]

1. Sustainable Membrane Technology Research Group (SMTRG), Faculty of Petroleum, Gas and Petrochemical Engineering (FPGPE), Persian Gulf University (PGU), Bushehr P.O. Box 75169-13817, Iran; benyaminbordbar@gmail.com
2. Department of Environmental Sciences, Faculty of Natural Resources, University of Guilan, Sowmeh Sara P.O. Box 43619-96196, Iran; orkomi@guilan.ac.ir
3. Department of Health Sciences and Technology, ETH Zurich, 8092 Zurich, Switzerland
* Correspondence: arash.khosravi@pgu.ac.ir (A.K.); mohammad.peydayesh@hest.ethz.ch (M.P.)

Abstract: Although emerging desalination technologies such as hybrid technologies are required to tackle water scarcity, the impacts of their application on the environment, resources, and human health, as prominent pillars of sustainability, should be evaluated in parallel. In the present study, the environmental footprint of five desalination plants, including multi-stage flash (MSF), hybrid reverse osmosis (RO)–MSF, hybrid nanofiltration (NF)–MSF, RO, and hybrid NF–RO, in the Persian Gulf region, have been analyzed using life cycle assessment (LCA) as an effective tool for policy making and opting sustainable technologies. The comparison was based on the impacts on climate change, ozone depletion, fossil depletion, human toxicity, and marine eutrophication. The LCA results revealed the superiority of the hybrid NF–RO plant in having the lowest environmental impact, although the RO process produces more desalinated water at the same feed and input flow rates. The hybrid NF–RO system achieves 1.74 kg CO_2 equivalent, 1.24×10^{-7} kg CFC-11 equivalent, 1.28×10^{-4} kg nitrogenous compounds, 0.16 kg 1,4-DB equivalent, and 0.56 kg oil equivalent in the mentioned impact indicators, which are 7.9 to 22.2% lower than the single-pass RO case. Furthermore, the sensitivity analysis showed the reliability of the results, which helps to provide an insight into the life cycle impacts of the desalination plants.

Keywords: life cycle assessment (LCA); hybrid desalination; multi-stage flash (MSF); reverse osmosis (RO); nanofiltration (NF)

Citation: Bordbar, B.; Khosravi, A.; Ahmadi Orkomi, A.; Peydayesh, M. Life Cycle Assessment of Hybrid Nanofiltration Desalination Plants in the Persian Gulf. *Membranes* 2022, 12, 467. https://doi.org/10.3390/membranes12050467

Academic Editor: Pei Sean Goh

Received: 30 March 2022
Accepted: 24 April 2022
Published: 26 April 2022

Publisher's Note: MDPI stays neutral with regard to jurisdictional claims in published maps and institutional affiliations.

Copyright: © 2022 by the authors. Licensee MDPI, Basel, Switzerland. This article is an open access article distributed under the terms and conditions of the Creative Commons Attribution (CC BY) license (https://creativecommons.org/licenses/by/4.0/).

1. Introduction

Global warming is one of the major conflicts in today's world, and freshwater supply is a prominent challenge of sustainability. Water shortage has increased due to global warming, population growth, industrialization, and pollution of freshwater resources due to anthropogenic activities. The world's population is estimated to increase by more than two billion by the next three decades [1]. Currently, more than one billion people in the world live in water-scarce areas. Water consumption has increased more than fivefold in the last century [2]. Water scarcity is affected by the supply and demand cycle. It is predicted that the average renewable water in Persian Gulf region is about 1000 cubic meters per capita per year, while the global average is more than 5000 cubic meters per capita per year. Additionally, the capacity of common water resources is endangered by increasing water demand and declining surface and groundwater quality. In order to solve this shortage, the countries located in the Persian Gulf region commenced the implementation of seawater desalination plants; however, the environmental impacts of seawater desalination have not been fully considered in development policies [3].

Seawater desalination is performed by a variety of processes and technologies. In general, desalination technologies can be classified into three categories: thermal, chemical, and membrane-based technologies [4]. Thermal processes include methods that use thermal energy to separate impurities from water, such as multi-stage flash (MSF), multi-effect distillation (MED), and thermal vapor compression (TVC). These technologies have high costs in addition to high thermal energy consumption. However, they have been prevalent in the past and are still used today. MSF is the most common thermal process [3,5]. The MSF water treatment process contributes significantly to the global capacity of the installed treatment plants and was the most common treatment technology in the Middle East [6]. Chemical methods desalinate seawater using chemical processes such as ion exchange resins. In membrane methods, water is purified and desalinated using a membrane. Some of these methods include RO, NF, microfiltration (MF), ultrafiltration (UF), electrodialysis (ED), etc. [5]. Membrane technology currently has an essential role in treating and desalinating seawater, and more than 60% of purified water is obtained using membrane technologies [5].

Economic constraints and technical specifications including the capacity, accessible technologies and specification of the feed are key factors in selecting the appropriate technology. However, desalination plants significantly impact the environment and natural resources and have direct and indirect emissions of pollutants into water, soil, and air through energy and chemical consumption [6,7]. Although seawater desalination is a well-established technology in the region, the assessment of environmental hazards and damages has not yet been adequately and thoroughly reviewed and considered by governments and industries [7,8].

Hybrid technologies are a new approach, which gain the benefits of two or more technologies simultaneously [5]. However, new technologies pose new challenges to ecosystems, resources, and human health. As per the proposal of Kloepffer, LCA is one of life cycle sustainability assessment concepts in sustainability studies [9,10]. LCA is used to assess and calculate the damage caused to the environment by a product or a process, which can be examined with several approaches, such as cradle to grave, gate to gate, cradle to gate, etc. [6,7,10–13].

There are several studies worldwide which have evaluated various aspects of desalination, including some the effects on the environment and human life. At first, pollutants' emissions of desalination were assessed qualitatively without LCA [7]. Basic LCA studies on desalination processes were started in the 1990s; however, in the 2000s, research processes improved, and new quantitative comparisons between desalination technologies were published by applying LCA [7]. In the early 2000s, Lundie et al. investigated the environmental impacts of the RO process [14]. Raluy et al. compared RO, MSF, and MED processes and integration with renewable energy resources [15–19]. At the same time, Stokes and Horvath analyzed environmental emissions of an RO plant in the United States [20]. Furthermore, they assessed the role of renewable energy on air pollution caused by water supply plants [21]. In 2008, two studies by Vince et al. investigated the midpoint impacts of several RO and UF processes located in France [7,22,23]. Muñoz et al. studied several life cycle impacts of RO processes in Spain, in four projects [7,24–26].

In the 2010s, more researchers around the world started investigating the environmental and economic life cycle impacts of desalination plants. Beery et al. assessed environmental emissions of several RO and hybrid RO–UF plants in Germany [7,27–29]. Analyzing midpoint and endpoint impacts of various RO configurations was the most common topic of LCA studies in the 2010s and 2020s [7]. Some studies focused only on the environmental impacts of operating RO plants [30–35]. Moreover, some analyzed and compared RO with other traditional processes [36–40]. Recently, more studies focused on investigating and comparing emerging technologies [5,7]. Hancock et al., Al-Sarkal and Arafat, and Linares et al., studied the environmental impacts of hybrid technologies and compared them with individual processes [41–43]. Furthermore, Antipova et al. and Cherif et al. assessed the role of renewable energy in desalination plants [44,45]. In addition, several researchers analyzed both environmental and economic impacts [46–49]. On the other hand, it should be considered that the sustainability studies might be somewhat region-based because the sources of energy and available technologies in the regions are different. For example, the impacts of using an electricity grid in the Middle East and the Europe are totally different due to different sources of power production in addition to the available capacity of renewable energies in these regions [7].

Recent literature in nanofiltration desalination plants focuses on the emerging technologies' environmental impact, such as hybrid processes, zero liquid discharge (ZLD) technologies, and the effect of renewable energy usage [5,7]. Ronquim et al. analyzed and compared the midpoint impacts of global warming, energy resources depletion, land use, and mineral resources depletion indicators for RO and ZLD processes [50]. Furthermore, Tsalidis et al. investigated ZLD plants in some European countries [51]. Recently, Khosravi et al. reviewed the LCA studies of emerging technologies in industrial wastewater treatment and desalination globally, and Figure 1 presented the distribution of LCA and sustainability studies of desalination plants in different territories and regions [7].

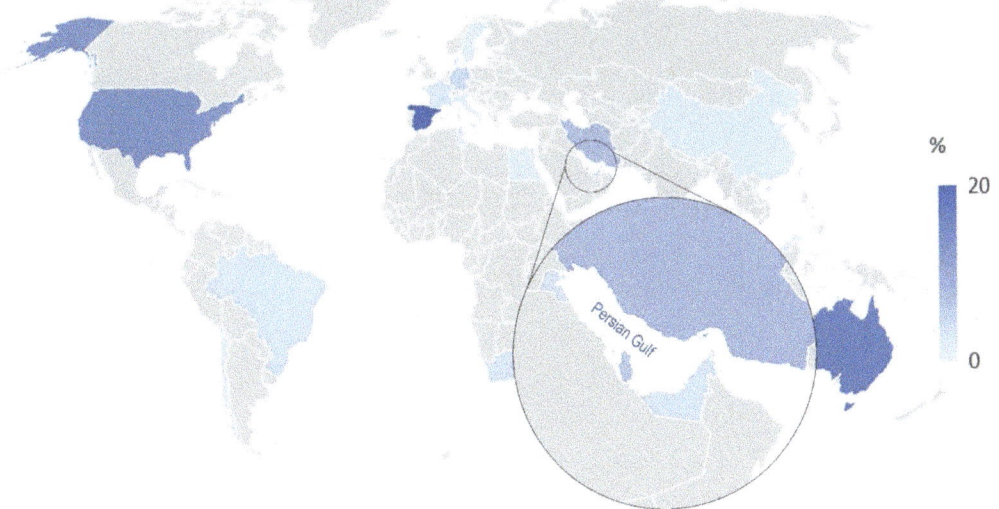

Figure 1. Distribution of water desalination LCA studies worldwide [7].

In this study, the LCA of five up-to-date real hybrid desalination plants in the Persian Gulf region, including recirculation multi-stage flash (R-MSF), hybrid RO/R-MSF, hybrid NF/R-MSF, single-pass RO, and hybrid NF/RO with a cradle-to-gate approach are assessed, and their emissions and environmental impacts are calculated, reviewed and compared. The contribution of the effective parameters (such as electricity, thermal energy, chemicals, and materials) on the impacts is also described. This study investigated the midpoint environmental impacts of emerging technologies from a long-term point of view and compared them with the traditional technologies, which helps to improve deep-seated plans to reduce their environmental impact and determine a policy for applying desalination technologies in the region.

2. Methodology

2.1. Goals and Scopes

LCA was implemented to investigate the life cycle impacts of five hybrid desalination plants. Case 1 is an R-MSF plant; case 2 is a hybrid RO/R-MSF plant; case 3 is a hybrid NF/R-MSF plant, case 4 is an RO plant, and case 5 is a hybrid NF/RO plant. The R-MSF desalination plant (case 1) is a real plant in operation in the Persian Gulf region as described and studied by Mannan et al. [8]. Case 2 is a pilot case where the R-MSF process is combined with RO (RO/R-MSF). In case 3, a pilot hybrid of R-MSF and NF technology (NF/R-MSF) is being conducted [8]. Case 4 is a single-stage RO plant, and case 5 is a hybrid NF/RO plant. Materials inventory, energy consumption, and membrane modules information have been extracted and used from industrial sources [8,52–56].

The cradle-to-gate approach of LCA is used to investigate the environmental impacts of the energy, chemicals, and materials used in the production of membranes and other parts of the aforementioned desalination plants, in addition to the impacts of energy and chemical consumption during the operation. SimaPro 9.3 software, Ecoinvent 3.8 cut-off database, and ReCipe 1.13 midpoint [57] method with global characterization factors were used to analyze the long-term impacts of the aforementioned cases.

2.2. Systems and Functional Unit

R-MSF was once the most common desalination method in Persian Gulf region. In the MSF process, seawater feed passes through pipes that are heated by the thermal energy of steam. The steam is in contact with the incoming saltwater pipe, leading to water evaporation [4]. In the chambers, the pressure decreases gradually in each stage, compared with the previous stage. When the heated seawater enters a low-pressure chamber, it suddenly evaporates. The vapor condenses by heat exchanging via the feed tubes on the top of the chamber and the condensate is collected by a vessel inside the chamber. The remaining saline water, called brine, exits the bottom of the chamber and re-enters the process in a cycle. High concentration brine is discharged into the sea. The inlet water pressure is less than 3 bar and the temperature is about 100 °C [8,58]. In addition, sodium hypochlorite is added to control biological growth in the desalination plant (chlorination process), and sodium bisulfite is added to control corrosion by removing dissolved gases (deaeration). The functional unit for LCA is 1 m^3 of produced freshwater. The schematic of case 1 (R-MSF) is depicted in Figure 2.

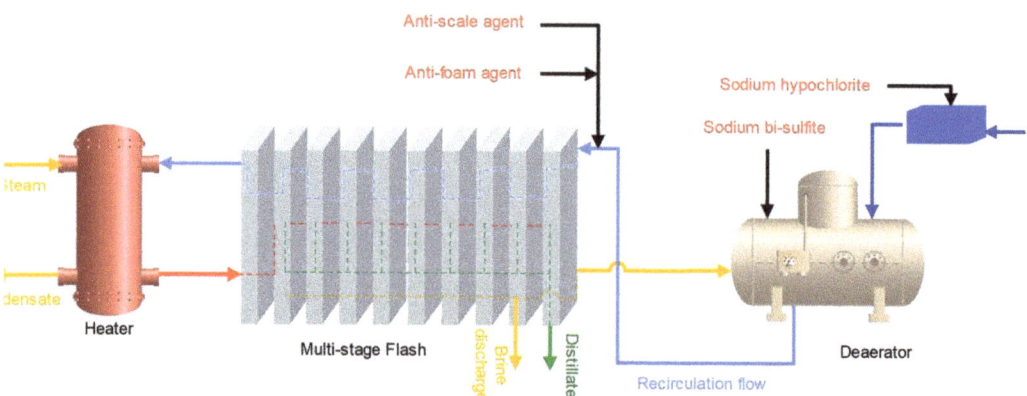

Figure 2. The schematic of the R-MSF system.

The required steam in case 1 is obtained from the steam returned from the turbine of a natural gas combined cycle power plant. This is undertaken to reduce energy consumption. The required thermal energy in the studied cases (cases 1–3) are 107, 64, and 64 MJ/m^3, respectively [8]. Cases 4 and 5 do not require any thermal energy.

The feed temperature in case 1 is kept below 112 °C to restrict the formation of scale due to the presence of calcium carbonate, calcium sulfate, and magnesium hydroxide in the feed. By adding RO and NF processes before the MSF process, these substances are reduced in the feed of MSF, and the MSF inlet temperature can be raised, which will ultimately increases plant productivity [8]. The schematic of the hybrid RO/R-MSF system is shown in Figure 3.

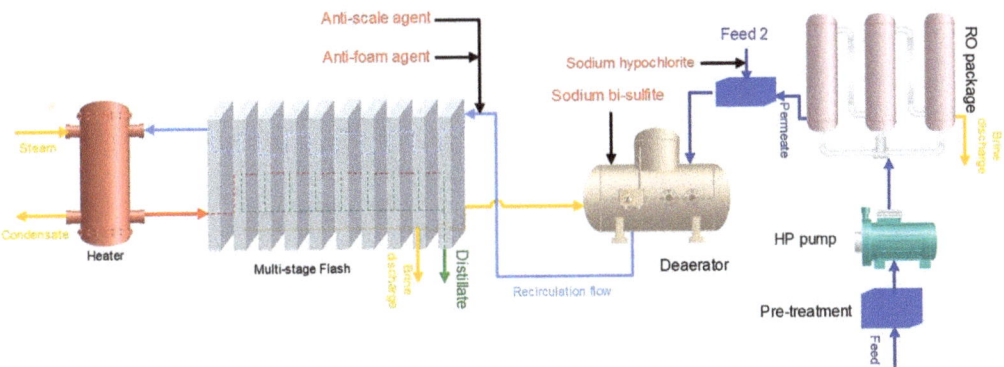

Figure 3. The schematic of case 1 (hybrid RO/R-MSF Flash).

Figure 4 shows the schematic of case 3 (hybrid NF/R-MSF). As shown in Figures 3 and 4, half the feed enters the membrane process, and the other half is mixed with permeate flow and enters MSF.

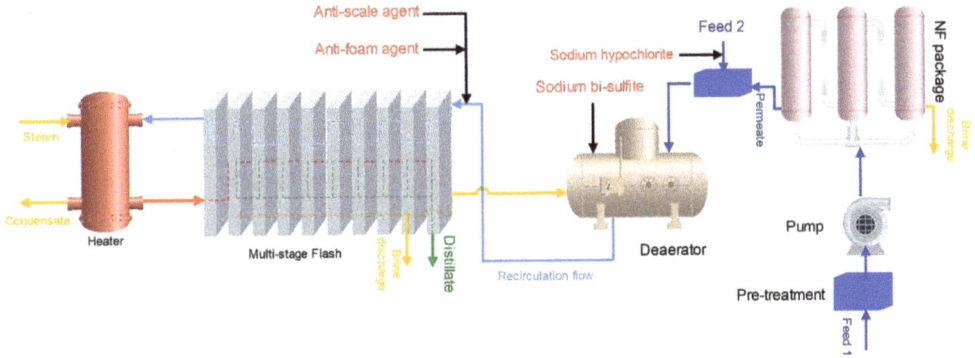

Figure 4. The schematic of case 3 (hybrid NF/R-MSF).

Case 4 is a single-stage RO plant. In this process, the seawater feed enters the RO system. At first, a pre-treatment process applied in order to remove substances that may cause membrane fouling, scaling and corrosion. Then the feed enters a high-pressure pump. The pump increases the pressure up to 16 bar to supply the transmembrane pressure (TMP) required for RO membrane modules. The required electrical energy for this process is 4.22 kWh for 1 cubic meter of desalted water [59]. The schematic of the RO plant is illustrated in Figure 5.

The last case is a hybrid NF–RO plant. After pre-treatment, the seawater feed enters a pump to supply the required TMP for NF membrane modules, then the feed enters the NF modules. The permeate goes to the high-pressure pumps to be treated by the RO process. Due to the pumping system at the NF process, the NF permeate pressure is more than the feed of RO in case 4; therefore, less electricity is needed for RO high-pressure pumps. The required electricity of this plant is 3.11 kWh for 1 cubic meter of desalted water [59]. The schematic of the hybrid NF–RO plant is drawn in Figure 6.

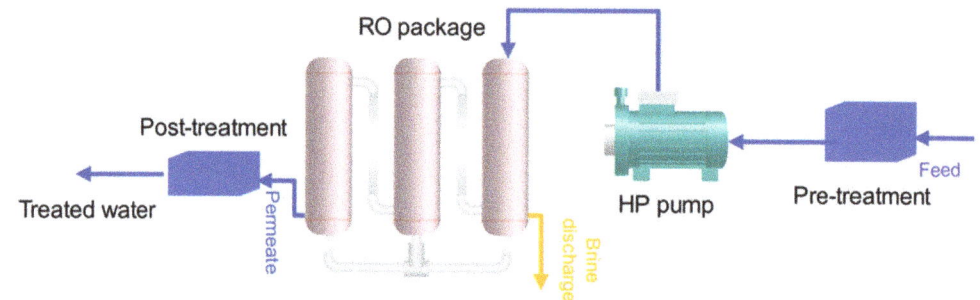

Figure 5. The schematic of the Reverse Osmosis system.

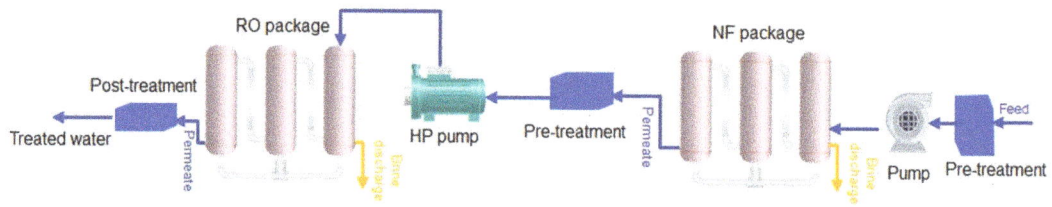

Figure 6. The schematic of case 5 (hybrid NF–RO).

The chemical additives used in pre-treatment and post-treatment of cases 1–5 are the same, except more chemicals are added to the feed to prevent fouling and corrosion of membranes in plants 2–5. Additionally, due to the high sensitivity of the RO membrane, additive chemicals such as citric acid and sodium sulfite are added exclusively to cases 2, 4 and 5. High-temperature antifouling is used in all cases to prevent the formation of scale due to the presence of calcium carbonate, calcium sulfate, and magnesium hydroxide in the feed at high temperatures. Applying high temperature increases the MSF performance. Other common pre-treatment methods include aeration and chlorination, the addition of sodium hydrogen sulfate, and ethanol to control descaling and foaming. To maintain distilled quality and control the growth of aquatic organisms, treatment in all plants is performed using calcium hydroxide and chlorination demineralizing agents. Chlorination, coagulation, and removal of the medium followed by the cartridge is used as a pre-treatment method for the NF system. In addition to common additives, citric acid and sodium sulfate are injected in the pre-treatment of RO in order to avoid membrane fouling and scaling, and hydrated lime and carbon dioxide are added to demineralize the water at the post-treatment stage [52,53]. Additionally, the environmental effects of the materials and solvents used for the membrane modules synthesis have been considered.

2.3. Life Cycle Inventory Analysis

In the life cycle inventory (LCI) stage, the data and information needed to estimate the emission rates of each process were collected [7]. In this study, the required data for cases 1 to 5 were collected from industrial data [8,52–56]. Table 1 shows the information for cases 1 to 5. More details about the data collection phase are presented in the supplementary section (Tables S1 to S5).

Table 1. Plant configuration and energy demand of the cases 1–5.

Plant Specification	Case 1	Case 2	Case 3	Case 4	Case 5
Technology	R-MSF	Hybrid RO/R-MSF	Hybrid NF/R-MSF	RO	Hybrid NF/RO
Configuration	Cross tube MSF	Cross tube MSF Single-pass RO	Cross tube MSF Single-pass NF	Single-pass RO	Single-pass NF Single-pass RO
Number of stages	21	35	35	-	-
Thermal energy (MJ)	107	64	64	-	-
Electrical energy (kWh/m^3)	4.19	4.6	3.42	4.22	3.11
Feed (seawater) Flowrate (m^3/h)	28,000	675	675	675	675
Treated water flowrate (m^3/h)	3430.57	378	288.1	405	303.75
Reference	[8]	[52,56]	[8]	[52,56]	[52,55,56]

Some chemicals are used in pre-treatment and post-treatment of the processes. Sodium hypochlorite is added to control microorganisms, bacteria, and other biological factors. To prevent corrosion, sodium bisulfite is used as a pre-treatment. Scaling is a challenging factor for desalination plants, where anticalins such as sulfuric acid and anti-foaming objects such as monoethylene oxide are added. The coagulant, iron chloride, is added to the NF and RO pre-treatment; however, in the post-treatment, sodium hypochlorite is added to the NF and RO process. Carbon dioxide and sodium hydroxide are used for the post-treatment of RO, and sodium sulfite is added as pre-treatment. The values of the additive chemicals are the dosages in water stream during the treatment, which are extracted from the industrial data [8,52,53] (Table 2).

Table 2. Chemical additives in pre-treatment and post-treatment of plants.

Stage	Chemicals	Amount (ppm)
Pre-treatment cases 1–5	Sodium hypochlorite	4
	Sodium bisulfite	0.5
	Sulfuric acid	2.4
	Monoethyleneoxide	0.1
Post-treatment for cases 1–5	Calcium hydroxide	0.5
	Sodium hypochlorite	0.5
Pre-treatment for NF and RO system	Ferric chloride	0.3
Post-treatment for NF and RO system	Chlorine	0.2
Pre-treatment for RO system	Citric acid	0.937
	Sodium sulfite	0.0739
Post-treatment for RO system	Lime	51.03
	Carbon dioxide	43

Components such as polyester, polysulfone, N,N-dimethylformamide (DMF), metaphenylene diamine (MPD), trimesoyl chloride (TMC), isopropanol (IPA), and phosphoric acid are the materials and solvents used to synthesize the membrane layer; and polypropylene as spacer, epoxy resin as glue and PVC as permeate tube are used. The data relating to the NF module were collected from the specifications of the commercial 8-inch NF membrane [6] and are listed in Table 3. The membrane lifetime was considered to be 4 years.

Table 3. Materials usage in fabrication of NF modules for their lifetime.

Component	Amount (kg/m^3)
Polyester	4.79452×10^{-11}
Polysulfone	1.0274×10^{-11}
DMF (N,N-dimethylformamide)	4.10959×10^{-11}
MPD (meta-phenylene diamine)	4.62329×10^{-13}
TMC (trimesoyl chloride)	1.19178×10^{-12}
Phosphoric acid	3.20548×10^{-12}
Polypropylene (spacers)	5.13699×10^{-11}
Epoxy resin (glue)	1.16438×10^{-11}
PVC (permeate tube)	1.78082×10^{-11}
IPA (isopropanol)	5.82192×10^{-12}

As for the NF modules, the environmental effects in the production phase of the RO modules were also considered. The data of the RO module were collected from SimaPro9.3 software database related to an 8-inch operating module [6]. The lifetime of membrane modules was considered to be 4 years. The inventory data of the RO module are been presented in Table 4.

Table 4. Materials usage in fabrication of RO modules for their lifetime.

Component	Amount (kg/m^3)
ABS (Acrylonitrile-butadiene-styrene copolymer)	2.90964×10^{-13}
Polyester	2.0077×10^{-13}
Polysulfone	2.12833×10^{-13}
DMF (N,N-dimethylformamide)	8.52231×10^{-12}
MPD (meta-phenylene diamine)	9.00922×10^{-16}
TMC (trimesoyl chloride)	2.69506×10^{-15}
Phosphoric acid	6.31159×10^{-14}
Polypropylene (spacers)	4.26565×10^{-13}
Epoxy resin (glue)	1.17299×10^{-14}
PVC (permeate tube)	1.3347×10^{-15}
IPA (isopropanol)	2.10215×10^{-14}

2.4. Life Cycle Impact Assessment (LCIA)

In the life cycle impact assessment (LCIA) phase, the environmental impacts of each process were calculated by using the emission inventories and the environmental impact potential of emitted materials [7]. The mentioned desalination processes release abundant pollutants and cause numerous environmental issues over their life cycle. The most common are CO_2, SO_X, NO_X, and different sized dust particles. Another important issue is fossil energy resources consumption [60]. The *ReCipe* method can assess 18 types of midpoint impacts; only 5 (climate change, ozone depletion, marine eutrophication, human toxicity, and fossil depletion) were examined in this study, in order to distinguish the cases regarding their environmental impacts. Climate change (CC) indicates the emission of equivalent carbon dioxide. Ozone depletion potential (ODP) represents the amount of CFC-11 equivalent released. Marine eutrophication potential (MEP) reveals the impacts of nitrogenous and phosphorous compounds. Human toxicity potential (HTP) indicates the degree of toxicity to humans, and fossil depletion potential (FDP) is also calculated based on consumed oil [7].

3. Results and Discussion

3.1. Life Cycle Impact Assessment (LCIA)

As shown in Figure 7, thermal energy and electricity consumption in case 1 had the largest contribution in all impacts. The thermal energy and electricity contributed to about 81.7% and 17.3% of CC index, respectively. This is due to the greenhouse gas emission from natural gas combustion in the gas boiler for steam and electricity production in a combined cycle power plant. Furthermore, chemicals and materials were responsible for 1.04%, 4.23%, 7.66%, 12.1%, and 0.749% share in CC, ODP, MEP, HTP and FDP indicators, respectively.

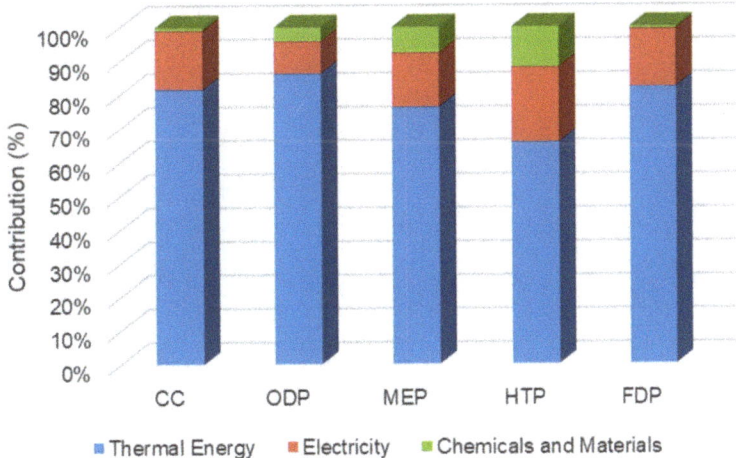

Figure 7. The contribution of inventories in midpoint impacts of case 1.

As shown in Figure 8, the share of electrical energy impacts in case 2 increased by comparison with case 1 because more electrical energy is consumed in high-pressure pumps of the RO process. The electrical energy contributions in the five mentioned impacts were 25.9%, 15.9%, 18.8%, 28.1%, and 26.7%, respectively. The thermal energy impact was still the highest in all indicators. In this case, the effect of chemicals and materials on MEP was higher than electricity, which contributed to a share of 32.4%, due to materials consumed for membrane fabrication and module production. In raw chemical and material production, membrane fabrication, and module packaging processes, a large amount of chemicals and materials are consumed or emitted, leading to changing the level of eutrophication.

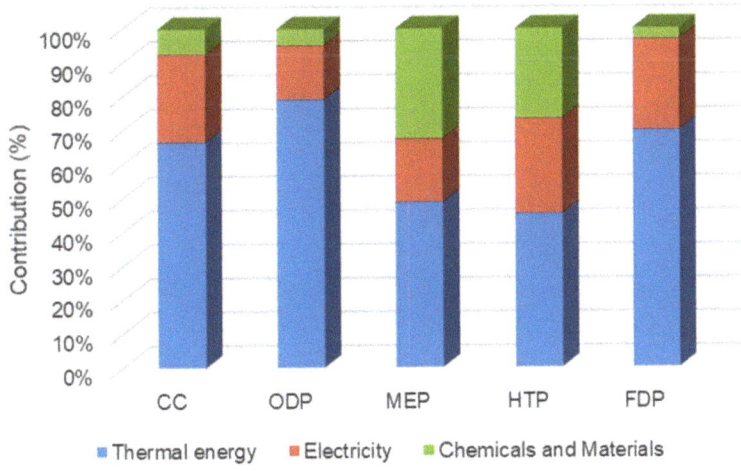

Figure 8. The contribution of inventories in midpoint impacts of case 2.

In case 3, the electricity consumption was lower than case 2. However, the consumption of thermal energy was as same as case 2. The relative contribution of thermal energy in midpoint impact indicators of case 3 (Figure 9) was higher compared with case 2 (Figure 8) due to lower electricity, chemicals, and materials consumptions. The contributions of thermal energy for the five mentioned impacts were 77.2%, 84.9%, 74.4%, 63.8% and 77.6%, respectively.

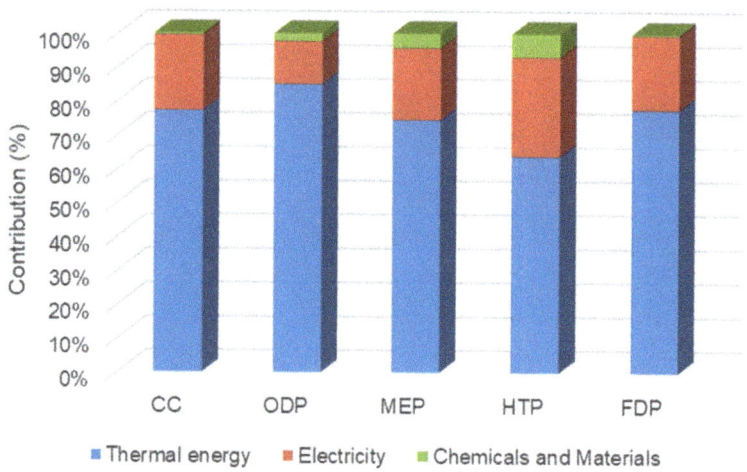

Figure 9. The contribution of inventories in midpoint impacts of case 3.

In case 4 (Figure 10), a single-pass RO process operates to desalinate feed seawater. In this case, thermal energy is not used; however, more electrical energy is applied for high-pressure pumps, the whole feed seawater enters the RO process, and the flowrate of pumps is twice than that of case 2. Electricity contributed to 81.6%, 77%, 41.7%, 55.6%, and 91% of CC, ODP, MEP, HTP, and FDP indices, respectively. Due to the absence of thermal energy, the operational and construction materials and chemicals had a major effect on the MEP indicator (58.3%).

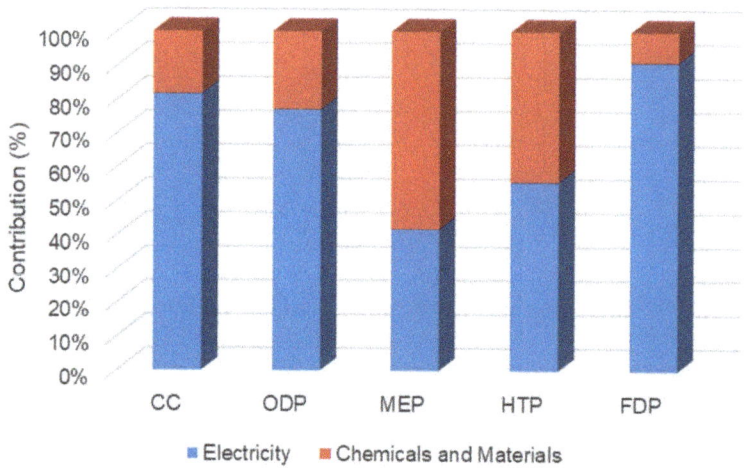

Figure 10. The contribution of inventories in midpoint impacts of case 4.

Figure 11 represents the contribution inventories in each midpoint impact in case 5. This case consists of a single-pass NF and single-pass RO in series. Due to the pressure applied by the pump before the NF process, less load is needed for the high-pressure pump; therefore, electricity consumption is less than that of case 4. On the other hand, more membrane modules are used due to applying the NF process, and more materials were manipulated rather than in a single RO case. The share of electricity was 74.2%, 71.9%, 33.3%, 46.6%, and 87.1% for the mentioned midpoint indicators, respectively. Moreover, MEP and HTP indicators were affected by chemicals and materials more than electricity.

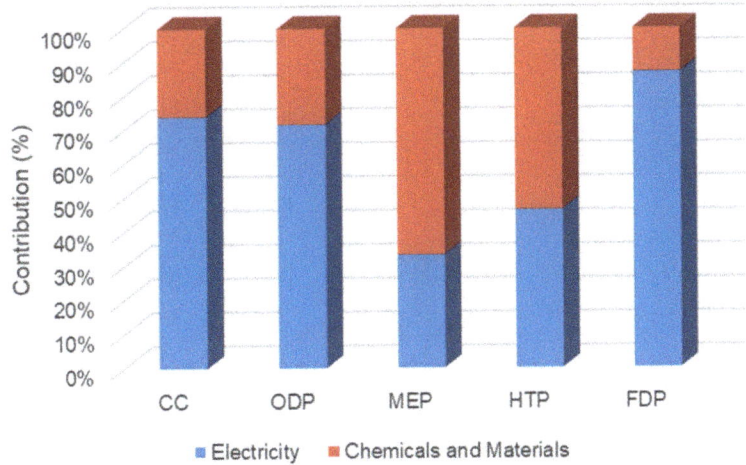

Figure 11. The contribution of inventories in midpoint impacts of case 5.

Figure 12 shows a comparison between the cases regarding the indicators. The analysis of the midpoint environmental factors for cases 1–5 revealed that case 1 (MSF) had the most impact on all five factors, which shows that the MSF process individually has the most environmental impact and emissions. The emission rate of combustion-induced gases such as CO_2, SO_X, and NO_X in case 1 was more than in other cases due to burning a large amount of natural gas to provide thermal energy. Case 2 (RO/R-MSF) was in the second place, due to lower thermal energy consumption compared with case 1. As a result, its environmental footprint was less than case 1. However, the high mechanical energy used in the high-pressure pumps caused high environmental impact.

Case 3 showed lower risks due to lower thermal energy usage than case 1 and lower electricity consumption than case 2. No thermal energy was applied in cases 4 and 5; therefore, they emitted less waste than other cases. The RO case (case 4) consumed more electrical energy than the hybrid NF/RO case (case 5) because of the greater load applied in the high-pressure pump. The last case achieved the lowest values in all the midpoint indicators, indicating than case 5 was superior to the other four cases regarding environmental footprint. It is noteworthy that the efficiency of case 5 was less than cases 2 and 4, and that the NF product quality cannot be as high as the RO technology. The RO system is able to treat 60% of the feed, versus 45% for the hybrid NF–RO process. The environmental impacts of all five conducted cases are shown in Table 5.

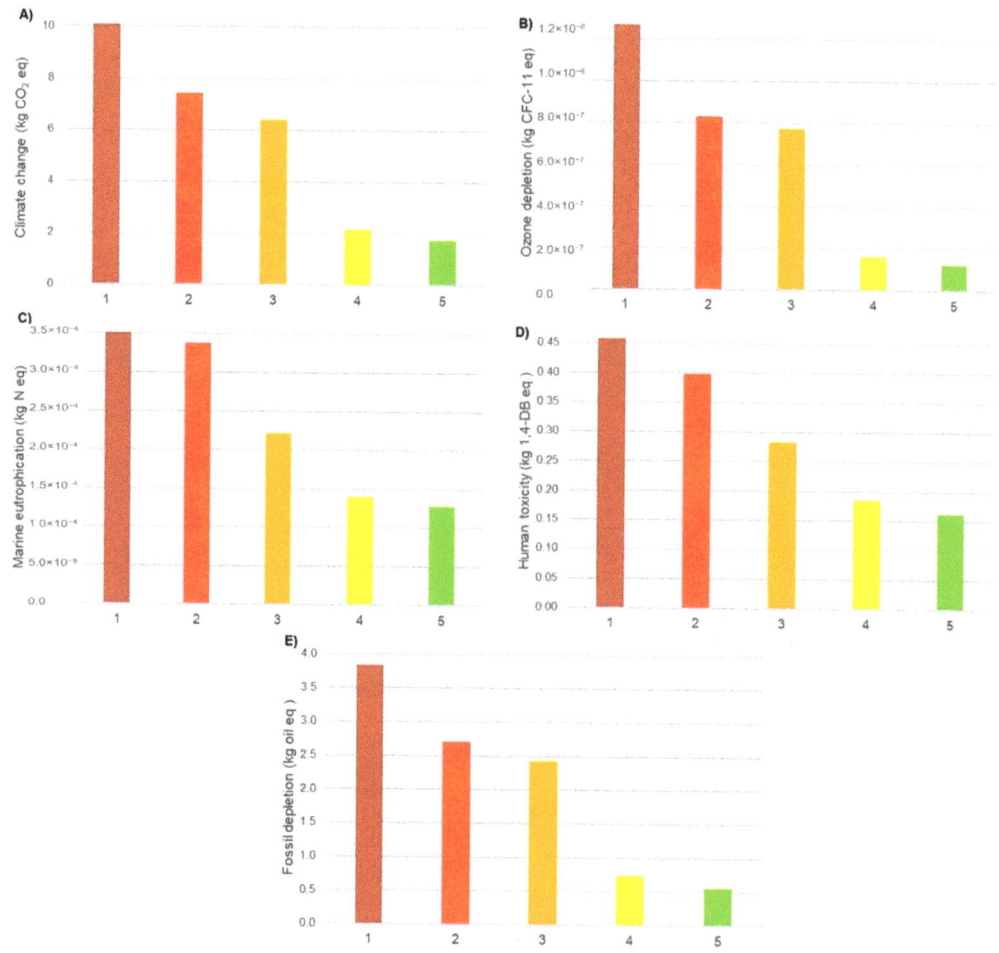

Figure 12. The impacts of desalination plants of cases 1 to 5 on (**A**) climate change, (**B**) ozone depletion, (**C**) marine eutrophication, (**D**) human toxicity, and (**E**) fossil depletion.

Table 5. Midpoint environmental indicator values for cases 1–5.

Impact	Unit	Case 1	Case 2	Case 3	Case 4	Case 5
Climate change	kg CO_2 eq	10.08	7.39	6.38	2.15	1.74
Ozone depletion	kg CFC-11 eq	1.27×10^{-6}	8.26×10^{-7}	7.70×10^{-7}	1.57×10^{-7}	1.24×10^{-7}
Marine eutrophication	kg N eq	3.60×10^{-4}	3.37×10^{-4}	2.21×10^{-4}	1.39×10^{-4}	1.28×10^{-4}
Human toxicity	kg 1,4-DB eq	0.46	0.40	0.28	0.18	0.16
Fossil depletion	kg oil eq	3.83	2.69	2.43	0.72	0.56

3.2. Sensitivity Analysis

Risk assessment is a scientific procedure to describe and determine the uncertainty and its characterizations to define or change decisions. A set of systematic, logical, analytical, evidence-based procedures were performed to find and measure the risks, probability, and possibility of a process and a way for decision-making [61]. Sensitivity analysis helps to improve the design and model by assessing the qualitative and quantitative responses of the studied analysis [62].

Due to the significant environmental impacts of electricity compared with the other inventories in the five studied cases, the sensitivity analysis was applied to the electrical energy with ±20% variation for cases 4 and 5. Electricity is the most effective parameter affecting CC, ODP, and FDP impact indicators due to burning natural gas in combined cycle power plants to supply electricity for mechanical equipment. The sensitivity analysis results showed that by 20% variation of electricity in case 4, the CC, ODP, MEP, HTP, and FDP indicators were varied at 16.31%, 15.4%, 8.34%, 11.12%, and 18.2%, respectively. Moreover, for the last case (hybrid NF–RO), the mentioned indicators were varied by 14.84%, 14.38%, 6.67%, 9.32%, and 17.42%, respectively (Figure 13). Generally, sensitivity analysis found that by ±20% changes in electrical energy, the environmental impacts changed from 8.34% to 18.2% for RO, and 6.67% to 17.42% for hybrid NF–RO processes by applying more efficient equipment or choosing more sustainable energy resources.

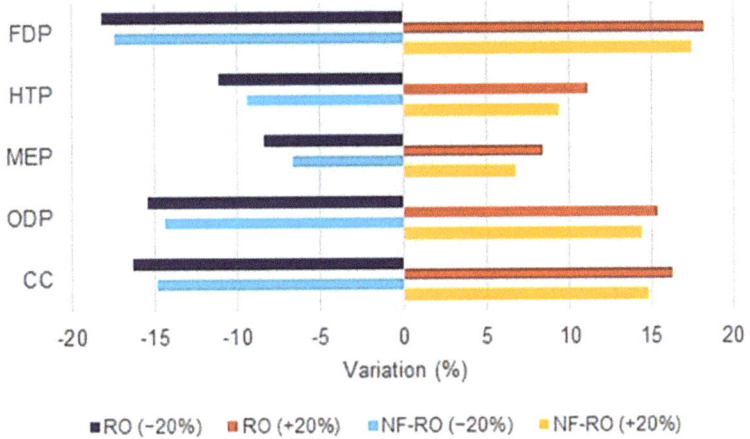

Figure 13. Sensitivity analysis for cases 4 and 5 based on the impacts of the electricity variation.

4. Conclusions

In this study, five hybrid desalination plants, including R-MSF, RO/R-MSF, NF/R-MSF, RO, and NF/RO in the Persian Gulf region, were examined by the LCA method. The results of this study may be helpful to reduce environmental impacts and determine a sustainable policy for developing desalination projects, considering the vital need for water resources and the growing development of seawater desalination technologies. There is a high potential to reduce the environmental impact of MSF desalination by increasing the efficiency of the process via flash stage addition or advanced feed water pre-treatment using NF or RO. Furthermore, membrane-based technologies such as NF and RO may be considered a good alternative for traditional processes such as thermal processes, which do not need thermal energy. This study revealed that by applying RO and hybrid NF/RO technologies, a notable reduction in environmental impact indicators might be expected compared with traditional technologies. The results showed that the hybrid NF/RO technology had the minimum environmental impact, which could guide the design of new plants in the future. However, it should be noted that the RO system can treat 60% of feed water, versus 45% for the hybrid NF/RO system, as studied in this work. The sensitivity analysis determined how the quantity of electrical energy would change the environmental impacts. It found that by ±20% changes in electrical energy, the impact indictors may change from 8.34% to 18.2% for RO, and from 6.67% to 17.42% for hybrid NF-RO processes, which means that by applying more efficient equipment or choosing clean energy resources, the environmental footprint can be reduced.

Supplementary Materials: The following supporting information can be downloaded at: https://www.mdpi.com/article/10.3390/membranes12050467/s1, Table S1: Life cycle inventory of case 1, Table S2: Life cycle inventory of case 2, Table S3: Life cycle inventory of case 3, Table S4: Life cycle inventory of case 4, Table S5: Life cycle inventory of case 5.

Author Contributions: Conceptualization, A.K.; methodology, B.B., A.K. and A.A.O.; software, B.B. and A.K.; validation, B.B., A.K. and A.A.O.; formal analysis, investigation, resources, data curation, B.B., A.K. and A.A.O.; writing—original draft preparation, B.B.; writing—review and editing, visualization, supervision, A.K., A.A.O. and M.P.; project administration, funding acquisition, A.K. All authors have read and agreed to the published version of the manuscript.

Funding: This research received no external funding.

Institutional Review Board Statement: Not applicable.

Informed Consent Statement: Not applicable.

Data Availability Statement: Not applicable.

Acknowledgments: The authors would like to acknowledge the kind support of Sh. Sharifi and SMTRG during the performing the research and preparation of the paper.

Conflicts of Interest: The authors declare no conflict of interest.

References

1. Nations, U. *World Population Prospects*; United Nations Department of Economic and Social Affairs: New York, NY, USA, 2019.
2. Water, U.W. *UN World Water Development Report 2020*; United Nation: New York, NY, USA, 2020.
3. Akbar, A. The Best Desalination Technology for the Persian Gulf. *Int. J. Soc. Ecol. Sustain. Dev. (IJSESD)* **2011**, *2*, 55–65. [CrossRef]
4. Younos, T.; Tulou, K.E. Overview of desalination techniques. *J. Contemp. Water Res. Educ.* **2005**, *132*, 3–10. [CrossRef]
5. Bordbar, B.; Khosravi, A.; Azin, R. A Review on Sustainable Hybrid Water Treatment Processes. In Proceedings of the 3rd Biennial Oil, Gas, and Petrochemical Conference (OGPC2020), Bushehr, Iran, 28–30 December 2020. Available online: https://www.researchgate.net/publication/355162533_A_Review_on_Sustainable_Hybrid_Water_Treatment_Processes (accessed on 28 December 2020).
6. Bonton, A.; Bouchard, C.; Barbeau, B.; Jedrzejak, S. Comparative life cycle assessment of water treatment plants. *Desalination* **2012**, *284*, 42–54. [CrossRef]
7. Khosravi, A.; Bordbar, B.; Orkomi, A.A. Life Cycle Assessment of Emerging Technologies in Industrial Wastewater Treatment and Desalination. In *Industrial Wastewater Treatment*; Sptringer Nature: Berlin, Germany, 2022. [CrossRef]
8. Mannan, M.; Alhaj, M.; Mabrouk, A.N.; Al-Ghamdi, S.G. Examining the life-cycle environmental impacts of desalination: A case study in the State of Qatar. *Desalination* **2019**, *452*, 238–246. [CrossRef]
9. Kloepffer, W. Life cycle sustainability assessment of products. *Int. J. Life Cycle Assess.* **2008**, *13*, 89. [CrossRef]
10. Hauschild, M.Z.; Rosenbaum, R.K.; Olsen, S.I. *Life Cycle Assessment*; Springer: Berlin, Germany, 2018; Volume 2018.
11. Jolliet, O.; Saadé, M.; Crettaz, P. *Analyse du Cycle de vie: Comprendre et Réaliser un Écobilan*; PPUR Presses Polytechniques: Lausanne, Switzerland, 2010; Volume 23.
12. Visentin, C.; da Silva Trentin, A.W.; Braun, A.B.; Thomé, A. Lifecycle assessment of environmental and economic impacts of nano-iron synthesis process for application in contaminated site remediation. *J. Clean. Prod.* **2019**, *231*, 307–319. [CrossRef]
13. Fernandes, S.; da Silva, J.C.E.; da Silva, L.P. Comparative life cycle assessment of high-yield synthesis routes for carbon dots. *NanoImpact* **2021**, *23*, 100332. [CrossRef]
14. Lundie, S.; Peters, G.M.; Beavis, P.C. Life cycle assessment for sustainable metropolitan water systems planning. *Environ. Sci. Technol.* **2004**, *38*, 3465–3473. [CrossRef]
15. Raluy, R.G.; Serra, L.; Uche, J.; Valero, A. Life-cycle assessment of desalination technologies integrated with energy production systems. *Desalination* **2004**, *167*, 445–458. [CrossRef]
16. Raluy, R.G.; Serra, L.; Uche, J. Life Cycle Assessment of Water Production Technologies—Part 1: Life Cycle Assessment of Different Commercial Desalination Technologies (MSF, MED, RO) (9 pp). *Int. J. Life Cycle Assess.* **2004**, *10*, 285–293. [CrossRef]
17. Raluy, R.G.; Serra, L.; Uche, J.; Valero, A. Life Cycle Assessment of Water Production Technologies—Part 2: Reverse Osmosis Desalination versus the Ebro River Water Transfer (9 pp). *Int. J. Life Cycle Assess.* **2005**, *10*, 346–354. [CrossRef]
18. Raluy, R.G.; Serra, L.; Uche, J. Life cycle assessment of desalination technologies integrated with renewable energies. *Desalination* **2005**, *183*, 81–93. [CrossRef]
19. Raluy, G.; Serra, L.; Uche, J. Life cycle assessment of MSF, MED and RO desalination technologies. *Energy* **2006**, *31*, 2361–2372. [CrossRef]
20. Stokes, J.; Horvath, A. Life Cycle Energy Assessment of Alternative Water Supply Systems (9 pp). *Int. J. Life Cycle Assess.* **2006**, *11*, 335–343. [CrossRef]
21. Stokes, J.R.; Horvath, A. Energy and Air Emission Effects of Water Supply. *Environ. Sci. Technol.* **2009**, *43*, 2680–2687. [CrossRef]
22. Vince, F.; Marechal, F.; Aoustin, E.; Bréant, P. Multi-objective optimization of RO desalination plants. *Desalination* **2008**, *222*, 96–118. [CrossRef]

23. Vince, F.; Aoustin, E.; Bréant, P.; Marechal, F. LCA tool for the environmental evaluation of potable water production. *Desalination* **2008**, *220*, 37–56. [CrossRef]
24. Muñoz, I.; Fernández-Alba, A.R. Reducing the environmental impacts of reverse osmosis desalination by using brackish groundwater resources. *Water Res.* **2008**, *42*, 801–811. [CrossRef]
25. Muñoz, I.; Rodríguez, A.; Rosal, R.; Fernández-Alba, A.R. Life Cycle Assessment of urban wastewater reuse with ozonation as tertiary treatment: A focus on toxicity-related impacts. *Sci. Total Environ.* **2009**, *407*, 1245–1256. [CrossRef]
26. Muñoz, I.; Milà-i-Canals, L.; Fernández-Alba, A.R. Life Cycle Assessment of Water Supply Plans in Mediterranean Spain. *J. Ind. Ecol.* **2010**, *14*, 902–918. [CrossRef]
27. Beery, M.; Wozny, G.; Repke, J.-U. Sustainable Design of Different Seawater Reverse Osmosis Desalination Pretreatment Processes. In *Computer Aided Chemical Engineering*; Pierucci, S., Ferraris, G.B., Eds.; Elsevier: Amsterdam, The Netherlands, 2010; Volume 28, pp. 1069–1074.
28. Beery, M.; Repke, J.-U. Sustainability analysis of different SWRO pre-treatment alternatives. *Desalin. Water Treat.* **2010**, *16*, 218–228. [CrossRef]
29. Beery, M.; Hortop, A.; Wozny, G.; Knops, F.; Repke, J.-U. Carbon footprint of seawater reverse osmosis desalination pre-treatment: Initial results from a new computational tool. *Desalin. Water Treat.* **2011**, *31*, 164–171. [CrossRef]
30. Meneses, M.; Pasqualino, J.C.; Céspedes-Sánchez, R.; Castells, F. Alternatives for Reducing the Environmental Impact of the Main Residue From a Desalination Plant. *J. Ind. Ecol.* **2010**, *14*, 512–527. [CrossRef]
31. Zhou, J.; Chang, V.W.C.; Fane, A.G. Environmental life cycle assessment of brackish water reverse osmosis desalination for different electricity production models. *Energy Environ. Sci.* **2011**, *4*, 2267–2278. [CrossRef]
32. Zhou, J.; Chang, V.W.C.; Fane, A.G. Environmental life cycle assessment of reverse osmosis desalination: The influence of different life cycle impact assessment methods on the characterization results. *Desalination* **2011**, *283*, 227–236. [CrossRef]
33. Zhou, J.; Chang, V.W.C.; Fane, A.G. An improved life cycle impact assessment (LCIA) approach for assessing aquatic eco-toxic impact of brine disposal from seawater desalination plants. *Desalination* **2013**, *308*, 233–241. [CrossRef]
34. Shahabi, M.P.; Anda, M.; Ho, G. Influence of site-specific parameters on environmental impacts of desalination. *Desalin. Water Treat.* **2015**, *55*, 2357–2363. [CrossRef]
35. Aleisa, E.; Al-Shayji, K. Ecological–economic modeling to optimize a desalination policy: Case study of an arid rentier state. *Desalination* **2018**, *430*, 64–73. [CrossRef]
36. Jijakli, K.; Arafat, H.; Kennedy, S.; Mande, P.; Theeyattuparampil, V.V. How green solar desalination really is? Environmental assessment using life-cycle analysis (LCA) approach. *Desalination* **2012**, *287*, 123–131. [CrossRef]
37. Tarnacki, K.; Meneses, M.; Melin, T.; van Medevoort, J.; Jansen, A. Environmental assessment of desalination processes: Reverse osmosis and Memstill. *Desalination* **2012**, *296*, 69–80. [CrossRef]
38. Ras, C.; Von Blottnitz, H. A comparative life cycle assessment of process water treatment technologies at the Secunda industrial complex, South Africa. *Water SA* **2012**, *38*, 549–554. [CrossRef]
39. Godskesen, B.; Hauschild, M.; Rygaard, M.; Zambrano, K.; Albrechtsen, H.J. Life-cycle and freshwater withdrawal impact assessment of water supply technologies. *Water Res.* **2013**, *47*, 2363–2374. [CrossRef] [PubMed]
40. Al-Shayji, K.; Aleisa, E. Characterizing the fossil fuel impacts in water desalination plants in Kuwait: A Life Cycle Assessment approach. *Energy* **2018**, *158*, 681–692. [CrossRef]
41. Hancock, N.T.; Black, N.D.; Cath, T.Y. A comparative life cycle assessment of hybrid osmotic dilution desalination and established seawater desalination and wastewater reclamation processes. *Water Res.* **2012**, *46*, 1145–1154. [CrossRef]
42. Al-Sarkal, T.; Arafat, H.A. Ultrafiltration versus sedimentation-based pretreatment in Fujairah-1 RO plant: Environmental impact study. *Desalination* **2013**, *317*, 55–66. [CrossRef]
43. Linares, R.V.; Li, Z.; Yangali-Quintanilla, V.; Ghaffour, N.; Amy, G.; Leiknes, T.; Vrouwenvelder, J.S. Life cycle cost of a hybrid forward osmosis–low pressure reverse osmosis system for seawater desalination and wastewater recovery. *Water Res.* **2016**, *88*, 225–234. [CrossRef]
44. Antipova, E.; Boer, D.; Cabeza, L.F.; Guillén-Gosálbez, G.; Jiménez, L. Uncovering relationships between environmental metrics in the multi-objective optimization of energy systems: A case study of a thermal solar Rankine reverse osmosis desalination plant. *Energy* **2013**, *51*, 50–60. [CrossRef]
45. Cherif, H.; Champenois, G.; Belhadj, J. Environmental life cycle analysis of a water pumping and desalination process powered by intermittent renewable energy sources. *Renew. Sustain. Energy Rev.* **2016**, *59*, 1504–1513. [CrossRef]
46. Karami, S.; Karami, E.; Zand-Parsa, S. Environmental and economic appraisal of agricultural water desalination use in South Iran: A comparative study of tomato production. *J. Appl. Water Eng. Res.* **2017**, *5*, 91–102. [CrossRef]
47. Shahabi, M.P.; McHugh, A.; Anda, M.; Ho, G. Comparative economic and environmental assessments of centralised and decentralised seawater desalination options. *Desalination* **2015**, *376*, 25–34. [CrossRef]
48. Shahabi, M.P.; McHugh, A.; Ho, G. Environmental and economic assessment of beach well intake versus open intake for seawater reverse osmosis desalination. *Desalination* **2015**, *357*, 259–266. [CrossRef]
49. Norwood, Z.; Kammen, D. Life cycle analysis of distributed concentrating solar combined heat and power: Economics, global warming potential and water. *Environ. Res. Lett.* **2012**, *7*, 044016. [CrossRef]
50. Ronquim, F.M.; Sakamoto, H.M.; Mierzwa, J.; Kulay, L.; Seckler, M.M. Eco-efficiency analysis of desalination by precipitation integrated with reverse osmosis for zero liquid discharge in oil refineries. *J. Clean. Prod.* **2020**, *250*, 119547. [CrossRef]

51. Tsalidis, G.A.; Gallart, J.J.E.; Corberá, J.B.; Blanco, F.C.; Harris, S.; Korevaar, G. Social life cycle assessment of brine treatment and recovery technology: A social hotspot and site-specific evaluation. *Sustain. Prod. Consum.* **2020**, *22*, 77–87. [CrossRef]
52. Biswas, W.K. Life cycle assessment of seawater desalinization in Western Australia. *World Acad. Sci. Eng. Technol.* **2009**, *56*, 369–375.
53. Goga, T.; Friedrich, E.; Buckley, C. Environmental life cycle assessment for potable water production—A case study of seawater desalination and mine-water reclamation in South Africa. *Water SA* **2019**, *45*, 700–709. [CrossRef]
54. Mabrouk, A.; Koc, M.; Abdala, A. Technoeconomic analysis of tri hybrid reverse osmosis-forward osmosis-multi stage flash desalination process. *Desalin. Water Treat.* **2017**, *98*, 1–15. [CrossRef]
55. Mabrouk, A.N.A.; Fath, H.E.-b.S. Techno-economic analysis of hybrid high performance MSF desalination plant with NF membrane. *Desalin. Water Treat.* **2013**, *51*, 844–856. [CrossRef]
56. Avlonitis, S.; Kouroumbas, K.; Vlachakis, N. Energy consumption and membrane replacement cost for seawater RO desalination plants. *Desalination* **2003**, *157*, 151–158. [CrossRef]
57. Huijbregts, M.; Steinmann, Z.; Elshout, P.; Stam, G.; Verones, F.; Vieira, M.; Hollander, A.; Zijp, M.; van Zelm, R. *ReCiPe 2016 v1. 1 A Harmonized Life Cycle Impact Assessment Method at Midpoint and Endpoint Level Report I: Characterization*; National Institute for Public Health and the Environment: Bilthoven, The Netherlands, 2017.
58. Darwish, M.A.; Abdulrahim, H.K.; Hassan, A.S. Realistic power and desalted water production costs in Qatar. *Desalin. Water Treat.* **2016**, *57*, 4296–4302. [CrossRef]
59. Elazhar, F.; Elazhar, M.; El Filali, N.; Belhamidi, S.; Elmidaoui, A.; Taky, M. Potential of hybrid NF-RO system to enhance chloride removal and reduce membrane fouling during surface water desalination. *Sep. Purif. Technol.* **2021**, *261*, 118299. [CrossRef]
60. Ghahremani, R.; Baheri, B.; Peydayesh, M.; Asarehpour, S.; Mohammadi, T. Novel crosslinked and zeolite-filled polyvinyl alcohol membrane adsorbents for dye removal. *Res. Chem. Intermed.* **2015**, *41*, 9845–9862. [CrossRef]
61. Gorris, L.G.M.; Yoe, C.E. Risk Analysis: Risk Assessment: Principles, Methods, and Applications. *Encycl. Food Saf.* **2014**. [CrossRef]
62. Pichery, C. Sensitivity Analysis. In *Encyclopedia of Toxicology*, 3rd ed.; Philip, W., Ed.; Academic Press: Oxford, UK, 2014; pp. 236–237. [CrossRef]

Article

Removal of MS2 and fr Bacteriophages Using MgAl$_2$O$_4$-Modified, Al$_2$O$_3$-Stabilized Porous Ceramic Granules for Drinking Water Treatment

Nur Sena Yüzbasi [1,*], Paweł A. Krawczyk [1,2], Kamila W. Domagała [1,2], Alexander Englert [3], Michael Burkhardt [3], Michael Stuer [1] and Thomas Graule [1]

1 Laboratory for High Performance Ceramics, Empa, Swiss Federal Laboratories for Materials Science and Technology, 8600 Dübendorf, Switzerland; pawel.krawczyk171@gmail.com (P.A.K.); domagalakamila11@gmail.com (K.W.D.); michael.stuer@empa.ch (M.S.); thomas.graule@empa.ch (T.G.)
2 Faculty of Materials Science and Ceramics, AGH University of Science and Technology, 30-059 Krakow, Poland
3 Department of Water and Wastewater Treatment, Institute of Environmental and Process Engineering, Eastern Switzerland University of Applied Sciences, 8640 Rapperswil, Switzerland; alexander.englert@ost.ch (A.E.); michael.burkhardt@ost.ch (M.B.)
* Correspondence: sena.yuezbasi@empa.ch

Citation: Yüzbasi, N.S.; Krawczyk, P.A.; Domagała, K.W.; Englert, A.; Burkhardt, M.; Stuer, M.; Graule, T. Removal of MS2 and fr Bacteriophages Using MgAl$_2$O$_4$-Modified, Al$_2$O$_3$-Stabilized Porous Ceramic Granules for Drinking Water Treatment. *Membranes* **2022**, *12*, 471. https://doi.org/10.3390/membranes12050471

Academic Editor: Mohammad Peydayesh

Received: 24 March 2022
Accepted: 25 April 2022
Published: 27 April 2022

Publisher's Note: MDPI stays neutral with regard to jurisdictional claims in published maps and institutional affiliations.

Copyright: © 2022 by the authors. Licensee MDPI, Basel, Switzerland. This article is an open access article distributed under the terms and conditions of the Creative Commons Attribution (CC BY) license (https://creativecommons.org/licenses/by/4.0/).

Abstract: Point-of-use ceramic filters are one of the strategies to address problems associated with waterborne diseases to remove harmful microorganisms in water sources prior to its consumption. In this study, development of adsorption-based ceramic depth filters composed of alumina platelets was achieved using spray granulation (calcined at 800 °C). Their virus retention performance was assessed using cartridges containing granular material (4 g) with two virus surrogates: MS2 and fr bacteriophages. Both materials showed complete removal, with a 7 log$_{10}$ reduction value (LRV) of MS2 up to 1 L. MgAl$_2$O$_4$-modified Al$_2$O$_3$ granules possessed a higher MS2 retention capacity, contrary to the shortcomings of retention limits in pure Al$_2$O$_3$ granules. No significant decline in the retention of fr occurred during filtration tests up to 2 L. The phase composition and morphology of the materials were preserved during filtration, with no magnesium or aluminum leakage during filtration, as confirmed by X-ray diffractograms, electron micrographs, and inductively coupled plasma-optical emission spectrometry. The proposed MgAl$_2$O$_4$-modified Al$_2$O$_3$ granular ceramic filter materials offer high virus retention, achieving the criterion for virus filtration as required by the World Health Organization (LRV $\geq$ 4). Owing to their high thermal and chemical stability, the developed materials are thus suitable for thermal and chemical-free regeneration treatments.

Keywords: drinking water; virus removal; MS2 bacteriophage; fr bacteriophage; granules; ceramic filters

1. Introduction

Safe and readily available drinking water is one of the major requirements of a healthy life. In 2017, 2.2 billion people had no access to safe drinking water, despite significant associated health risks [1]. Waterborne diseases can be readily transmitted through bacteria (e.g., *Vibrio cholerae*, *Legionella pneumophilia*, *Salmonella typhi*) and viruses (e.g., poliovirus, rotaviruses A–F, hepatitis A virus) and cause severe illnesses and deaths of millions of people [2].

Access to clean drinking water has been achieved based on several chemical, physical, and mechanical processes (such as heat treatment, chlorination, ozonation, chemical precipitation, or coagulation and flocculation and photochemical inactivation with UV irradiation) and filtration technologies, which have been proven to effectively remove or inactivate viruses or other microorganisms [3–5]. Application of some of these processes can sometimes be highly challenging in developing countries or rural areas in developed countries due to the high costs of treatment and distribution systems and a lack of or limited infrastructure [6,7].

One of the main strategies to address problems associated with waterborne diseases worldwide is to apply on-site water treatment systems, i.e., point-of-use (POU) or household water treatment (HWT) technologies, to reduce harmful microorganisms in water sources prior to consumption. Porous ceramic filters are widespread and increasingly used as drinking water treatment technologies [8–11], particularly in rural areas and developing countries [10,12].

Ceramic filters are highly advantageous as they are compatible with regeneration processes such as steam sterilization, calcination, backflushing, or chemical agents [2,13,14]. Such filters are generally effective in the removal of pathogens in the microporous range, such as bacteria and protozoa. This approach, however, fails in virus filtration due to rapid fouling of the nanometric pores (to trap viruses with typical dimensions of <100 nm), among other considerations [15–17]. The application of filters with nanopores may be limited due to the high pressure drop (high energy consumption), their low throughput, and especially the risk of fast blocking by colloidal fouling.

Viruses are nano-sized amphoteric microbes with a varying surface charge depending on the individual virus type and strain [12,18–20]. The net surface charge of viruses is pH- and surface chemistry-dependent [12,19,21,22]. An increase in the pH of the medium can lead to an increase in the ionization of carboxyl and sulfhydryl groups, and a decrease in ionization of amine groups at the surface of viruses [12,17]. Typically, isoelectric points (IEP) of viruses vary between pH 3 to 9, leading to the presence of both positively and negatively charged viruses in natural waters, depending on the virus type [17,20,23]. Surface characteristics can play a significant role in virus removal/inactivation in porous media [9,20,21,24]. Attempts have been made to predict the adsorption characteristics of viruses in porous media using the DLVO theory to model the electrostatic and van der Waals forces [12,17,21]. Additionally, previous studies have demonstrated the effect of non-DLVO factors on virus-media sorption and/or inactivation, such as hydrophobicity [23,25,26] effects arising from structural incompatibility between viruses and sorbents [20,24], roughness of the deposition surface/sorbent [27], and water chemistry [28,29].

Metal oxide surfaces are expected to possess a positive surface charge at pH values below the isoelectric point [12,30,31], which can in turn promote the attraction of viruses. In fact, functionalization of conventional depth filter surfaces, e.g., sand filters or diatomaceous earth or fiber structures, by metal oxides such as iron oxide [32–35], aluminum oxide [15,33,36], copper oxide [15,35,37], magnesium oxide [38], hydrated oxides of yttrium [29], and zirconium [2,29], resulted in enhanced virus retention.

In addition to the metal oxides mentioned above, $MgAl_2O_4$ is also known for its high IEP (~pH = 11.8) [39,40]. Even if Al_2O_3 or MgO were previously studied for virus retention applications, to the best of our knowledge, $MgAl_2O_4$ has not been implemented as an adsorbent for virus removal. Utilization of $MgAl_2O_4$ for water filtration applications was only performed by Kamato et al., for the removal of submicron-sized colloidal particles (simulating bacteria) from a suspension [41]. $MgAl_2O_4$ can be a suitable material for ceramic filters due to its non-toxicity, low cost, and excellent chemical stability. The latter property enables easy regeneration of the filter materials by backflushing [8,36], thermal [42], acidic, or basic treatment [43], without any phase transition. Such phase changes were previously observed in Cu_xO_y-based granules upon thermal treatment, which makes these materials less functional for potential applications in water filtration [15].

To this end, this study investigates the development of granular ceramic filter materials through the modification of Al_2O_3 granules with $MgAl_2O_4$ nanoparticles (Mg-NP), where the granular structures were developed by the spray granulation technique. The granules were calcined at 800 °C for further consolidation. The granular materials were tested in flow tests to determine the retention capacity of two different bacteriophages (MS2 and fr bacteriophages), serving as surrogates for human pathogenic waterborne viruses. MS2 bacteriophage is often used and was chosen as a surrogate for apolar and negatively charged human enteric viruses [16,44], while fr bacteriophage was selected due its electropositive surface in water at pH in the range from 3 to 9 [45,46].

2. Materials and Methods

Materials. The synthesis of filter materials was achieved using the spray granulation technique. Commercially available $MgAl_2O_4$ nanoparticles (d_{v50} = 0.2–0.3 µm, spinel S25CR, Baikowski SA, France, purity $\geq$ 99%, surface area of 21–24 $m^2 \cdot g^{-1}$) and plate-like Al_2O_3 (d_{v50} = 6–12 µm, white sapphire alumina, Merck Group, Germany, purity > 99.0%, surface area of 1–2 $m^2 \cdot g^{-1}$) were selected as starting materials. PAA5 (Polyacrylic acid, 50% soln. in water (MW ~ 5000), Polyscience, Inc., Warrington, PA, USA) and polyvinyl alcohol (PVA, MW 31,000–50,000, 98–99% hydrolyzed, Sigma Aldrich, St. Louis, MO, USA) were used as a dispersant and a binder, respectively.

Spray granulation. Materials were developed using the Büchi Mini Spray Dryer B290 (Büchi Labortechnik AG, Flawil, Switzerland) [39,40]. The details regarding slurry preparation and synthesis parameters are provided in the Supplementary Materials. To describe the materials, the following nomenclature is used throughout this paper: MgAl, Al, Al-Pl, Mg-NP, for $MgAl_2O_4$-modified Al_2O_3 granules, Al_2O_3 granules, plate-like Al_2O_3 powder (white sapphire), and $MgAl_2O_4$ nanoparticles (spinel S25CR), respectively, and summarized in Table 1. In order to consolidate the granules, remove the polymer binder matrix, and achieve strong bonding between Mg-NP and Al-Pl, granules were calcined in air at 800 °C with a heating (and cooling) rate of 5 °C·min^{-1} and a 1 h dwell time in PY 12 H (Pyrotec Brennofenbau GmbH, Osnabrück, Germany).

Table 1. Material nomenclature.

	Starting Materials	Spray-Dried Granules
Al-Pl	plate-like Al_2O_3 powder (white sapphire)	-
Mg-NP	$MgAl_2O_4$ nanoparticles (spinel S25CR)	-
MgAl	-	$MgAl_2O_4$-modified Al_2O_3 granules
Al	-	Al_2O_3 granules

Characterization. Synthesized materials were characterized using X-ray diffraction (XRD), N_2 physisorption, energy-dispersive X-ray spectroscopy (EDX), zeta potential, laser diffraction (LD), and helium pycnometry. The details of the characterization measurements are provided in the Supplementary Materials.

Bacteriophages and filtration tests. Two different bacteriophages, *Escherichia* phage MS2 (MS2; diameter = 25 nm, DSMZ 13767, Braunschweig Germany, IEP ~3.5–3.9 [17]) and *Escherichia coli* bacteriophage fr (fr, diameter = 19 nm, ATCC 15767-B1, Virginia, USA, IEP ~8.9–9.0 [45,46]), were used as virus surrogates. The associated host organisms for MS2 and fr were *Escherichia coli* strain W1485 (DSM-5695, Braunschweig, Germany) and *Escherichia coli* strain 3300-141 (ATCC 19853, Manassas, VA, USA), respectively.

Virus solutions with a concentration of 10^8 PFU·mL^{-1} in TRIS buffer (0.02 M tris (hydroxymethyl)-aminomethane (Merck Group, Darmstadt, Germany) and 5 mM magnesium sulfate (Merck Group, Darmstadt, Germany), pH = 7.3) and their host bacteria were purchased from the Culture Collection of Switzerland (CCOS, Wädenswil, Switzerland). For enumeration of phages, the double agar layer (DAL) method was applied according to the US EPA Method 1602, 2001 [47] (as described in the Supplementary Materials in detail). The phage concentration (I_f) was calculated, accounting for the dilution (D), using the following Equation (1):

$$I_f = \text{Number of plaques} \cdot D \ \left[\text{PFU mL}^{-1}\right] \quad (1)$$

Additionally, $\log_{10}$ virus removal (LVR) efficiency was determined for each tested material based on Equation (2), where I_0 is the initial phage concentration:

$$LVR = \log \frac{I_0}{I_f} \qquad (2)$$

Virus retention tests were performed in a laboratory-scale filtration setup (Figure S1), where 4 g of granular material was placed over glass fiber filter paper (pore size 0.4 µm, binder free, Macherey-Nagel filters) in a 70 mm-long cartridge with a diameter of 15 mm. The flow rate of the solution was adjusted to 300 mL·h^{-1} and the pressure inside the cartridge at the beginning of the tests was measured as 0.4 bars.

To assess the virus retention of the materials, continuous-flow filtration tests were conducted using MS2 or fr bacteriophages, where the initial concentration of the virus solutions was fixed to 10^7 PFU·mL^{-1} in TRIS buffer. The filtration characteristics of the developed granular ceramic filter materials were assessed based on dead-end filtration tests. Approximately 20 mL of permeate was collected after 250, 650, 1000, 1250, 1650, and 2000 mL of virus solution passed through the filtration medium to follow the decline in the LRV and thus the saturation trend of the granular ceramic filter materials. For each material, two cartridges containing 4 g of granular ceramic filter material were prepared to verify the reproducibility of the results.

Furthermore, to assess the solubility of the materials in contact with the filtration medium, magnesium and aluminum content of permeates were measured by inductively coupled optical emission spectrometry (ICP-OES, Acros, Spectro Analytical Instruments GmbH, Kleve, Germany).

3. Results and Discussion

3.1. Characterization of Starting Materials

Figure 1 shows the surface morphology of the raw powders. Al-Pl (Figure 1a) is composed of micron-sized plate-like particles with various irregular shapes, which tend to pile on each (i.e., agglomeration). Similarly, Mg-NP (Figure 1b) possesses a rough surface texture, as a result of irregularly shaped nanoparticle aggregate formation. The specific surface area (SSA) of raw materials was determined as 1.7 and 24.6 m^2·g^{-1} (Table 2), for Al-Pl and Mg-NP, respectively.

Table 2. Particle size distribution (d_{v90}, d_{v50}, and d_{v10}), surface area, density, and IEP of starting powders and synthesized granules.

	Starting Materials			Granules	
	Al-Pl	Mg-NP	Mg-NP/1 wt.% PAA	Al Granules	MgAl Granules
Particle size					
d_{v90} (µm)	25.7	31.7	1.1	98.20	123.5
d_{v50} (µm)	11.6	2.7	0.8	51.69	52.02
d_{v10} (µm)	1.1	0.5	0.6	28.20	26.37
Surface area (m^2·g^{-1})	1.7	24.6	-	1.7	7.6
Cumulative pore volume (cm^3·g^{-1})	-	-	-	1.01	0.91
Density (g·cm^{-3})	3.94	3.80	-	-	-
IEP	9.06	11.84	-	-	-

Crystalline phases of the commercial raw powders were determined using XRD and are shown in Figure 1c. The powders were crystalline without the presence of impurity phases. The diffraction peaks of Al-Pl are characteristic of trigonal α-Al$_2$O$_3$ with a rhombohedral (corundum) structure and R-3 c (167) space group (PDF: 43-1484). The diffraction pattern of Mg-NP corresponds to that of cubic MgAl$_2$O$_4$ with a spinel structure and $F\,d\,\text{-}3\,m$ (227) space group (PDF: 21-1152).

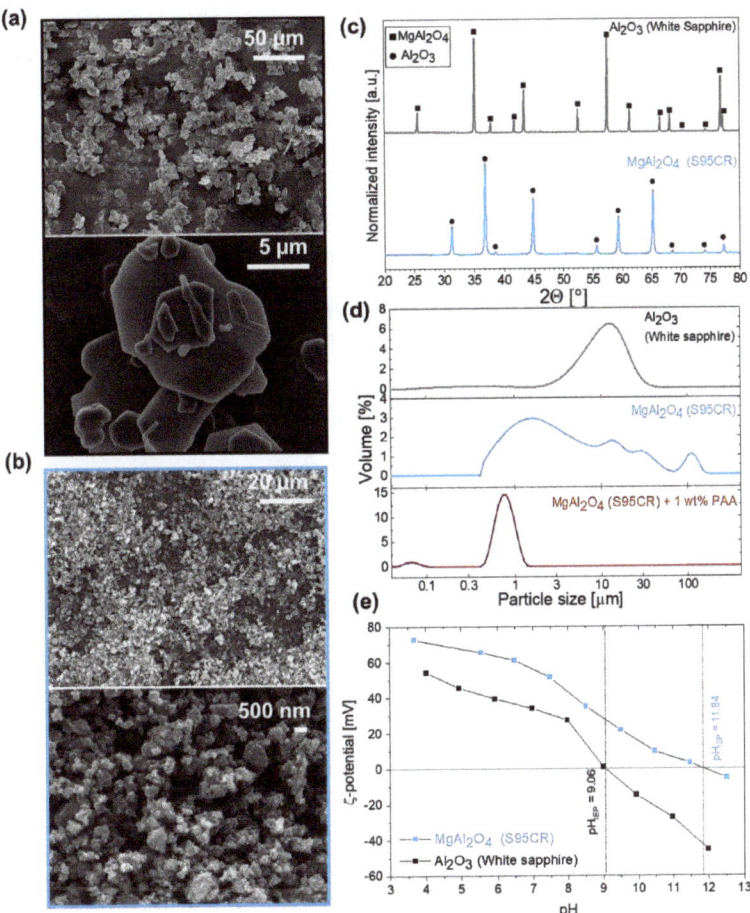

Figure 1. Characterization of the starting materials, MgAl$_2$O$_4$ nanoparticles and Al$_2$O$_3$ (white sapphire), as represented by blue and gray colors, respectively. Electron micrographs of (**a**) Al$_2$O$_3$ (white sapphire, Al-Pl) and (**b**) MgAl$_2$O$_4$ nanoparticles (Mg-NP), (**c**) X-ray diffractogram, (**d**) particle size distribution, and (**e**) zeta potential as a function of pH.

The particle size distribution of the starting materials is shown in Figure 1b, and d_{v10}, d_{v50}, and d_{v90} values are summarized in Table 2. The volume-based LD measurements for alumina platelets represented a monomodal and relatively broad particle size distribution. d_{v10} of 1 µm, d_{v50} of 12 µm, and d_{v90} of 26 µm were determined for alumina platelets, in agreement with the company-provided values. Note that the shape of particles has a strong impact on the particle size measurements, since the particle size distribution of non-spherical particles is calculated on the basis of equivalent spherical diameters [48,49].

On the other hand, spinel nanoparticles showed a broad and polymodal particle size distribution with a d_{v50} of 2.7 µm, in disagreement with the company-provided values (0.2–0.3 µm). The volume-based LD measurements thus indicate strong agglomeration of spinel nanoparticles, as seen on the electron micrographs, requiring extensive milling to re-disperse Mg-NP prior to granulation. Therefore, PAA5 was selected as a dispersant to stabilize Mg-NP. The influence of dispersant concentration and milling time on particle size distribution was critically assessed and further explained in the Supplementary Materials (Figure S2). The addition of 1 wt.% dispersant and 10 h of milling allowed to reduce the particle size distribution of Mg-NP (d_{v10} of 0.6 µm, d_{v50} of 0.8 µm, and d_{v90} of 1.1 µm), as shown in Figure 1b.

The zeta potential of alumina platelets and spinel nanoparticles was measured as a function of pH, as represented in Figure 1c. Both starting materials showed a positive zeta potential (above 30 mV) in the pH range typical for drinking water (pH 6 to 8). The IEP of AL-Pl and Mg-NP was determined at 9.06 and 11.84, respectively (Table 2), in agreement with the literature [39,40,50].

3.2. Granulation

Spray-drying allows changing the granule morphology by tuning the slurry properties and granulation parameters. Preliminary tests pre-established an appropriate solid load of the dispersion, nozzle type, and binder fraction to optimize the granulation yield, material size, porosity, and surface area. Only a marginal difference in particle size distribution and morphology of the granules could be observed depending on the nozzle type, i.e., ultrasonic or two-fluid nozzle (Figure S3). Due to the ease of handling, the two-fluid nozzle was selected for further experiments. On the contrary, the binder content had a distinct impact on the granule morphology and size, as represented in Figure S5. The PVA content was varied between 0 and 5 wt.% (referring to the total amount of all powders in the slurry). At least 2 wt.% of binder was found to be required to form well-defined granules rather than a mix of broken granules and loose powder. Higher amounts of binder (5 wt.% PVA), however, caused the formation of large granule agglomerates and thus heterogeneous, bimodal size distributions with diameters up to 1 mm. The optimal fraction of PVA was thus found to be 2 wt.% (6.2 vol.%) in order to obtain homogenous, spherical-shaped granular material. Finally, high SSA, as one of the important prerequisites for successful virus adsorption, could be achieved by lowering the solid loading within the powder slurry for spray granulation. Lowering the solid load from 20 to 10 vol.% almost doubled the SSA of MgAl granules from 4.0 to 7.6 $m^2 \cdot g^{-1}$.

Figure 2a represents the surface morphology of the MgAl and Al granules with a solid loading of 10 vol.% in the presence of 2 wt.% PVA (and 1 wt.% PAA5 in the case of MgAl for spinel dispersion) after their calcination at 800 °C. Electron micrographs reveal the presence of spherical granules in both materials, which have been collected in the coarse collector of the spray-dryer (as illustrated in Figure S4). However, observing the granules collected in the fine collector of the spray-dryer revealed that they had been broken and that the granulation was especially poor in the case of Al compared to MgAl. This can be attributed to an additional binding effect from $MgAl_2O_4$ nanoparticles that tend to adhere strongly to each other, as previously described by Kendall et al. [51,52]. SEM images of the broken or polished granules in Figure 2b display the sub-surface morphology of the materials. Both materials possessed a highly porous internal structure (as previously also confirmed by mercury intrusion porosimetry [15]), a critical feature to ensure effective water flow during filtration, as a result of randomly oriented alumina platelets and the low solid load of the ceramic slurry. To evaluate the compositional homogeneity between Al_2O_3 and $MgAl_2O_4$, EDX measurements were carried out. The elemental maps of aluminum and magnesium provided in Figure 2c revealed that Mg-NP were homogeneously distributed on the alumina platelets and within the granule volume (surface and sub-surface), as also confirmed in Figure S6. There was no phase change during granulation and calcination steps, as confirmed by the diffraction patterns of Al and MgAl in Figure 2d that show the presence of α-Al_2O_3 and cubic $MgAl_2O_4$.

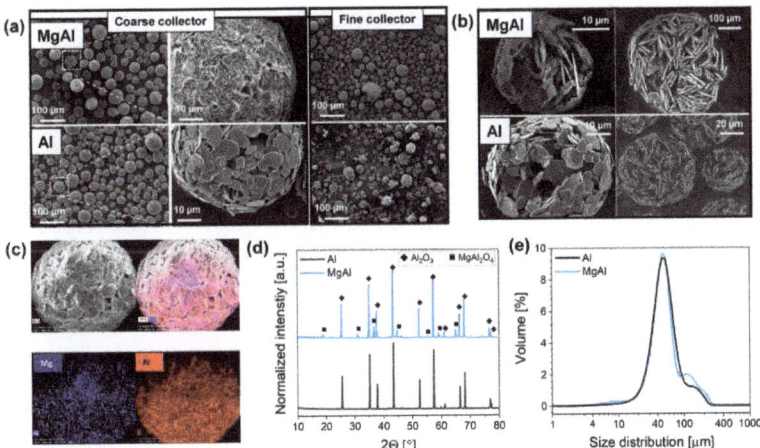

Figure 2. Characterization of the spray-dried Al (—) and MgAl (—) granules that were calcined at 800 °C. (**a**) Electron micrographs, (**b**) sub-surface morphology of the broken (left) or polished (right) granules, (**c**) elemental mapping of MgAl granules, (**d**) X-ray diffractograms, and (**e**) particle size distribution of the granules.

The particle size distribution of the granules (Figure 2e and Table 2) was affected only marginally by the presence of Mg-NP. On the contrary, the introduction of Mg-NP resulted in a four times larger specific surface area for MgAl (7.6 m$^2 \cdot$g^{-1}) granules compared to Al granules (1.7 m$^2 \cdot$g^{-1}) when alumina was partially substituted by a material with a larger specific surface area, such as Mg-NP (24.6 m$^2 \cdot$g^{-1}). The porous structure of the granules was further characterized by using mercury intrusion porosimetry (MIP). The cumulative pore volume porosity and pore size distribution of the granules are represented in Figure S7. The cumulative pore volume of Al and MgAl was only marginally different, detected as 1.01 and 0.91 cm^3/g, respectively, indicating that the presence of Mg-NP did not block the pores of the granules. In both materials, two distinct pore size ranges were noticeable between 0.5 to 2 μm and 8 to 20 μm.

3.3. Filtration Tests and Virus Removal Performance

Prior to the tests with the granular material, control tests (with cartridges containing only glass fiber filter paper) were performed to examine bacteriophage binding on the filter paper. These tests confirmed that the presence of glass fiber filter paper did not have any contribution in the removal of bacteriophages. LRV results obtained during dead-end filtration tests performed using 4 g of ceramic filter materials are presented in Figure 3. Additionally, LRV was also plotted against normalized filtrate volume per filter bed volume, where bed volumes were calculated as 4.8 and 6.2 mL for granules and Al-Pl, respectively, as shown in Figure S8. Figure 3 shows that the MS2 phages can be effectively retained and completely removed from water up to 1 L with 4 g of both synthesized granules. The granular structure, in a similar context with depth filtration, enabled higher contact time between the adsorbent and MS2-contaminated water, which permitted effective physisorption and chemisorption of the contaminants, when compared to surface filtration [53]. However, there was a sharp decline in LRV of Al granules after 1.25 L, and ultimately no virus removal after 1.65 L, indicating that saturation of granules was reached. On the other hand, virus retention of MgAl slightly decreased after 1.65 L from LRV 7 to 5, while still meeting the WHO standards for drinking water (LRV ≥ 4) over 2 L [54].

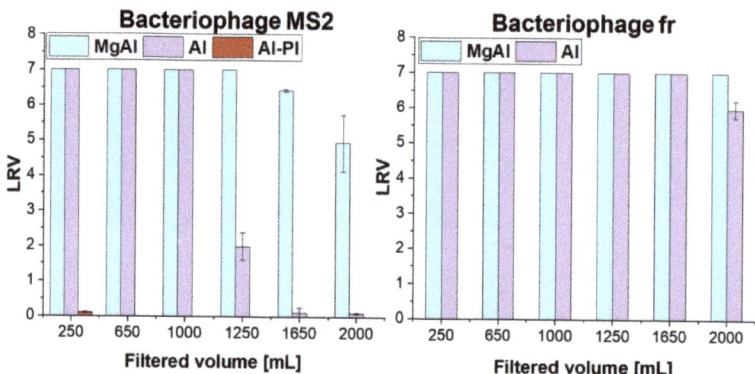

Figure 3. Retention performance of the granules based on MS2 and fr $\log_{10}$ removal as a function of filtered volume. All filter media contained the same amount of material (4 g).

As a point of comparison, non-granulated alumina platelets showed significantly poorer virus retention capacity and there was no removal of negatively charged MS2 bacteriophages (Figure 3). Due to the high powder packing density, filtration with Al-Pl was significantly more challenging compared to filtration with the granules and accompanied by a severe pressure drop in the cartridges. This clearly illustrates the importance of the microstructure (e.g., tortuosity) and porosity of the filter media during filtration. Due to the filtration challenges in Al-Pl filter media, they were only evaluated up to 250 mL, and the tests continued thereafter with granular materials only. Filtration tests revealed that despite a high positive surface charge of Al-Pl (IEP = 9.1), there was no removal of negatively charged MS2 bacteriophages (IEP $\approx$ 3.5 [17]) at pH 7.3. Such a high charge difference between the filter and bacteriophage surface has been reported to lead to a virion sorption through electrostatic forces [20,38,55]; however, it is insufficient in explaining our experimental data. Indeed, despite attractive electrostatic forces, the poor retention performance of Al-Pl may be a result of: (i) a significantly low surface area due to platelet agglomeration via the basal planes, and thus a small number of adsorption sites in Al-Pl, and/or (ii) the formation of preferential flow paths, e.g., short-circuits of the filter. The latter leads to an insufficient contact of contaminated water with the filter surface for adsorption of bacteriophages, quickly saturating the little-exposed filter surface area on the flow paths. The former prevents the fulfilment of a key prerequisite for effective virus retention: having a high number of adsorption sites. Indeed, a higher virus removal capacity is often not solely correlated with the IEP of viruses and sorption surface, but also with the surface area and roughness of the sorption media [27]. Dika et al. demonstrated that substrate roughness has an impact on the adhesion of bacteriophages, where weaker adhesion was observed on a low-roughness surface (glass) when compared to substrates with a higher roughness (polypropylene or stainless steel). Our results strongly suggest that the slower saturation of MgAl granules observed in filtration tests with MS2 bacteriophage can be linked to the larger surface area, surface roughness, and higher adsorption sites for viruses provided by spinel nanoparticles, when compared to Al granules.

One important characteristic of a virus is its IEP, which represents the pH value at which the surface charge of a virus is zero. Usual IEPs of viruses range from 3 to 7 [17]. To cover the IEPs of relevant viruses found in water in a wider range and have a better understanding on the potential contribution of electrostatic forces, MS2 filtration tests were complemented by separate filtration tests using fr bacteriophages. The filtration tests with fr bacteriophages revealed successful retention of up to 2 L of contaminated water for Al and MgAl granules. There was a small decline only in LRV of Al granules from LRV 7 to 6, after 1.65 L, while MgAl granules could successfully achieve complete removal of fr bacteriophages even after 2 L.

fr bacteriophage has a high IEP of 8.9 according to literature-reported values [56–58]. The filtration tests with fr bacteriophages revealed successful retention up to 2 L of contaminated water, in spite of the low zeta-potential difference between the filter material and virus surrogate and the resulting low electrostatic interaction forces' contributions. A more detailed literature review, however, shows that the reported IEP of fr bacteriophage varies widely, from 3.5 to 9.0 [55]. Recent studies experimentally validated by light scattering and electrophoretic mobility measurements show that fr bacteriophages have mostly a negative surface charge [20,21,55]. Armanious et al. theoretically estimated the surface charge of fr bacteriophage based on the ionizable amino-acids and the tertiary structure of fr capsid protein and reported its surface charge and IEP as -2.5×10^2 C·m^{-2} and 4.5, respectively [20]. Due to the contradictory findings in the literature with respect to the surface charge of fr bacteriophages, the present retention results need to be assessed considering two scenarios: fr bacteriophages have a (i) negative surface charge and (ii) positive surface charge. In the first scenario, short-range attraction forces, i.e., van der Waals forces or hydrophobic effects, may dominate or replace the electrostatic forces in virion sorption. Following the second scenario, fr adsorption may occur similarly to MS2, driven by longer range electrostatic forces which can be further complemented with attractive van der Waals forces and hydrophobic effects. In the latter scenario, the retention performance difference obtained in filtration with fr bacteriophages and MS2 bacteriophages can be linked to the level of hydrophobicity caused by differences in the surface polarities. Armanious et al. calculated the relative hydrophobicity of MS2 and fr bacteriophages and suggested that fr bacteriophage experienced larger contributions from the hydrophobic effect due to its higher apolarity [20].

Ongoing efforts clearly demonstrate the complexity of the interactions that play a critical role in the virus trapping, and further investigations are necessary to validate the exact mechanisms as well as exclude or quantify additional contributions that may result from filter aging. According to the WHO International Scheme to Evaluate Household Water Treatment Technologies report [54], a typical virus concentration of challenge water and a minimum test water volume is indicated as 10^5 PFU·mL^{-1} and 20 L·day^{-1} for laboratory verification tests of POU water filtration technologies, e.g., granular media and porous or membrane filters [59]. Test waters used in this study were spiked two orders of magnitude higher than suggested challenge concentrations (in the absence of humic acids or natural organic materials). Filtration tests with 2 L of MS2- or fr-contaminated water in controlled systems show outstanding adsorption properties of MgAl granules (with 4 g) and bring the material system closer to testing in real water systems. These materials need to be tested for longer periods with higher volumes of challenge test waters based on guideline values of the WHO [54,60], prior to the application stage.

3.4. Characterization of Materials after Filtration

Both materials were characterized following filtration in order to investigate the influence of water exposure on the morphology and phase composition of the granules. The phase composition of the granules was preserved during filtration according to X-ray diffractograms, as shown in Figure 4a.

The electron micrographs presented in Figure 4b indicate that the granular structure was preserved after their exposure to 2 L of contaminated water. Further tests on the stability of the granules were performed using ICP-OES to evaluate the dissolution of alumina and magnesium aluminate spinel to the form of Al^{3+} and Mg^{2+} cations in the permeate after filtration tests with fr bacteriophage. Aluminum was not detected after 2 L of filtration (where the detection limit of Al is 50 µg·L^{-1}), and only a small quantity of magnesium release (<60 µg·L^{-1}) was observed in the case of MgAl granules (Figure 4c). The magnesium concentration in the permeate, which is known to cause hardness in drinking water, was three orders of magnitude lower than the taste threshold value recommended by the WHO for drinking water (<25–50 mg·L^{-1}) [59].

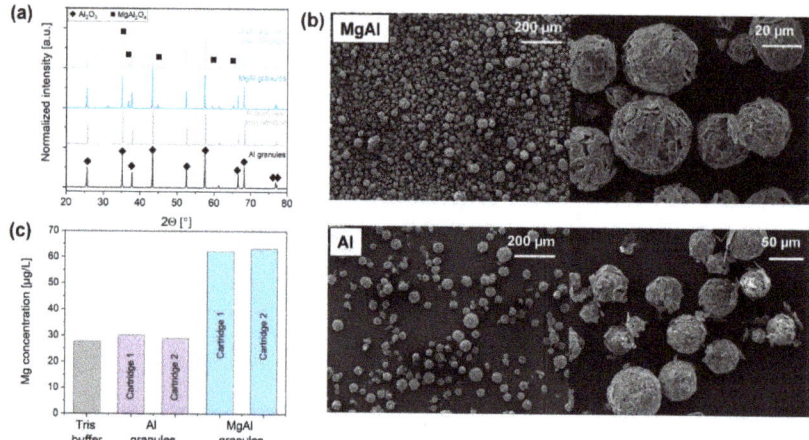

Figure 4. Characterization of the granules after filtration. (**a**) X-ray diffractograms, (**b**) electron micrographs, and (**c**) associated magnesium concentrations detected in permeate, as determined by ICP-OES.

As the limited lifetime is a severe technical challenge within an adsorption-based, dead-end filtration process, the thermal regeneration will be a beneficial alternative even after the filter itself starts losing its efficiency due to clogging of its pores and the occupation of all adsorption sites by the virus contaminants (and by concomitant humic acid and other competitive adsorbing water contaminants in real water systems). One of the main advantages of granular ceramic filter materials is the possibility of regenerating the filter media by thermal means [11,29]. Preliminary tests revealed that after heat treatment of the saturated MgAl filter at 400 °C in air, the virus retention capacity was recovered (Figure S9). However, it is important to mention here that the filters were not fully saturated and filtration after regeneration was performed with only 250 mL of MS2 solution (10^7 PFU /mL). Therefore, optimized regeneration conditions need be developed by assessing the process parameters such as temperature and regeneration cycles in the presence of competitive adsorbing water contaminants, as envisaged to be conducted in the future.

4. Conclusions

In this work, we shed light on the structure–performance relationship of spray-dried granules as a ceramic filter material for virus removal in drinking water applications. The following conclusions were reached under the observations of this study:

1. The presence of homogenously distributed Mg-NP in Al_2O_3 granules offers effective means to enhance adsorption sites of virus surrogates (MS2 and fr bacteriophages).
2. $MgAl_2O_4$-modified Al_2O_3 granules exceeded the retention performance of pristine Al_2O_3 granules, as revealed through flow tests.
3. $MgAl_2O_4$-modified Al_2O_3 granules possess promising adsorption properties, and could successfully achieve a $\log_{10}$ reduction of 5 and 7 of MS2 and fr bacteriophages, respectively, with 4 g of MgAl after 2 L of filtration.
4. There was no degradation in phase composition and morphology of the granules upon filtration.
5. No aluminum nor significant magnesium leakage was detected during the filtration, suggesting a high stability of the developed materials as a result of consolidation at 800 °C.
6. Preliminary regeneration tests indicated that the developed granular ceramic filter materials can be potentially reused after thermal treatment.

The underlying mechanism of successful virus retention is still not clear; however, experimental findings suggest that highly porous granular structures play a key role in the removal of bacteriophages. It is suggested that it enables a good permeability and thus contact between the material and the influent.

The current study therefore highlights the potential of $MgAl_2O_4$-Al_2O_3 granules for drinking water treatment. Prior to real applications, however, the filter materials developed in this study need to be tested with more complex water chemistries, such as the presence of complexing factors (e.g., natural organic matter, different pH) and regeneration options (e.g., by thermal means) need to be evaluated to increase the lifetime and reuse the absorber material.

Supplementary Materials: The following supporting information can be downloaded at: https://www.mdpi.com/article/10.3390/membranes12050471/s1. Details regarding characterization techniques, spray-drying, and virus removal experimental set-up descriptions, additional material characterization and retention results (PDF). References [61–65] are cited in the supplementary materials.

Author Contributions: Conceptualization, N.S.Y. and T.G.; data curation, N.S.Y. and P.A.K.; formal analysis, N.S.Y., P.A.K. and K.W.D.; funding acquisition, T.G.; investigation, N.S.Y. and P.A.K.; methodology, N.S.Y., P.A.K., A.E. and M.S.; project administration, N.S.Y. and T.G.; resources, M.B.; supervision, M.S. and T.G.; validation, N.S.Y., P.A.K., K.W.D. and A.E.; visualization, N.S.Y. and P.A.K.; writing—original draft, N.S.Y.; writing—review and editing, N.S.Y., P.A.K., A.E., M.B., M.S. and T.G. All authors have read and agreed to the published version of the manuscript.

Funding: This work was funded by a private donation, with a project acronym of "MultiCarboVir" and project number of 5211.01432.100.01.

Institutional Review Board Statement: Not applicable.

Informed Consent Statement: Not applicable.

Data Availability Statement: Not applicable.

Acknowledgments: This work was funded by a private donation and our donor is highly acknowledged. We thank Brian Sinnet from Eawag, Switzerland, for ICP-OES measurements and Laura Conti from Empa, Switzerland, for MIP measurements.

Conflicts of Interest: The authors declare no conflict of interest.

References

1. WHO. Drinking-Water, Key Facts. 2019. Available online: https://www.who.int/news-room/fact-sheets/detail/drinking-water (accessed on 26 April 2022).
2. Kroll, S.; de Moura, M.O.C.; Meder, F.; Grathwohl, G.; Rezwan, K. High virus retention mediated by zirconia microtubes with tailored porosity. *J. Eur. Ceram. Soc.* **2012**, *32*, 4111–4120. [CrossRef]
3. Peter-Varbanets, M.; Zurbrügg, C.; Swartz, C.; Pronk, W. Decentralized systems for potable water and the potential of membrane technology. *Water Res.* **2009**, *43*, 245–265. [CrossRef] [PubMed]
4. Chen, C.; Guo, L.; Yang, Y.; Oguma, K.; Hou, L.-a. Comparative effectiveness of membrane technologies and disinfection methods for virus elimination in water: A review. *Sci. Total Environ.* **2021**, *801*, 149678. [CrossRef] [PubMed]
5. Sobsey, M.D.; Stauber, C.E.; Casanova, L.M.; Brown, J.M.; Elliott, M.A. Point of use household drinking water filtration: A practical, effective solution for providing sustained access to safe drinking water in the developing world. *Environ. Sci. Technol.* **2008**, *42*, 4261–4267. [CrossRef] [PubMed]
6. Mankad, A.; Tapsuwan, S. Review of socio-economic drivers of community acceptance and adoption of decentralised water systems. *J. Environ. Manag.* **2011**, *92*, 380–391. [CrossRef] [PubMed]
7. Pooi, C.K.; Ng, H.Y. Review of low-cost point-of-use water treatment systems for developing communities. *NPJ Clean Water* **2018**, *1*, 11. [CrossRef]
8. Michen, B.; Diatta, A.; Fritsch, J.; Aneziris, C.; Graule, T. Removal of colloidal particles in ceramic depth filters based on diatomaceous earth. *Sep. Purif. Technol.* **2011**, *81*, 77–87. [CrossRef]
9. Sellaoui, L.; Badawi, M.; Monari, A.; Tatarchuk, T.; Jemli, S.; Dotto, G.L.; Bonilla-Petriciolet, A.; Chen, Z. Make it clean, make it safe: A review on virus elimination via adsorption. *Chem. Eng. J.* **2021**, *412*, 128682. [CrossRef]
10. Van Halem, D.; Van der Laan, H.; Heijman, S.; Van Dijk, J.; Amy, G. Assessing the sustainability of the silver-impregnated ceramic pot filter for low-cost household drinking water treatment. *Phys. Chem. Earth Parts A/B/C* **2009**, *34*, 36–42. [CrossRef]

11. Wegmann, M.; Michen, B.; Graule, T. Nanostructured surface modification of microporous ceramics for efficient virus filtration. *J. Eur. Ceram. Soc.* **2008**, *28*, 1603–1612. [CrossRef]
12. Brown, J.; Sobsey, M.D. Microbiological effectiveness of locally produced ceramic filters for drinking water treatment in Cambodia. *J. Water Health* **2010**, *8*, 1–10. [CrossRef]
13. Mugnier, N.; Howell, J.A.; Ruf, M. Optimisation of a back-flush sequence for zeolite microfiltration. *J. Membr. Sci.* **2000**, *175*, 149–161. [CrossRef]
14. Willemse, R.; Brekvoort, Y. Full-scale recycling of backwash water from sand filters using dead-end membrane filtration. *Water Res.* **1999**, *33*, 3379–3385. [CrossRef]
15. Mazurkow, J.M.; Yuüzbasi, N.S.; Domagala, K.W.; Pfeiffer, S.; Kata, D.; Graule, T. Nano-sized copper (oxide) on alumina granules for water filtration: Effect of copper oxidation state on virus removal performance. *Environ. Sci. Technol.* **2019**, *54*, 1214–1222. [CrossRef]
16. Michen, B. Virus Removal in Ceramic Depth Filters: The Electrostatic Enhanced Adsorption Approach. Ph.D. Thesis, Technical University Bergakademie Freiberg, Freiberg, Germany, 2010.
17. Michen, B.; Graule, T. Isoelectric points of viruses. *J. Appl. Microbiol.* **2010**, *109*, 388–397. [CrossRef]
18. Dowd, S.E.; Pillai, S.D.; Wang, S.; Corapcioglu, M.Y. Delineating the specific influence of virus isoelectric point and size on virus adsorption and transport through sandy soils. *Appl. Environ. Microbiol.* **1998**, *64*, 405–410. [CrossRef] [PubMed]
19. Sobsey, M.; Shields, P.; Hauchman, F.; Hazard, R.; Caton Iii, L. Survival and transport of hepatitis A virus in soils, groundwater and wastewater. *Water Sci. Technol.* **1986**, *18*, 97–106. [CrossRef]
20. Armanious, A.; Aeppli, M.; Jacak, R.; Refardt, D.; Sigstam, T.; Kohn, T.; Sander, M. Viruses at solid–water interfaces: A systematic assessment of interactions driving adsorption. *Environ. Sci. Technol.* **2016**, *50*, 732–743. [CrossRef]
21. Armanious, A.; Münch, M.; Kohn, T.; Sander, M. Competitive Coadsorption Dynamics of Viruses and Dissolved Organic Matter to Positively Charged Sorbent Surfaces. *Environ. Sci. Technol.* **2016**, *50*, 3597–3606. [CrossRef] [PubMed]
22. Domagała, K.; Bell, J.; Yüzbasi, N.S.; Sinnet, B.; Kata, D.; Graule, T. Virus removal from drinking water using modified activated carbon fibers. *RSC Adv.* **2021**, *11*, 31547–31556. [CrossRef]
23. Gerba, C.P. Applied and theoretical aspects of virus adsorption to surfaces. *Adv. Appl. Microbiol.* **1984**, *30*, 133–168. [PubMed]
24. Heffron, J.; Mayer, B.K. Emerging investigators series: Virus mitigation by coagulation: Recent discoveries and future directions. *Environ. Sci. Water Res. Technol.* **2016**, *2*, 443–459. [CrossRef]
25. Dika, C.; Duval, J.F.; Francius, G.; Perrin, A.; Gantzer, C. Isoelectric point is an inadequate descriptor of MS2, Phi X 174 and PRD1 phages adhesion on abiotic surfaces. *J. Colloid Interface Sci.* **2015**, *446*, 327–334. [CrossRef] [PubMed]
26. Farrah, S.R.; Shah, D.O.; Ingram, L.O. Effects of chaotropic and antichaotropic agents on elution of poliovirus adsorbed on membrane filters. *Proc. Natl. Acad. Sci. USA* **1981**, *78*, 1229–1232. [CrossRef] [PubMed]
27. Dika, C.; Ly-Chatain, M.; Francius, G.; Duval, J.; Gantzer, C. Non-DLVO adhesion of F-specific RNA bacteriophages to abiotic surfaces: Importance of surface roughness, hydrophobic and electrostatic interactions. *Colloids Surf. A Physicochem. Eng. Asp.* **2013**, *435*, 178–187. [CrossRef]
28. Watts, S.; Julian, T.R.; Maniura-Weber, K.; Graule, T.; Salentinig, S. Colloidal Transformations in MS2 Virus Particles: Driven by pH, Influenced by Natural Organic Matter. *ACS Nano* **2020**, *14*, 1879–1887. [CrossRef]
29. Wegmann, M.; Michen, B.; Luxbacher, T.; Fritsch, J.; Graule, T. Modification of ceramic microfilters with colloidal zirconia to promote the adsorption of viruses from water. *Water Res.* **2008**, *42*, 1726–1734. [CrossRef]
30. Michen, B.; Meder, F.; Rust, A.; Fritsch, J.; Aneziris, C.; Graule, T. Virus removal in ceramic depth filters based on diatomaceous earth. *Environ. Sci. Technol.* **2012**, *46*, 1170–1177. [CrossRef]
31. Parks, G.A. The isoelectric points of solid oxides, solid hydroxides, and aqueous hydroxo complex systems. *Chem. Rev.* **1965**, *65*, 177–198. [CrossRef]
32. Gutierrez, L.; Li, X.; Wang, J.; Nangmenyi, G.; Economy, J.; Kuhlenschmidt, T.B.; Kuhlenschmidt, M.S.; Nguyen, T.H. Adsorption of rotavirus and bacteriophage MS2 using glass fiber coated with hematite nanoparticles. *Water Res.* **2009**, *43*, 5198–5208. [CrossRef]
33. Lukasik, J.; Cheng, Y.-F.; Lu, F.; Tamplin, M.; Farrah, S.R. Removal of microorganisms from water by columns containing sand coated with ferric and aluminum hydroxides. *Water Res.* **1999**, *33*, 769–777. [CrossRef]
34. Noubactep, C. Metallic iron for environmental remediation: A review of reviews. *Water Res.* **2015**, *85*, 114–123. [CrossRef]
35. Schabikowski, M.; Cichoń, A.; Németh, Z.; Kubiak, W.; Kata, D.; Graule, T. Electrospun iron and copper oxide fibers for virus retention applications. *Text. Res. J.* **2019**, *89*, 4373–4382. [CrossRef]
36. Ke, X.B.; Zhu, H.Y.; Gao, X.P.; Liu, J.W.; Zheng, Z.F. High-performance ceramic membranes with a separation layer of metal oxide nanofibers. *Adv. Mater.* **2007**, *19*, 785–790. [CrossRef]
37. Szekeres, G.P.; Németh, Z.n.; Schrantz, K.; Németh, K.n.; Schabikowski, M.; Traber, J.; Pronk, W.; Hernádi, K.r.; Graule, T. Copper-coated cellulose-based water filters for virus retention. *ACS Omega* **2018**, *3*, 446–454. [CrossRef]
38. Michen, B.; Fritsch, J.; Aneziris, C.; Graule, T. Improved virus removal in ceramic depth filters modified with MgO. *Environ. Sci. Technol.* **2013**, *47*, 1526–1533. [CrossRef]
39. Haijun, Z.; Xiaolin, J.; Zhanjie, L.; Zhenzhen, L. The low temperature preparation of nanocrystalline $MgAl_2O_4$ spinel by citrate sol–gel process. *Mater. Lett.* **2004**, *58*, 1625–1628. [CrossRef]

40. Kadosh, T.; Cohen, Y.; Talmon, Y.; Kaplan, W.D.; Krell, A. In Situ Characterization of Spinel Nanoceramic Suspensions. *J. Am. Ceram. Soc.* **2012**, *95*, 3103–3108. [CrossRef]
41. Kamato, Y.; Suzuki, Y. Reactively sintered porous MgAl$_2$O$_4$ for water-purification filter with controlled particle morphology. *Ceram. Int.* **2017**, *43*, 14090–14095. [CrossRef]
42. Finley, J. Ceramic membranes: A robust filtration alternative. *Filtr. Sep.* **2005**, *42*, 34–37. [CrossRef]
43. Wang, Z.; Meng, F.; He, X.; Zhou, Z.; Huang, L.-N.; Liang, S. Optimisation and performance of NaClO-assisted maintenance cleaning for fouling control in membrane bioreactors. *Water Res.* **2014**, *53*, 1–11. [CrossRef]
44. Jacquin, C.; Yu, D.; Sander, M.; Domagala, K.; Traber, J.; Morgenroth, E.; Julian, T.R. Competitive co-adsorption of bacteriophage MS2 and natural organic matter onto multiwalled carbon nanotubes. *Water Res. X* **2020**, *9*, 100058. [CrossRef]
45. Herath, G.; Yamamoto, K.; Urase, T. Removal of Viruses by Microfiltration Membranes at Different Solution Environments. *Water Sci. Technol.* **1999**, *40*, 331–338. [CrossRef]
46. Sakoda, A.; Sakai, Y.; Hayakawa, K.; Suzuki, M. Adsorption of viruses in water environment onto solid surfaces. *Water Sci. Technol.* **1997**, *35*, 107–114. [CrossRef]
47. USEPA. *Method 1602: Male-Specific (F+) and Somatic Coliphage in Water by Single Agar Layer (SAL) Procedure*; EPA 821-R-01-029; USEPA: Washington, DC, USA, 2001.
48. Mishchenko, M.I.; Travis, L.D. Light scattering by polydisperse, rotationally symmetric nonspherical particles: Linear polarization. *J. Quant. Spectrosc. Radiat. Transf.* **1994**, *51*, 759–778. [CrossRef]
49. Scott, D.M.; Matsuyama, T. Laser diffraction of acicular particles: Practical applications. In Proceedings of the International Conference on Optical Particle Characterization (OPC 2014), Tokyo, Japan, 10–14 March 2014; International Society for Optics and Photonics: Bellingham, WA, USA, 2014; p. 923210.
50. Pfeiffer, S.; Florio, K.; Makowska, M.; Ferreira Sanchez, D.; Van Swygenhoven, H.; Aneziris, C.G.; Wegener, K.; Graule, T. Iron oxide doped spray dried aluminum oxide granules for selective laser sintering and melting of ceramic parts. *Adv. Eng. Mater.* **2019**, *21*, 1801351. [CrossRef]
51. Kendall, K. Adhesion: Molecules and mechanics. *Science* **1994**, *263*, 1720–1725. [CrossRef]
52. Kendall, K.; Weihs, T. Adhesion of nanoparticles within spray dried agglomerates. *J. Phys. D Appl. Phys.* **1992**, *25*, A3. [CrossRef]
53. Herzig, J.; Leclerc, D.; Goff, P.L. Flow of suspensions through porous media—application to deep filtration. *Ind. Eng. Chem.* **1970**, *62*, 8–35. [CrossRef]
54. WHO. *Evaluating Household Water Treatment Options: Health-Based Targets and Microbiological Performance Specifications*; World Health Organization: Geneva, Switzerland, 2011.
55. Heffron, J.; McDermid, B.; Maher, E.; McNamara, P.J.; Mayer, B.K. Mechanisms of virus mitigation and suitability of bacteriophages as surrogates in drinking water treatment by iron electrocoagulation. *Water Res.* **2019**, *163*, 114877. [CrossRef]
56. Abbaszadegan, M.; Monteiro, P.; Nwachuku, N.; Alum, A.; Ryu, H. Removal of adenovirus, calicivirus, and bacteriophages by conventional drinking water treatment. *J. Environ. Sci. Health Part A* **2008**, *43*, 171–177. [CrossRef] [PubMed]
57. Gerba, C.P.; Riley, K.R.; Nwachuku, N.; Ryu, H.; Abbaszadegan, M. Removal of Encephalitozoon intestinalis, Calicivirus, and Coliphages by Conventional Drinking Water Treatment. *J. Environ. Sci. Health Part A* **2003**, *38*, 1259–1268. [CrossRef] [PubMed]
58. Overby, L.R.; Barlow, G.H.; Doi, R.H.; Jacob, M.; Spiegelman, S. Comparison of Two Serologically Distinct Ribonucleic Acid Bacteriophages, I. Properties of the Viral Particles. *J. Bacteriol.* **1966**, *91*, 442–448. [CrossRef] [PubMed]
59. WHO. *Calcium and Magnesium in Drinking Water: Public Health Significance*; World Health Organization: Geneva, Switzerland, 2009.
60. Thompson, M.; Wood, R. The international harmonized protocol for the proficiency testing of (chemical) analytical laboratories (Technical Report). *Pure Appl. Chem.* **1993**, *65*, 2123–2144. [CrossRef]
61. Asakura, S.; Oosawa, F. Interaction between particles suspended in solutions of macromolecules. *J. Polym. Sci.* **1958**, *33*, 183–192. [CrossRef]
62. Lu, K.; Kessler, C. Colloidal dispersion and rheology study of nanoparticles. *J. Mater. Sci.* **2006**, *41*, 5613–5618. [CrossRef]
63. Brunauer, S.; Emmett, P.H.; Teller, E. Adsorption of gases in multimolecular layers. *J. Am. Chem. Soc.* **1938**, *60*, 309–319. [CrossRef]
64. Carvalho, C.; Susano, M.; Fernandes, E.; Santos, S.; Gannon, B.; Nicolau, A.; Gibbs, P.; Teixeira, P.; Azeredo, J. Method for bacteriophage isolation against target Campylobacter strains. *Lett. Appl. Microbiol.* **2010**, *50*, 192–197. [CrossRef]
65. Chu, Y.; Jin, Y.; Yates, M.V. *Virus Transport through Saturated Sand Columns as Affected by Different Buffer Solutions*; Wiley Online Library: New York, NY, USA, 2000. [CrossRef]

MDPI
St. Alban-Anlage 66
4052 Basel
Switzerland
Tel. +41 61 683 77 34
Fax +41 61 302 89 18
www.mdpi.com

Membranes Editorial Office
E-mail: membranes@mdpi.com
www.mdpi.com/journal/membranes

www.ingramcontent.com/pod-product-compliance
Lightning Source LLC
LaVergne TN
LVHW070734100526
838202LV00013B/1230